AF906266

Collagen-Derived Materials

Collagen-Derived Materials

Synthesis and Applications in Electrochemical Energy Storage and Conversion

Feng Wang, Yaqin Huang, and Jin Niu

WILEY-VCH

Authors

Prof. Feng Wang
Beijing University of Chemical Technology
Key Laboratory of Chemical Resource Engineering
Laboratory of Electrochemical Process and Technology for materials
15 Beisanhuan East Road
100029 Beijing
China

Prof. Yaqin Huang
Beijing University of Chemical Technology
Key Laboratory of Chemical Resource Engineering
Laboratory of Electrochemical Process and Technology for materials
15 Beisanhuan East Road
100029 Beijing
China

Associate Professor Jin Niu
Beijing University of Chemical Technology
Key Laboratory of Chemical Resource Engineering
Laboratory of Electrochemical Process and Technology for materials
15 Beisanhuan East Road
100029 Beijing
China

Cover Image: © halepak/Getty images

All books published by **WILEY-VCH** are carefully produced. Nevertheless, authors, editors, and publisher do not warrant the information contained in these books, including this book, to be free of errors. Readers are advised to keep in mind that statements, data, illustrations, procedural details or other items may inadvertently be inaccurate.

Library of Congress Card No.: applied for

British Library Cataloguing-in-Publication Data
A catalogue record for this book is available from the British Library.

Bibliographic information published by the Deutsche Nationalbibliothek
The Deutsche Nationalbibliothek lists this publication in the Deutsche Nationalbibliografie; detailed bibliographic data are available on the Internet at <http://dnb.d-nb.de>.

© 2022 WILEY-VCH GmbH, Boschstr. 12, 69469 Weinheim, Germany

All rights reserved (including those of translation into other languages). No part of this book may be reproduced in any form – by photoprinting, microfilm, or any other means – nor transmitted or translated into a machine language without written permission from the publishers. Registered names, trademarks, etc. used in this book, even when not specifically marked as such, are not to be considered unprotected by law.

Print ISBN: 978-3-527-34962-3
ePDF ISBN: 978-3-527-83450-1
ePub ISBN: 978-3-527-83451-8
oBook ISBN: 978-3-527-83452-5

Typesetting Straive, Chennai, India
Printing and Binding CPI Group (UK) Ltd, Croydon, CR0 4YY

Printed on acid-free paper

C116599_030222

Contents

Preface

Global energy demand and environmental degradation have resulted in a severe crisis in the sustainable development of the international community. In this context, it is necessary to develop high-efficiency electrochemical energy storage and conversion devices to address the challenges. The assembly of the devices needs materials (especially electrode materials) that should not only show high performance but also have sustainability and low cost.

As one kind of natural biomass, collagen exists in the connective tissues of animals, comprising 23 kinds of amino acids. As the hydrolysate of collagen, gelatin has unique water solubility, rheological property, and gel property. By virtue of rich carbon atoms and non-carbon atoms (e.g. nitrogen, oxygen, and sulfur), collagen and gelatin can be converted into a new kind of carbon material. The collagen-derived carbons have heteroatom doping and unique morphologies, which are derived from the natural structural compositions of the collagen-enriched precursors. By virtue of their structural compositions and unique properties, the collagen derived materials (including gelatin and collagen-derived carbons) show promising applications in electrochemical energy storage and conversion.

Our group has been devoted to collagen and collagen-derived materials since 1997. We started from the gelatin extraction from collagen and gradually expanded the applications of collagen-derived materials. So far, gelatin with and without modification has been used as different modifying agents for electrodes, separators, and electrolytes. Collagen-derived carbons with different morphology structures and chemical compositions have also been used as electrodes and support materials for electrochemical energy storage and conversion. Based on the related works, we have published over 100 peer-reviewed journal papers and authored more than 40 patents.

This book is based on our published works and also includes typical works by other researchers. It is divided into ten chapters, and written by Prof. Feng Wang, Prof. Yaqin Huang, and Prof. Jin Niu. This book covers basic knowledge of collagen and collagen-derived materials, and their typical synthesis and applications. Moreover, it also presents the up-to-date ideas and strategies for the synthesis and characterization of advanced collagen-derived materials, as well as multifunctional applications in electrochemical energy storage and conversion. Therefore, this book can not only enable students and new researchers to obtain easy understanding of these materials and their applications but also serve as an important reference book for scientists and engineers in other research fields.

I would like to thank the other group members for their assistance during the book writing, including Mengyue Liu, Yongzheng Shi, Feng Zhu, Junhui Sun, Binke Li, Na Yang, Rong Shao, Jiaxing Liu, and Weihao Song. I would also like to thank my current and former group members, as well as collaborators who have contributed to the study on collagen-derived materials. Finally, we would like to acknowledge the financial support from the National Natural Science Foundation of China (52130206, U20A20337, 51432003, 22178016, and 52003025) and Fundamental Research Funds for the Central Universities (buctrc202024).

Beijing, China *Feng Wang*
10 August 2021

1

Introduction

1.1 Electrochemical Energy Storage and Conversion

Electric energy plays a very important role in our daily life and industrial production. Fossil fuel, nuclear heat, and renewable energies (e.g. solar energy, wind energy, and biomass energy) all can be converted into electric energy [1]. Unfortunately, the energy-conversion processes are always accompanied by a lot of energy loss. For instance, the conversion efficiency of nuclear heat to electric energy is only ~30%. Moreover, the electric energy derived from renewable energies depends highly on weather, season, and territory, which cannot meet the actual demand in time. Thus, it is an urgent need to address the issue of electric energy storage and conversion.

Developing advanced energy storage and conversion technologies is critical to improving the energy utilization efficiency and expanding the energy application field. Secondary batteries, supercapacitors, water electrolyzers, and fuel cells are some of the typical electrochemical energy storage and conversion devices. Figure 1.1 shows the schematics of these electrochemical energy storage and conversion systems [2]. Electric energy can be converted into chemical energy by water electrolyzers, producing hydrogen (with a conversion efficiency of ~70%) for further use in fuel cells. In the opposite process, fuel cells convert chemical energy into electric energy. The energy-conversion process is reversible in secondary batteries (e.g. lithium-ion battery). Electric energy can be converted into chemical energy in the charging process [3]. In the discharging process, the chemical energy is converted back into electric energy. The conversion speed determines the system power, while the storage capacity is relevant to the system energy. Generally, the active materials for energy conversion and storage are integrated into secondary batteries because of the internal system. Different from secondary batteries, the electrolyzer and fuel cell systems are suitable for separated converters and storage. This kind of electrochemical storage and conversion systems commonly supplies higher energy than that with integrated storage and converters. Therefore, electrolyzers and fuel cells have also attracted much attention [4]. Herein, a brief overview of typical secondary batteries, supercapacitors, fuel cells, and water electrolyzers is presented.

Collagen-Derived Materials: Synthesis and Applications in Electrochemical Energy Storage and Conversion, First Edition. Feng Wang, Yaqin Huang, and Jin Niu.
© 2022 WILEY-VCH GmbH. Published 2022 by WILEY-VCH GmbH.

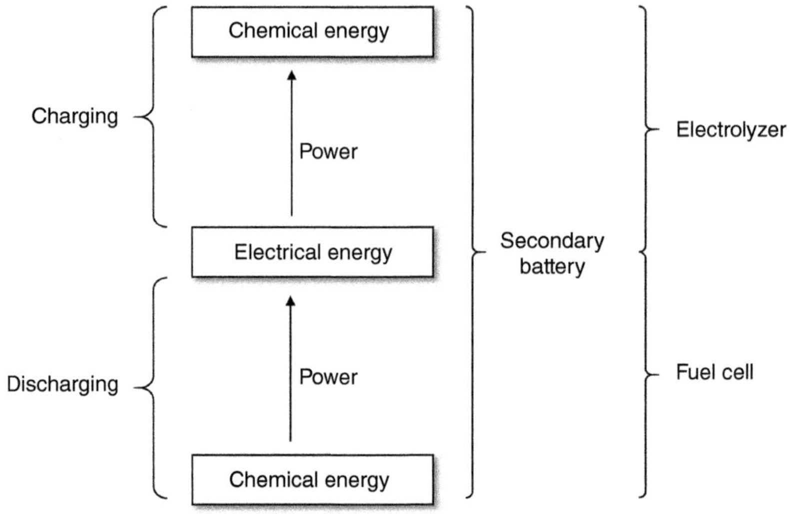

Figure 1.1 Schematics of typical systems for electrochemical energy storage and conversion. Source: Modified from Shen et al. [2]; © 2016, Taylor & Francis Group.

1.1.1 Secondary Batteries and Supercapacitors

Currently, secondary batteries and supercapacitors are regarded as the most efficient electrochemical systems for electric energy storage and conversion. As a typical secondary battery, lithium-ion battery has been widely studied and commercialized. The working mechanism of lithium-ion battery is shown in Figure 1.2. It involves a reversible intercalation/deintercalation of Li ions into/from host materials, combined with electron flow through the external circuit in the charging/discharging process [5]. Taking a commercial lithium-ion battery as an example, graphite and layered $LiCoO_2$ are usually employed as anode host and cathode host, respectively. A mixture of lithium-containing salt (e.g. $LiPF_6$) and organic solvents (e.g. diethyl carbonate and ethylene carbonate) is commonly used as an electrolyte. The chemical reactions during the charging and discharging process are shown as below:

$$LiCoO_2 + C_6 \xrightarrow{\text{Charging}} Li_{1-x}CoO_2 + Li_xC_6 \quad (1.1)$$

$$LiCoO_2 + C_6 \xleftarrow{\text{Discharging}} Li_{1-x}CoO_2 + Li_xC_6 \quad (1.2)$$

As a "rocking chair" battery, the energy storage and release of the lithium-ion battery primarily depend on the "rock" of Li ions between cathode and anode. As compared with Ni-based battery and lead–acid battery, the lithium-ion battery possesses characteristics of higher energy density and longer lifetime. Moreover, the lithium-ion battery is easier to be integrated into portable devices.

Based on the energy storage mechanisms, supercapacitors are primarily classified into electric double-layer capacitors (EDLCs) and pseudocapacitors (Figure 1.3) [7]. No matter what kind of mechanism, the charge storage occurs only on the electrode surface without ion diffusion within the electrode bulk, which endows

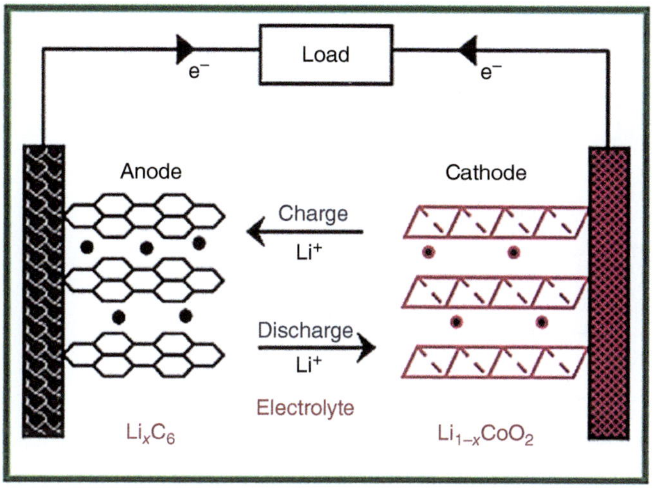

Figure 1.2 Illustration of the charging/discharging process within a lithium-ion battery, composed of a graphite anode and a $LiCoO_2$ cathode. Source: Reproduced with permission from Manthiram et al. [5]; © 2008, The Royal Society of Chemistry.

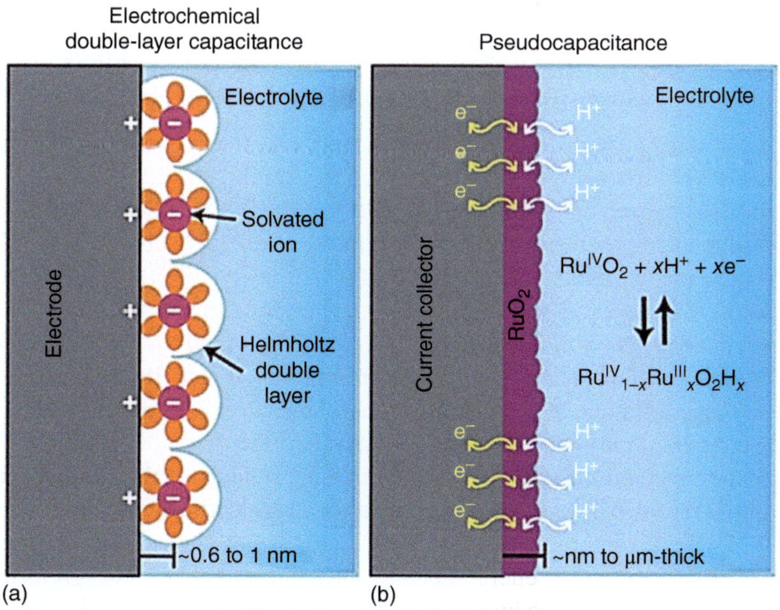

Figure 1.3 Schematic of energy storage mechanisms of (a) an EDLC and (b) a pseudocapacitor. Source: Reproduced with permission from Simon et al. [6]; © 2011, Materials Research Society.

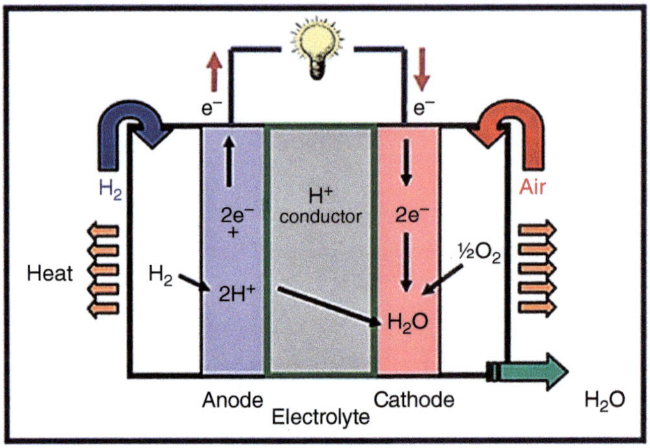

Figure 1.4 Schematic of a PEMFC. Source: Reproduced with permission from Manthiram et al. [5]; © 2008, The Royal Society of Chemistry.

supercapacitors with a higher power density (~10 kW kg^{-1}) but lower energy density (<10 Wh kg^{-1}) than lithium-ion battery [6]. The EDLC stores energy through the fast adsorption and desorption of electrolyte ions at the electrode/electrolyte interface, which forms the electric double layer during the charging/discharging process. The pseudocapacitor reversibly stores energy via Faradaic reactions or surface redox reactions, which also occur at the electrode surface [8].

1.1.2 Fuel Cells and Electrolyzers

Fuel cells are promising energy-conversion devices because they can show high conversion efficiency besides being environment friendly. Among the different kinds of fuel cells, the proton exchange membrane fuel cell (PEMFC) is widely studied by virtue of its high energy delivery and low working temperature (<100 °C) [9, 10]. Figure 1.4 shows the working process of a PEMFC that uses oxygen and hydrogen as oxidant and fuel, respectively [5]. H^+ ions are generated through the electrocatalytic oxidation of H_2 gas on the surface of an anode with Pt catalyst, which further migrates through a Nafion membrane with high proton conductivity. Meanwhile, the electrons move from anode to cathode through an external circuit and electrocatalytically reduce O_2 gas to O^{2-} ions on the cathode surface with the assistance of Pt catalyst. Water is produced by the combination of the H^+ and O^{2-} ions. Thus, the chemical energy involved in reaction 1.4 is converted into electricity:

$$2H_2 + O_2 \rightarrow 2H_2O \tag{1.3}$$

Contrary to fuel cells, electrolyzers convert electric energy to chemical energy. Figure 1.5 displays a schematic representation of a water electrolyzer, which mainly consists of a cathode, an anode, an electrolyte, and an external power. At a critical voltage, electrons flow from the power source to the cathode and reduce H^+ ions to H_2 gas. At the same time, OH^- ions move to the anode through the electrolyte to

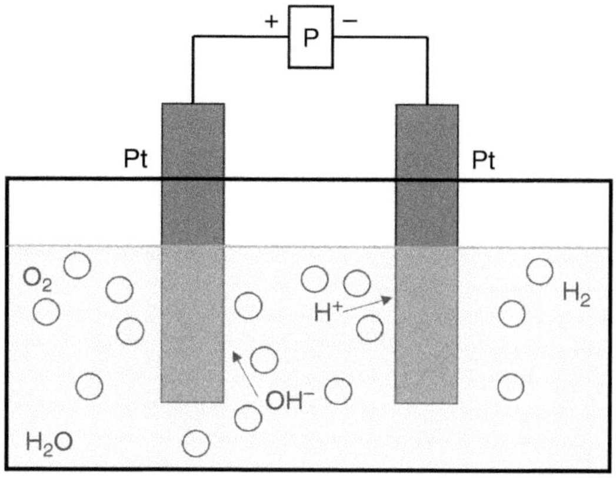

Figure 1.5 Schematic of a water electrolyzer. Source: Reproduced with permission from Ma et al. [11]; © 2008, Elsevier Ltd.

keep an electrical charge balance. O_2 gas is then generated by oxidization of OH^- ions by giving away electrons. Then, the electrons flow back to the power source. The half-reactions occurring on the electrodes are represented below:

$$\text{Overall: } H_2O \rightarrow H_2 + \frac{1}{2}O_2 \tag{1.4}$$

1.2 Materials for Electrochemical Energy Storage and Conversion

Materials (especially electrode materials) play very important roles in the performance of the aforementioned devices. Great efforts have been devoted to developing novel materials to improve the device's performance. Herein, we review typical materials for lithium-ion batteries, supercapacitors, and electrocatalysis (especially for oxygen reduction and evolution reactions).

1.2.1 Materials for Lithium-Ion Batteries

Based on the lithium-ion storage mechanism, the anode materials for the lithium-ion battery can be classified into intercalated-type materials (e.g. graphite, carbon nanotubes, and $Li_4Ti_5O_{12}$), conversion-type materials (e.g. SnO_2, V_2O_5, and MoS_2), alloy-type materials (e.g. Si and Sn), and organic materials (e.g. carboxylates) [12] (Figure 1.6). To date, graphite is the most used anode material owing to its low cost, good thermal stability, and low working potential. Although conversion-type and alloy-type anode materials have a high theoretical capacity, they commonly have poor electrical conductivity and show large volume change during the lithiation process, which results in low coulombic efficiency and short lifetime in practical cells. Layered transition metal oxides are generally used as cathode materials for lithium-ion battery, such as $LiFePO_4$, $LiCoO_2$, $LiMn_2O_4$, and $Li[Ni_xCo_yMn_z]O_2$ [13]. The electrolytes for lithium-ion battery are mainly composed of lithium-containing salt (e.g. $LiPF_6$) and organic carbonate solvents

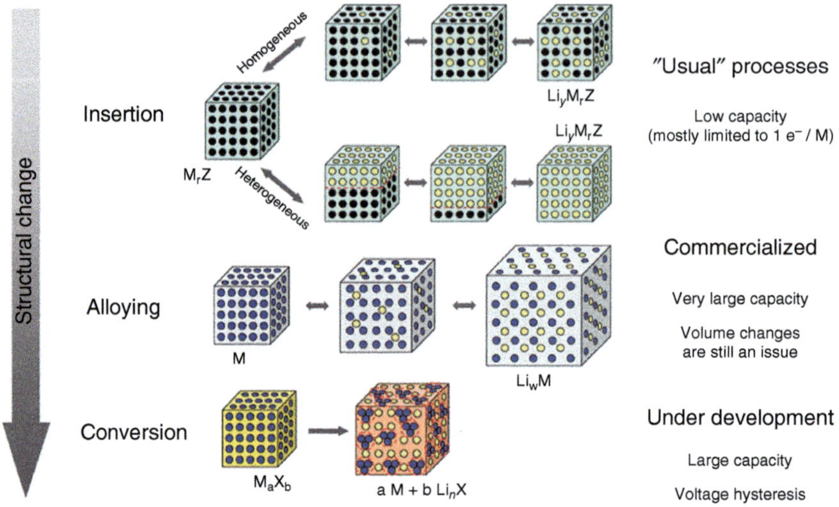

Figure 1.6 Different types of anode materials for lithium-ion batteries. Source: Reproduced with permission from Long et al. [12]; © 2009, The Royal Society of Chemistry.

(e.g. ethylene carbonate), which are used for the balance of fluidity and dielectric constant. As separator, electronically insulated polymer film (e.g. polypropylene) with high porosity is widely used to avoid the direct contact between cathode and anode as well as ensure fast lithium-ion transport.

1.2.2 Materials for Supercapacitors

Carbon materials (e.g. activated carbons, graphene, carbon nanotubes, and porous carbons) have been employed as electrode materials for EDLC owing to their easy processing, large specific surface area, high electronic conductivity, and good chemical stability. Generally, the capacitance of EDLCs can be enlarged when the specific surface area of carbon electrode material is increased. In addition to the high specific surface area, suitable porosity for accessible ions (e.g. pore shape, pore size, and pore size distribution) also significantly influences the capacitance. For instance, the hierarchical porous structure can facilitate ion insertion and diffusion, thus endowing carbon electrodes with high capacitance and good rate capability. The electrode materials for pseudocapacitor contribute capacitance by reversible and rapid redox reactions, occurring on the electrode surface. Metal oxides (e.g. RuO_2), metal carbides (e.g. MXene), electronically conducting polymers (e.g. polyaniline), and functionalized nanocarbons (e.g. heteroatom-doped graphene) are widely used as electrode materials for pseudocapacitor [6]. In addition, some biomaterials with functional groups (e.g. protein) are also electrode materials that can deliver pseudocapacitance [14]. The electrolytes for supercapacitors include aqueous and nonaqueous (including organic and ionic liquid) systems [15]. The two standards for choosing electrolytes are stable electrochemical window and high ionic conductivity, which determine the energy and power densities of the supercapacitors, respectively. In comparison with organic electrolytes (with wide electrochemical windows ranging from 2.5 to 3.5 V), alkali- (e.g. KOH) and acid-based (e.g. H_2SO_4) aqueous electrolytes provide a higher ionic conductivity (up to ~1 S cm^{-1}) but a lower electrochemical window (1.0–1.8 V). Ionic liquid electrolytes show the widest electrochemical window (3.0–5.5 V) but the lowest ionic conductivity among these three electrolytes.

1.2.3 Materials for Oxygen Electrochemistry

The performance of water electrolyzers and fuel cells is primarily restricted by oxygen reduction reactions (ORRs) and oxygen evolution reactions (OERs) performance [16]. Although oxygen is a kind of clean and abundant resource, its efficient and controllable use is difficult because the sluggish kinetic processes of ORR and OER lead to the low device efficiency in acidic and alkaline electrolytes [17]. The electrocatalysts for ORR and OER can be divided into precious metallic catalysts, non-noble metallic catalysts, and carbon-based catalysts (Figure 1.7). Pt catalyst has been used as an ORR electrocatalyst since the 1960s, and it is still widely studied because it not only shows a superior ORR performance but also provides a relatively simple system for investigating the ORR mechanism [19]. Pt-based alloy materials show more excellent ORR activity and stability than a pure Pt electrocatalyst. The selected secondary metals and the alloy architectures are of great importance to the performance of Pt-based alloy electrocatalysts. Non-noble metals (e.g. Ni-, Fe-, Co-, and Mn-based oxides) are the most promising candidates to Pt because of their better electrocatalysis activities in alkaline solutions [20, 21]. By virtue of low cost and high electric conductivity, carbon materials with heteroatom doping (e.g. nitrogen, oxygen, sulfur, phosphor, boron) have attracted extensive attention as electrocatalysts for ORR and OER. However, precious metallic catalysts and non-noble metallic catalysts generally show better activity and practical performance than carbon materials.

1.2.4 Biomass-Derived Materials for Electrochemical Energy Storage and Conversion

It is well established that the ever-increasing global energy demand and environmental degradation have resulted in a severe crisis in the sustainable development of the international community [22]. In this context, it is imperative to develop high-performance energy storage and conversion devices to address the challenges. The assembly of the devices needs materials (e.g. electrode materials) that should not only show high performance but also have sustainability and low cost. Biomass-derived materials are such kind of materials for the energy storage and conversion devices because of their excellent characteristics, which include: (i) As the most abundant carbon source in nature, biomass is derived from all life species (including animals, plants, and microorganisms); (ii) Biomass possesses inherently unique microarchitecture, which is beneficial for the design of microstructured materials (especially electrode materials). (iii) Some biomass also contains non-carbon elements (e.g. nitrogen, oxygen, sulfur, and phosphor). These

Reduced cost and enhanced stability

Category	Pure Pt	Pt-based alloys	Non-Pt compounds	Metal-free materials
Materials	Pt/C	$Fe_{50}Pt_{50}$, CoPt, Au-Pt, Pt_3Ni, Pd-Pt, etc.	$Fe\text{-}N_4$, $Co\text{-}N_4$, Co-PPY, etc.	N-Carbon, $G\text{-}C_3N_4$, N-Graphene
Featured examples	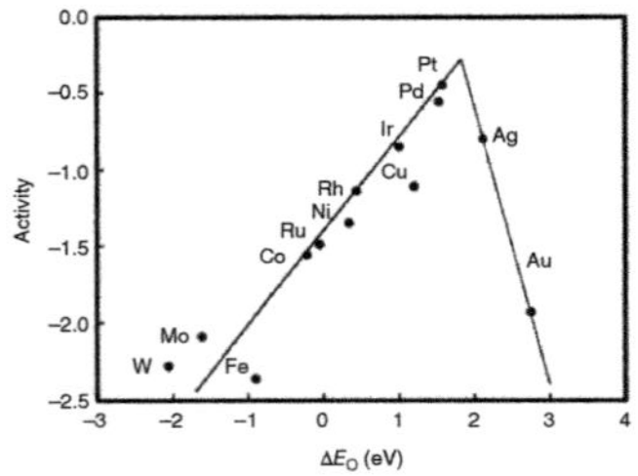	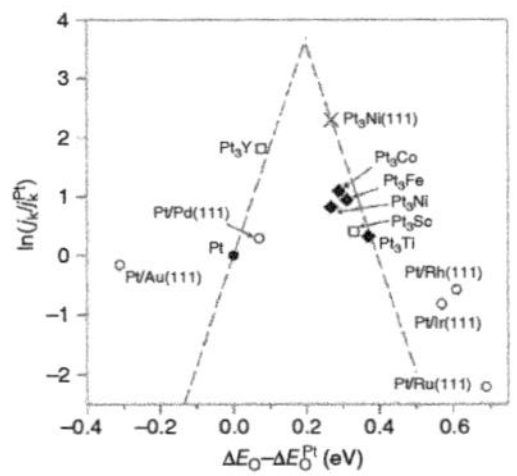	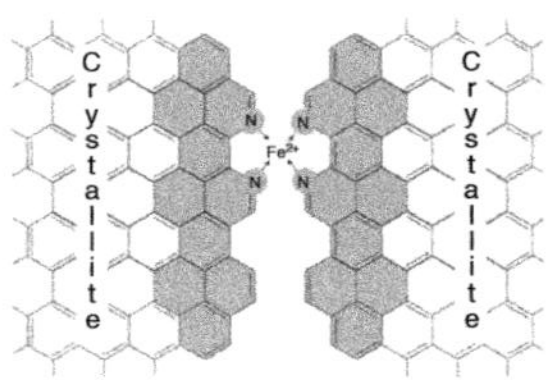	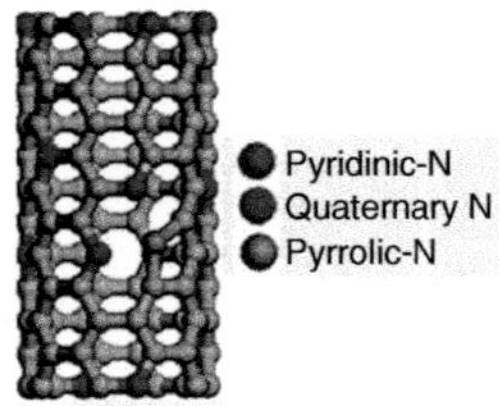

Figure 1.7 Progress in the development of ORR electrochemical catalysts. Source: Reproduced with permission from Chen et al. [18]; © 2012, John Wiley & Sons.

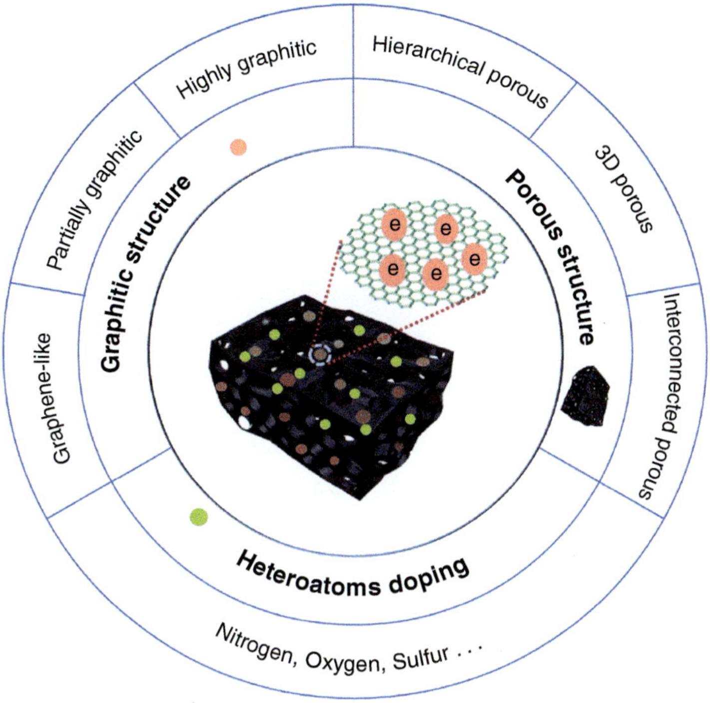

Figure 1.8 Schematic diagram showing the structure and properties of biomass-derived materials. Source: Reproduced with permission from Zheng et al. [23]; © 2020, The Royal Society of Chemistry.

heteroatoms can be in situ doped into biomass-derived carbons during the carbonization process, introducing additional active sites to enhance the performance of biomass-derived carbon electrodes. Therefore, biomass-derived materials have been widely prepared and used for electrochemical energy storage and conversion devices in recent years (Figure 1.8). Table 1.1 shows the properties and applications of the biomass-derived materials for different devices [3].

1.3 Collagen-Derived Materials for Electrochemical Energy Storage and Conversion

As one kind of natural fibrous protein, collagen exists in the connective tissue of animals. It is composed of 23 kinds of amino acids (mainly including glycine, proline, and alanine). Collagen consists of three polypeptide chains that assemble in a triple helix by hydrogen and hydrophobic bonds [24]. The polypeptide chains contain two identical chains (α1) and an additional chain (α2), which have slightly different chemical compositions. Figure 1.9 illustrates the hierarchical structure of a collagen triple helix. A single-collagen fiber is called tropocollagen, which has a diameter of ~1.6 nm and a length of ~300 nm. So far, 28 different types of collagen have been

Table 1.1 Biomass-derived materials for electrochemical energy storage and conversion [3].

Electrochemical energy storage and conversion devices	Roles of biomass-derived materials	Biomass-derived materials	Desired features of the biomass-derived materials
Supercapacitor	Electrode	Porous/activated carbon	Porous, conducting, high specific surface area, double-layer capacitance, functional groups, heteroatom doping, engineered pore structure
Lithium-ion battery	Anode electrode	Hard/soft carbon	Micrographite domain for Li/Na intercalation, conducting, porosity
PEMFC	Electrocatalyst or catalyst support	Porous/activated carbon	Conducting, high specific surface area, heteroatom doping catalytic property for ORR
PEMFC	Gas diffusion layer	Porous/activated carbon, carbon cloth/felt	Conducting, high porosity, hydrophobic
Lithium-ion battery; Supercapacitor	Electrode binder	Cellulose, chitosan, other polysaccharides	Polymer chain and binding ability, chemical stability, functional groups
Supercapacitor	Electrolyte	Organic acids and salts	Ion conductivity, stability over a reasonably wide potential window
Lithium-ion battery; supercapacitor	Separator	Cellulose, fiber	Electron insulator, thin film, wettability, flexibility, mechanical strength, chemical stability

Source: Yu et al. [3]/John Wiley & Sons.

discovered, which are divided into fibrillar and non-fibrillar collagen based on their structures [26].

As a soluble protein compound, gelatin is produced by partially hydrolyzing collagen within animals. During the hydrolysis process, three polypeptide chains are disentangled and the collagen molecules are dissociated into smaller molecules, after breaking the hydrogen bonds and hydrophobic bonds [27]. Thus, the structure, composition, and properties of gelatin are dependent on the collagen sources and types. It contains 18 kinds of amino acids. Different from collagen, gelatin consists of a relatively high content of nonpolar amino acids (>80%, such as glycine, proline,

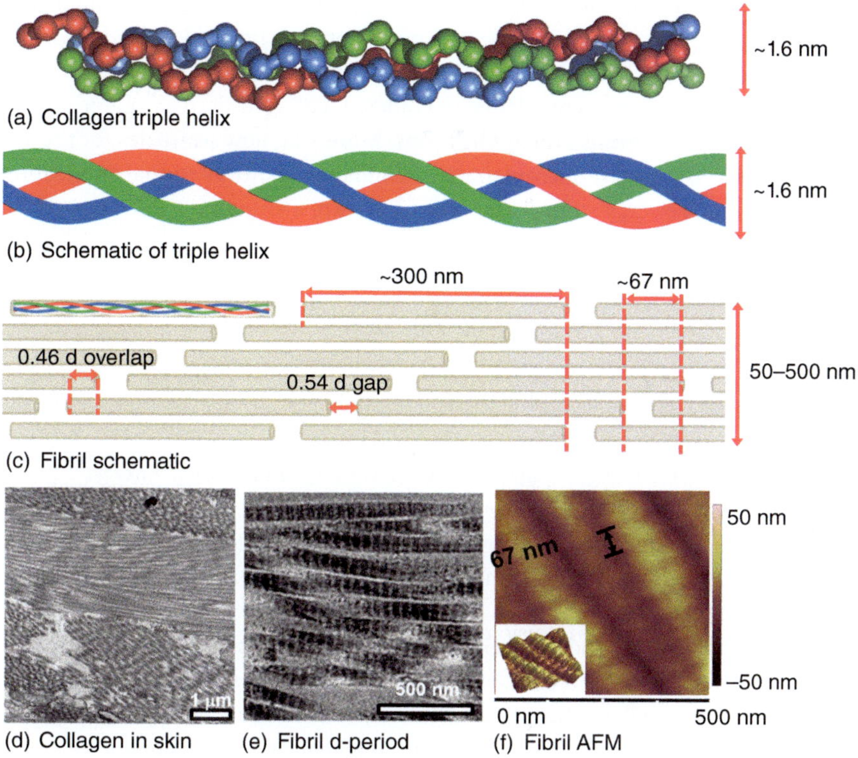

Figure 1.9 Overview of the hierarchical structure for collagen triple helix. (a, b) Schematic illustration of a collagen triple helix. (c) Schematic illustration of collagen fibrils. (d) Cross-section image of collagen fibrils in skin. (e) Transmission electron microscopy image of collagen fibrils. (f) Atomic force microscope image of collagen fibrils. Source: Sherman et al. [25]/with permission of © Elsevier Ltd.

and alanine). In addition to essential physicochemical characteristics (e.g. solubility, transparency, and color), gelatin has unique rheological property, gel strength, and thermal stability, which are mainly dependent on its molecular structures [28]. Since the polypeptide chains are composed of various amino acids with rich carbon, nitrogen, oxygen, and sulfur atoms, collagen and gelatin can be converted into carbon materials by controllable carbonization. The collagen-derived carbons have heteroatom doping and unique morphologies, which are derived from the structural compositions of collagen.

By virtue of their structures and compositions, the collagen-derived materials (including gelatin and collagen-derived carbons) show promising applications in electrochemical energy storage and conversion. Gelatin materials with and without modification have been used as various modifying agents for electrodes [29], separators [30], and electrolytes [31]. Collagen-derived carbons with different morphology structures and chemical compositions have also been prepared by different methods. They have been used as electrodes or support materials for electrochemical energy storage (e.g. supercapacitors [32], alkali-ion batteries [33], and lithium–sulfur batteries [34]) and electrochemical energy conversion (e.g. oxygen reduction reactions [35], oxygen evolution reactions [36], and hydrogen evolution reactions [37]), showing good performance and practical prospect.

This book presents basic knowledge of collagen, collagen-derived materials, and their conventional synthesis and applications. Moreover, new strategies for the synthesis of advanced collagen-derived materials and their multifunctional applications in electrochemical energy storage and conversion are also summarized.

References

1 Yoo, H.D., Markevich, E., Salitra, G. et al. (2014). On the challenge of developing advanced technologies for electrochemical energy storage and conversion. *Materials Today* 17: 110–121.

2 Shen, J., Fang, D., Hong, Z., and Mei, C. (2015). *Electrochemical Energy: Advanced Materials and Technologies*. Florida: CRC Press.

3 Yu, F., Li, S., Chen, W. et al. (2019). Biomass-derived materials for electrochemical energy storage and conversion: overview and perspectives. *Energy & Environmental Materials* 2: 55–67.

4 Kalyanasundaram, K. and Grätzel, M. (2012). Themed issue: nanomaterials for energy conversion and storage. *Journal of Materials Chemistry* 22: 24190–24194.

5 Manthiram, A., Vadivel Murugan, A., Sarkar, A., and Muraliganth, T. (2008). Nanostructured electrode materials for electrochemical energy storage and conversion. *Energy & Environmental Science* 1: 621–638.

6 Simon, P. and Gogotsi, Y. (2008). Materials for electrochemical capacitors. *Nature Materials* 7: 845–854.

7 Long, J.W., Bélanger, D., Brousse, T. et al. (2011). Asymmetric electrochemical capacitors-stretching the limits of aqueous electrolytes. *MRS Bulletin* 36: 513–522.

8 Conway, B.E. (1999). *Electrochemical Supercapacitors*. New York: Plenum Press.

9 Appleby, A.J. (1988). *Fuel cell handbook*. New York: Van Nostrand Reinhold Co. Inc.

10 Acres, G.J.K. (2001). Recent advances in fuel cell technology and its applications. *Journal of Power Sources* 100: 60–66.

11 Ma, L., Sui, S., and Zhai, Y. (2009). Investigations on high performance proton exchange membrane water electrolyzer. *International Journal of Hydrogen Energy* 34: 678–684.

12 Palacín, M.R. (2009). Recent advances in rechargeable battery materials: a chemist's perspective. *Chemical Society Reviews* 38: 2565–2575.

13 Whittingham, M.S. (2004). Lithium batteries and cathode materials. *Chemical Reviews* 104: 4271–4302.

14 Khairy, M. and El-Safty, S.A. (2015). Promising supercapacitor electrodes based immobilization of proteins onto macroporous Ni foam materials. *Journal of Energy Chemistry* 24: 31–38.

15 Béguin, F., Presser, V., Balducci, A., and Frackowiak, E. (2014). Carbons and electrolytes for advanced supercapacitors. *Advanced Materials* 26: 2219–2251.

16 Jiao, Y., Zheng, Y., Jaroniec, M., and Qiao, S.Z. (2015). Design of electrocatalysts for oxygen- and hydrogen-involving energy conversion reactions. *Chemical Society Reviews* 44: 2060–2086.

17 Katsounaros, I., Cherevko, S., Zeradjanin, A.R., and Mayrhofer, K.J.J. (2014). Oxygen electrochemistry as a cornerstone for sustainable energy conversion. *Angewandte Chemie International Edition* 53: 102–121.

18 Chen, Q., Tan, X., Liu, Y. et al. (2020). Biomass-derived porous graphitic carbon materials for energy and environmental applications. *Journal of Materials Chemistry A* 8: 5773–5811.

19 Gómez-Marín, A.M., Rizo, R., and Feliu, J.M. (2014). Oxygen reduction reaction at Pt single crystals: a critical overview. *Catalysis Science & Technology* 4: 1685–1698.

20 Frydendal, R., Paoli, E.A., Knudsen, B.P. et al. (2014). Benchmarking the stability of oxygen evolution reaction catalysts: the importance of monitoring mass losses. *ChemElectroChem* 1: 2075–2081.

21 Gorlin, Y. and Jaramillo, T.F. (2010). A bifunctional nonprecious metal catalyst for oxygen reduction and water oxidation. *Journal of the American Chemical Society* 132: 13612–13614.

22 Larcher, D. and Tarascon, J.M. (2015). Towards greener and more sustainable batteries for electric energy storage. *Nature Chemistry* 7: 19–29.

23 Zheng, Y., Jiao, Y., Jaroniec, M. et al. (2012). Nanostructured metal-free electrochemical catalysts for highly efficient oxygen reduction. *Small* 8: 3550–3566.

24 Rich, A. and Crick, F.H.C. (1955). The structure of collagen. *Nature* 176: 915–916.

25 Sherman, V.R., Yang, W., and Meyers, M.A. (2015). The materials science of collagen. *Journal of the Mechanical Behavior of Biomedical Materials* 52: 22–50.

26 Shamhart, P.E. and Meszaros, J.G. (2010). Non-fibrillar collagens: key mediators of post-infarction cardiac remodeling? *Journal of Molecular and Cellular Cardiology* 48: 530–537.

27 Alipal, J., Mohd Pu'ad, N.A.S., Lee, T.C. et al. (2021). A review of gelatin: properties, sources, process, applications, and commercialisation. *Materials Today: Proceedings* 42: 240–250.

28 Kozlov, P.V. and Burdygina, G.I. (1983). The structure and properties of solid gelatin and the principles of their modification. *Polymer* 24: 651–666.

29 Chen, M., Chen, Z., Fu, X., and Zhong, W.-H. (2020). A Janus protein-based nanofabric for trapping polysulfides and stabilizing lithium metal in lithium–sulfur batteries. *Journal of Materials Chemistry A* 8: 7377–7389.

30 Liu, N., Huang, B., Wang, W. et al. (2016). Modified separator using thin carbon layer obtained from its cathode for advanced lithium sulfur batteries. *ACS Applied Materials & Interfaces* 8: 16101–16107.

31 Liu, Y., Shi, Q., Wu, Y. et al. (2021). Highly efficient dendrite suppressor and corrosion inhibitor based on gelatin/Mn^{2+} co-additives for aqueous rechargeable zinc-manganese dioxide battery. *Chemical Engineering Journal* 407, 127189.

32 Chen, W., Zhang, H., Huang, Y., and Wang, W. (2010). A fish scale based hierarchical lamellar porous carbon material obtained using a natural template for high performance electrochemical capacitors. *Journal of Materials Chemistry* 20: 4773–4775.

33 Niu, J., Shao, R., Liang, J. et al. (2017). Biomass-derived mesopore-dominant porous carbons with large specific surface area and high defect density as high performance electrode materials for Li-ion batteries and supercapacitors. *Nano Energy* 36: 322–330.

34 Wei, S., Zhang, H., Huang, Y. et al. (2011). Pig bone derived hierarchical porous carbon and its enhanced cycling performance of lithium–sulfur batteries. *Energy & Environmental Science* 4: 736–740.

35 Liu, H., Cao, Y., Wang, F. et al. (2014). Pig bone derived hierarchical porous carbon-supported platinum nanoparticles with superior electrocatalytic activity towards oxygen reduction reaction. *Electroanalysis* 26: 1831–1839.

36 Guan, J., Zhang, Z., Ji, J. et al. (2017). Hydrothermal synthesis of highly dispersed Co_3O_4 nanoparticles on biomass-derived nitrogen-doped hierarchically porous carbon networks as an efficient bifunctional electrocatalyst for oxygen reduction and evolution reactions. *ACS Applied Materials & Interfaces* 9: 30662–30669.

37 Li, T., Liu, J., Song, Y., and Wang, F. (2018). Photochemical solid-phase synthesis of platinum single atoms on nitrogen-doped carbon with high loading as bifunctional catalysts for hydrogen evolution and oxygen reduction reactions. *ACS Catalysis* 8: 8450–8458.

2

Collagen

2.1 History of Collagen

The word "collagen" comes from Greek, which means "glue producer." Collagen can be gradually hydrolyzed into water-soluble gelatin (also referred to as animal glue) when it is treated with hot water. As the richest protein in animals, collagen is the primary composition for the connective tissues and extracellular matrices. By virtue of its unique structure, collagen can form insoluble fibers that endow animal tissues with high strength [1].

In the 1950s, modern studies on collagen started, especially at a molecular level. As typical researchers and works, Highberger et al. [2] and Schmitt et al. [3] used an electron microscope to characterize the structure of collagen molecule. Then, Ramachandran and Kartha [4] and Rich and Crick [5] developed the triple-helical structure models of the collagen. Finally, Boedtker and Doty [6] measured the physical and chemical properties of the collagen in solution [7].

In the 1960s, detailed information about collagen and its structure was further revealed by Miller et al. through the investigation of protein chemistry and biochemistry [8, 9]. Based on the electron microscope results, it was found that band-shape-like collagen fibrils with a length of 70 nm are composed of three different but homologous collagen types, including types I, II, and III. Type I collagen is the most abundant among them; type II is the main collagenous component within cartilage; type III coexists with type I with different content in most tissues. After the discovery of type II and III collagen, Kefalides reported that a specific collagen is also present within basement membranes, which was categorized as the type IV collagen [10]. Then, Kuhn et al. [11] and Timpl et al. [12] showed that the structure of the type IV collagen is obviously different from that of the collagen, which can form fibers.

In addition to the above collagens, the type VI, VII, IX, and X collagens were also reported subsequently. Their structures are all different from each other [13]. The detailed structural information of the collagens was gradually provided by different approaches, including electron microscopy, biosynthesis, immunology, and molecular biology methods [7]. Generally, all collagens have a triple-helical domain. The formation of the triple-helical structure is different from that of globular and nonhelical structures. The collagen molecules are synthesized intracellularly and then secreted

Collagen-Derived Materials: Synthesis and Applications in Electrochemical Energy Storage and Conversion, First Edition. Feng Wang, Yaqin Huang, and Jin Niu.
© 2022 WILEY-VCH GmbH. Published 2022 by WILEY-VCH GmbH.

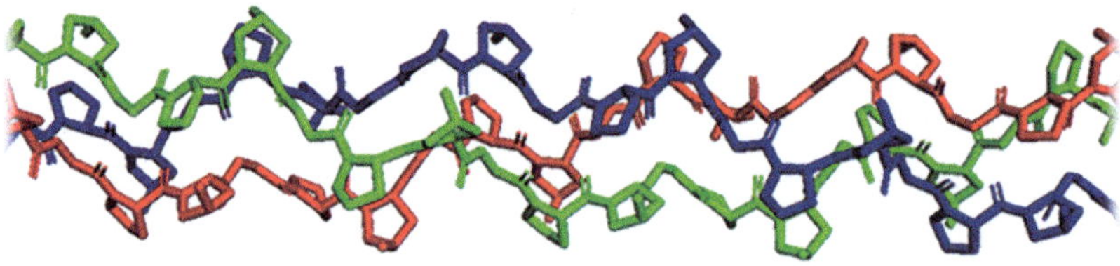

Figure 2.1 Collagen triple-helix structure. Source: Reproduced with permission from Chmielewski [19]; © 2020, John Wiley and Sons.

into the extracellular space. After aggregation with other molecules, supramolecular structures are formed and stabilized by intermolecular covalent bonds [7].

2.2 Structure and Composition of Collagen

Collagens are the major components of connective tissues. They also exist in the interstitial tissue of parenchymal organs. The collagen ensures the stability and structural integrity of the tissues and organs. The role played by collagen in a specific organ determines its type and organization [14]. The collagen-enriched tissues include cartilage, tendons, cornea, and the organic component of bone. In tendons, collagen forms parallel fibers that support high strength. In the cornea, collagen is stacked to ensure the least scattering of transmitting light. In extracellular matrix, collagen exists in the form of a gel that maintains the tissue structure. In bone, collagen forms a framework with the hydroxyapatites, which show high mechanical performance as a composite material [15].

In terms of composition, collagen has a repeated motif of Gly–X–Y. Gly is glycine while X and Y are other amino acids. The positions of X and Y are mostly occupied by proline and 4-hydroxyproline, respectively [16, 17]. Although collagens with different types have different compositions, all of them have the same characteristics, such as a right-handed triple-helix structure containing three polypeptide chains [18]. As shown in Figure 2.1, the collagen molecule has a long and rigid structure composed of three α chains that intertwine with each other to form a triple helix [19]. The collagen molecule shows a length of ~300 nm and a diameter of 1.5 nm. The collagen fibers have high tensile strength because of the tight coupling α chains, which is higher than that of a steel wire.

Based on the genetic type, the triple-helical motif can further form a major or minor part of the collagen molecule. In addition to collagenous domains, different noncollagenous parts also exist in the other regions of the collagen [20]. Specifically, three left-handed helices form a supercoiled right-handed structure in a rope-like shape [21]. It should be noted that the three chains are also staggered by one residue relative to one another [22]. On the whole, the collagens are composed of continuous triple helix and short-helical telopeptides. The telopeptides do not have triple-helical structures without repeating Gly–X–Y units. Although the telopeptides only account for 2% of the collagen molecule, they are essential to the formation of collagen fibrils [23].

2.2.1 Amino-Acid Residue and Triple-Helix Structure

As mentioned above, the composition units of the collagen are amino-acid residues, which were found early in the 1940s. Early studies have already shown that the general nature of the amino-acid composition has a high uniformity although the collagen comes from different sources.

Each peptide of the collagen has ~1050 amino-acid units, including glycine, proline, hydroxyproline, and lysine residues [24]. Among these amino-acid residues, glycine and proline residues play important roles in the formation of the triple-helical structure. As the smallest amino-acid residue, glycine residue occupies every third position in the polypeptide chain. The specific position for the glycine is because that glycine residue is small enough to stably occupy the narrow core within the triple-helical structure. As a result, glycine residue accounted for about one-third of the total number of the chain residues, including mammalian, amphibian, and avian collagens [25]. In addition to glycine residue, proline and hydroxyproline residues are also important components of collagen, which account for ~25% of the amino-acid residues within the polypeptide chain. These two imino acid residues have a unique characteristic in that the C^{β} atom is linked to the peptide nitrogen by the side chain, resulting in a 5-membered ring. The ring structures of proline and hydroxyproline residues act as "kinks" in the polypeptide chain, which are beneficial to the formation of the helical conformation. Specifically, the peptide unit links the amino-acid residue with the imino acid residue, leading to its low freedom degree of rotation about the $N—C^{\alpha}$ bond. The restriction imposed on the orientation of the peptide unit can stabilize the collagen structure, endowing the collagen chain with high rigidity. As another amino-acid residue of collagen, alanine residue accounts for ~10% of all residues within the polypeptide chain. The other compositions (e.g. arginine, lysine, glutamic acid residues) in polar side chains account for ~20% of the amino-acid residues. Although these amino-acid residues are not very essential to the triple-helical structure, they play important roles in linkages of the triple chains to form the collagen fibrils [26].

It was known in early times that glycyl residues account for one-third of the total number of residues in the polypeptide chain. However, there was no specific structure to explain the content of glycine residue until 1954 [4]. Ramachandran and Reddi proposed the structure that glycine residue occupies every third position in the polypeptide chain. The position of glycine residue cannot be substituted by other residues that have β-carbon atoms. It is also assumed that all the peptide units in this structure should have their trans conformations, which is different from the previous structures. Figure 2.2a displays the projection of the triple-chain structure down the central axis [26]. The structure is composed of three polypeptide chains (marked as A, B, and C). Every polypeptide is in the shape of a threefold helix with a pitch of 9 Å. The helical chains are related to each other by a threefold screw axis, having a left-handed twist.

As shown in Figure 2.2a, the three chains are held together by hydrogen bonds [26]. In addition to the backbone of chain A, the atoms connected to the α-carbon atoms are also marked in Figure 2.2b [26]. It is shown that the bonds connecting

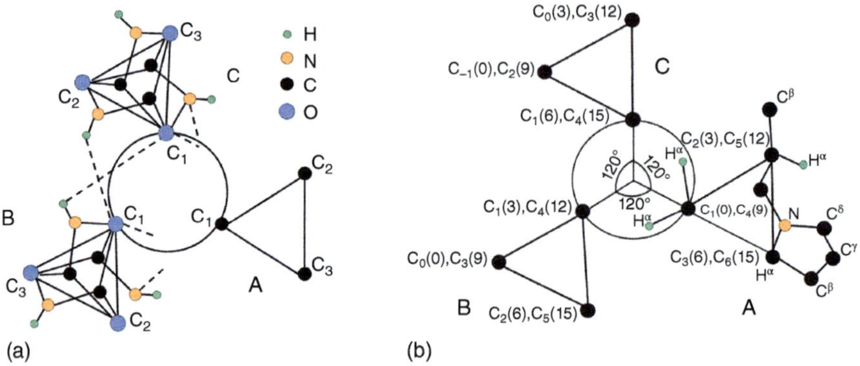

Figure 2.2 Projection of the triple-helical structure of collagen. (a) Indication of the postulated hydrogen bonds between chains B and C. (b) Indication of the angles and heights of the α-carbon atoms in the chains. Source: Modified from Ramachandran and Reddi [26]; © 1976, Springer Nature.

the β-carbon atoms to C_2^α and C_3^α point away from the triple-helix center, while the bonds connecting two hydrogen atoms to C_1^α (as part of a glycine residue) point to the triple-helix center. It should be noted that there is no space for β-carbon atoms to form in this central region. If C^β occurred at position 1, the three chains would be separated from one another. In this case, the hydrogen bonds linking the chains could not be formed since the relevant bond length was too large to maintain stability. Therefore, it can be concluded that only the side chain of glycyl residue can be located at C_1^α in the chains. Also, the residue at every symmetry-related atom (e.g. C_4^α, C_7^α, etc.) can only be glycine. Thus, glycine residue must occur at every third position in the collagen chain to maintain the triple-helical structure. Different from position 1, it is easy to fix a prolyl side chain at position 2 or 3. The dihedral angle for proline residue has a limited range, which is perfectly satisfactory in the above structure. It should be noted that the positions 2 and 3 are not always occupied by imino acid residues (proline and hydroxyproline residues) although they can be fittingly located at these positions without distorting the main structure of the chains. Experimental evidence further confirmed the above speculation. A complete amino-acid sequence of one peptide chain showed that proline and hydroxyproline residues occur only in the positions 2 and 3. Moreover, it is revealed that hydroxyproline residue occurs only in position 3 while proline residue can occur in both positions. The unique occupation of hydroxyproline residue in the chain is very important to understand the hydrogen bonding within the triple-helical structure and the stability of collagen structure [26].

Engel and Bächinger further illustrate the triple-helical structure of collagen in detail. As shown in Figure 2.3, the three helices are staggered by one residue relative to each other, supercoiling along a common axis into a right-handed triple helix. The repeated sequence of Gly–X–Y is essential to the superhelical formation. There are three residues in the straight and non-supercoiled polyproline-II-helix every turn, while the residues of glycine are in the face of each other in the interior of the helix structure (Figure 2.3b) [27]. Since the side chain at the C^α is larger than that at H,

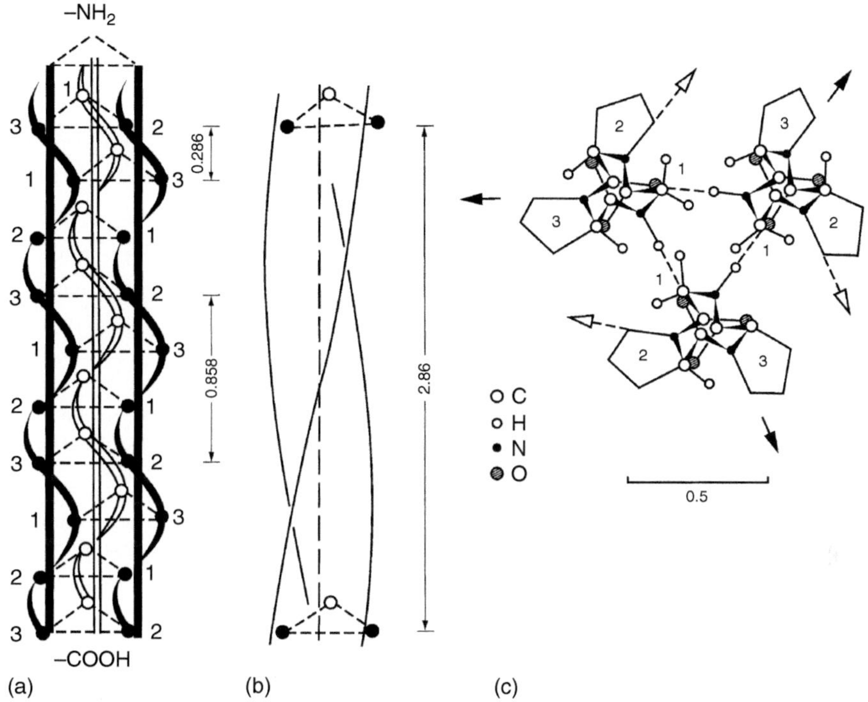

Figure 2.3 Collagen triple-helix structure in which glycine occupies position 1, while proline occupies positions 2 and 3: (a) Side view, three left-handed polyproline-II-type helices; (b) side view, the right-handed triple helix (dashed lines show the C^{α} positions); (c) top view (dashed lines show the hydrogen bond positions). The scale bars are indicated in nm. Source: Reproduced with permission from Brinckmann et al. [27]; © 2005, Springer Nature.

the hydrogen bond would not be formed between the backbone NH-group of glycine and the backbone CO-group of the residue at X position within the adjacent chain. These hydrogen bonds play important roles in the structural stability. If there are other residues (in addition to proline and hydroxyproline) in the chains, the isolated polyproline-II-helices would be unstable [27].

As shown in Figure 2.3a, ten tripeptide units form three turns when they are viewed at each left-handed helix. As a result, there are 3.33 residues per turn. In the right-handed triple helix, an identical structural element repeats every ten residues, resulting in the axial repeat of 2.86 and 0.286 nm per residue (Figure 2.3b). It should be noted that the lateral association of triple helices is very important in fibril formation. The interaction edges are critically determined by the chain sequence and helical symmetry. The difference between the collagen triple helix and other multi-stranded helices (e.g. DNA double helices) is the staggered arrangement of the three chains. As displayed in Figure 2.3a, a glycine residue in one chain faces a proline residue at position X within another chain. Meanwhile, the proline residue faces a proline residue at position Y within the third chain. Due to the staggered arrangement of the chains, six alignments are possible for three different chains in collagen.

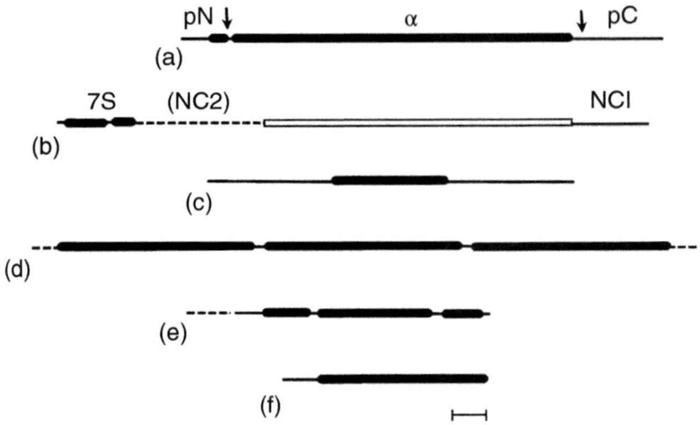

Figure 2.4 Illustrations of the structural features for collagen chains. (a) A pro-α chain of the fiber-forming collagens; (b) the α1(IV) chain; (c) a chain of type VI collagen; (d) a chain of type VIII or VII collagen; (e) the α1(IX) chain; (f) a chain of type X collagen. The scale bar at the bottom is equivalent to 100 amino-acid residues. Solid lines and filled bars represent the regions of nontriplet structure and triplet structures, respectively. The stripped bar in (b) represents a region that has interrupted triplet structure. Broken lines represent the regions that have undetermined sequences. Source: Reproduced with permission from Nimni [32]; © 1988, CRC Press, Inc.

Therefore, the topoisomerism of the triple helices for natural collagens is difficult to know. However, the stagger of the chains is important for binding processes, which could be used to recognize the residue arrays in different chains [28].

2.2.2 Primary Structure of Chains

Table 2.1 summarizes the chains and corresponding collagen types. There are three kinds of collagens according to the chain properties [13]. The chains within type I, II, III, and V collagens have a relative molecular mass (M_r) ≥95 000. In addition, these chains have a continuous Gly–X–Y repetitive triplet structure with over one thousand amino-acid residues [29]. Moreover, the molecules derived from these collagen chains also have similar structures. These collagens could be categorized as fibril collagens and fiber-forming collagens [30, 31]. Although the chains associated with type IV, VI, VII, and VIII collagens also have M_r over 95 000, they have smaller stretches of the repetitive triplet structures of Gly–X–Y than the chains within type I, II, III, and V collagens. For the chains of type IX and X collagens, they not only have smaller stretches of the repetitive triplet structures but also show a M_r <95 000.

The model in Figure 2.4a is obviously suitable for the pro-α chains associated with type I, II, and III collagens [32]. However, there are some differences between different α chains for the same type of collagens. The differences primarily exist in the pN domain. For example, the pN domain of pro-α1(I) contains 86 amino-acid residues. Nevertheless, the pN is deleted in pro-α2(I). As a result, the length of

Table 2.1 Chains of collagen family.

Chains	Molecular assembly	Molecular weight (kDa)/α chain	Collagen type
$[\alpha1(I)]_2\alpha2(I)]$	Monomers staggered by 67 nm	95	I (heterotrimer)
$[\alpha1(I)]_3$			I (homotrimer)
$[\alpha1(II)]_3$	Monomers staggered by 67 nm	95	II
$[\alpha1(III)]_3$	Monomers staggered by 67 nm	95	III
$[\alpha1(IV)_2\alpha2(IV)]$; α3(IV), α4(IV), α5(IV), α6(IV)	Association of 4N- and 2C-termini	170–180	IV
$[\alpha1(V)]_2\alpha2(V)[\alpha1(V)\ \alpha2(V)\alpha3(V)]\ [\alpha1(V)]_3$	Monomers staggered by 67 nm	120–145	V
[α1(VI) α2(VI) αα3(VI)]	Association into tetramers that aggregate end to end	αl(VI) 140 α2(VI) 140 α(VI) 340	VI
$[\alpha1(VII)]_3$	Lateral aggregation of antiparallel dimers	170	VII
$[\alpha1(VIII)]_2\ \alpha2(VIII)$	Interrupted helical structure	61	VIII
[α1(IX) α2(IX) αα3(IX)]	Covalently cross-linked to surface of collagen II fibrils	68–115	IX
$[\alpha1(X)]_3$	Assemble a mat-like structure	59s	X

Source: Modified from Sorushanova et al. [13].

pro-α2(I) is shorter than that of pro-αl(I). Moreover, the pN of pro-α1(II) contains numerous basic amino-acid residues, while the other nontriplet sequences of these chains are generally composed of numerous acidic and hydrophobic residues (Figure 2.5) [33].

Apart from the different amino-acid numbers within the pN region, the pro-α chains associated with type I, II, and III collagens are highly homologous [30]. Not considering the glycyl residues within the triplet-sequence regions, the homologous level for pro-α1(I) and pro-α1(II) is ~65% while the homologous level for pro-α2(I) and pro-α1(III) is ~45%. It is also revealed that the pro-α1(III) is most different from the other chains. The sequence data of pro-α2(V) confirm that pro-α2(V) also shows a high level of homology to the chains within type I, II, and III collagens, indicating that the chains of type V collagen have similar structural features to the chains of type I, II, and III collagens due to the homology [34, 35].

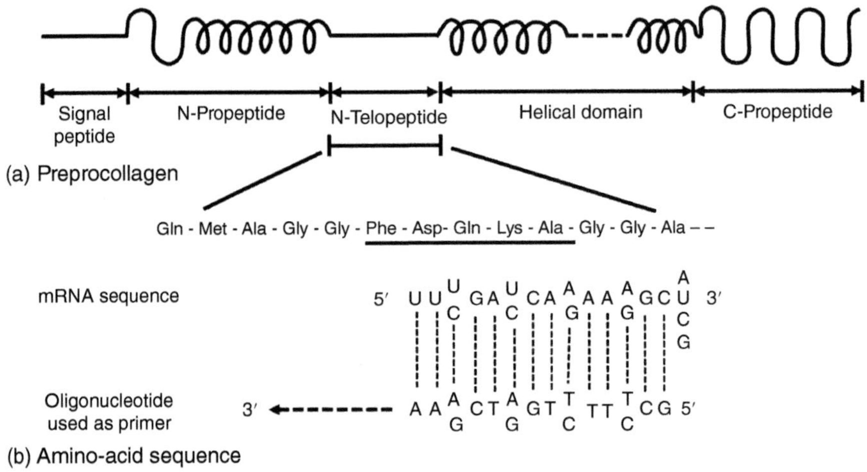

Figure 2.5 (a) Diagrammatic representation of entire α(II) chain. (b) The amino-acid sequence of the amino-terminal telopeptide. Source: Kohno et al. [33]/Elsevier/Licensed under CC BY 4.0.

As displayed in Table 2.1, there are more than seven unique chains that compose the type IV, VI, VII, and VIII collagens [36]. Generally, these collagens are less common than the fiber-forming collagens (type I, II, III, and V collagens). Moreover, it is very difficult to recover the monomer molecular species of the proteins since the yield of tissue extraction is relatively low. Therefore, the sequence data of the chains for these collagens still need to be further determined. The α1(IV) chain is one of the known chains. The structural features of the α1(IV) chain have been determined by electron microscopy, and the molecule compositions have been confirmed by limited proteolysis.

For the type IV collagenits rotary shadowing shows that there is a globular knob zone (referred to as NC1 domain) at one end and a compact helical zone (referred to as 7S domain) at the other end of the thread-like molecule [12]. Moreover, the flexibility pattern of the molecules indicates that there is also a nonhelical zone (referred to as the NC2 domain) next to the 7S domain. The molecule length is observed to be ~400 nm, implying that there are 1700–1800 amino-acid residues within each chain. Further research on the NC1 domain showed that this zone also includes the C-terminal regions of the constituent chains [37]. The primary structure of the αl(IV) chains within this zone has been determined by nucleotide sequencing [38]. It is shown that the C-terminal globular region of the α1(IV) chain is shorter than the corresponding region of pro-α chains associated with fiber-forming collagens. Notably, the nontriplet region of the α1(IV) chain shows a twofold symmetry since the N-terminal half and C-terminal half are highly homologous [39]. In addition, the repetitive Gly–X–Y triplet structure of the α1(IV) chain is frequently interrupted, which contributes to the thread-like region preceding the globular domain. The above results construct the structural basis for the good

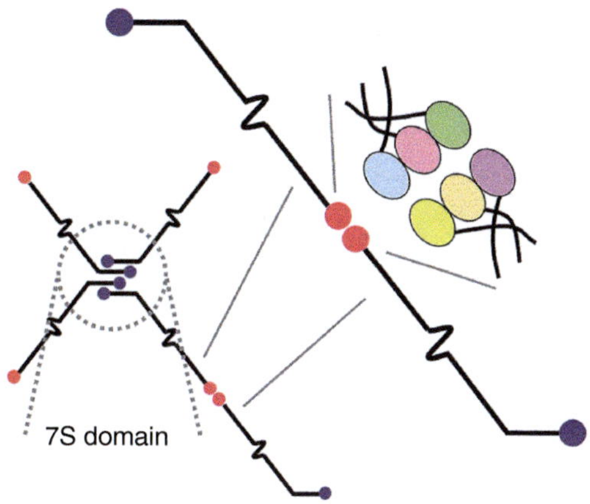

Figure 2.6 Schematic diagram of type IV collagen and chains in the basement membrane. Source: Reproduced with permission from Kadler et al. [40]; © 2007, Journal of Cell Science.

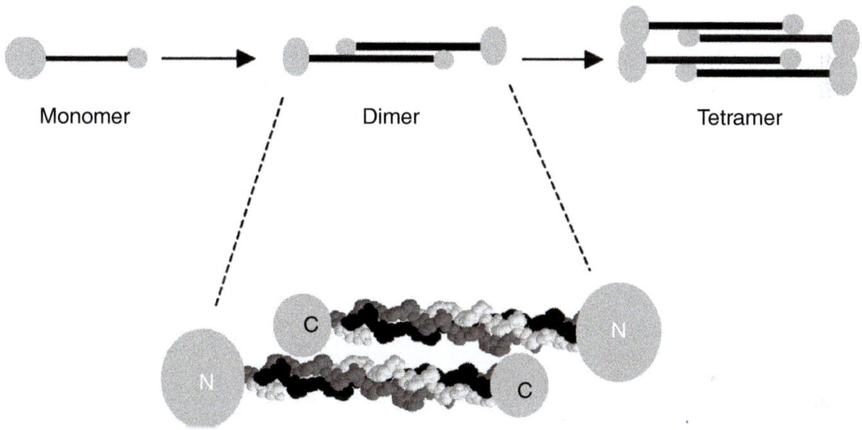

Figure 2.7 Schematic diagram of type VI collagen and chains. Source: Ball et al. [42]/Elsevier/Licensed under CC BY 4.0.

flexibility and susceptibility to proteolysis of the type IV molecules. The related data are summarized and illustrated in Figures 2.4b and 2.6 [40].

The primary structure of the chains associated with type VI collagen is partially determined. The N-terminal sequences of chains derived from peptic fragments of the molecules have been confirmed. It is seen that the corresponding molecules consist of three chains [41]. The rotary shadowing of the peptic fragments has indicated that there are molecules that contain a triple-helical zone at the center and a globular zone at the end (Figure 2.7) [42, 43]. These results about the characterization of the chains associated with type VI have been proved in the latter research, thus realizing the isolation and characterization of the intact composition and structure for the protein [44, 45]. In the subsequent studies, the molecules that were confirmed by rotary shadowing were revealed to show the same characteristics as the molecules that were obtained following limited pepsin digestion. However, the former has larger

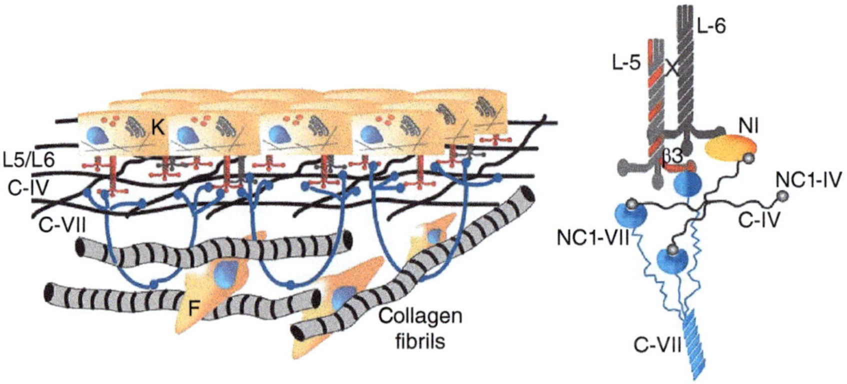

Figure 2.8 Schematic diagram of type VII collagen and chains in the dermis. Source: Reproduced with permission from Bentz et al. [46]; © 2006, Elsevier.

globular ends. Moreover, the components of the molecule dissociation showed an apparent M_r of 140 000 when compared to noncollagenous standard proteins. As compared to collagenous standards proteins, the components had an apparent M_r of 110 000. Considering the relatively small size of the collagenous sequence for each chain, the triplet structures might account for only one-third or less of the total weight for each chain. These structural features are schematically displayed in Figure 2.4c. A chain with ~1100 amino-acid residues, containing a triplet-sequence domain with 345 residues at the center and two nontriplet-sequence domains with 378 residues at both ends, is shown.

Type VII collagens derived from placental membranes and epidermal basement membranes have been isolated by limited pepsin digestion (Figure 2.8) [46]. The molecules isolated in this way show an unusually long triple-helical domain with a length of 450 nm. Therefore, the pepsin-solubilized type VII molecules have a similar size to the type IV molecules. Different from the type IV molecules, there are fewer interruptions of triple-helical conformation within the type VII molecules. It is shown that the molecules of type VII are cleaved in only one locus, which produces two fragments of similar size after exposure to pepsin for a long time. Based on the above results, it can be assumed that the integrated α chains of the type VII molecules contain a high proportion of triplet structures with few interrupted sites.

As one of the collagenous products, type VIII collagen has been isolated from the secreta of cultured endothelial cells [47, 48]. The structure feature of the chain associated with type VIII collagen is its unusual length with a M_r of ~180 000. After exposure to pepsin, the yielded fragments of the molecules have a M_r of 50 000–60 000. The structure features of the chains associated with the type VIII collagen have been determined based on the result [49]. The corresponding model shows a primary chain with a M_r of 180 000. The triplet structure is interrupted twice and separated into three equational parts. This structure model could well correspond to the experimental data of the chains within the type VIII collagen. It is also indicated that the size of structural features of the chains for the type VIII collagen is similar to the chains for the type VII collagen. However, the chains of type VII and VIII collagens are different in the placement of interruptions in the triplet domain. The structure of the chains associated with types VII and VIII is illustrated in Figure 2.4d.

Type IX and X collagens are the minor components of the extracellular matrix for hyaline cartilage. The difference is that type X collagen primarily exists in the regions of chondrocyte hypertrophy. In general, these two types of collagens are composed of more than four unique chains (Table 2.1) [13].

By limited pepsin digestion of hyaline cartilages, type IX molecules are generally cleaved into several small triple-helical fragments [29]. The experimental results of the fragments for chick cartilage showed that type IX collagen was assembled by three unique polypeptides. The result revealed that the type IX molecule is a heterotrimer that is composed of three various chains [50]. The experimental results indicate that the α1(IX) chain contains at least 738 amino-acid residues. There are four nontriplet regions alternate with three triplet regions in it [51, 52]. The N-terminal triplet regions contain one intercept point, while the C-terminal triplet regions contain two intercept points. As a result, the repetitive triplet structure is discontinuous. The structure of the chains associated with type IX is illustrated in Figure 2.4e. In addition to the above unusual structural features, polymers of chondroitin sulfate and/or dermatan sulfate are found to be covalently attached to more than one chain of type IX collagen [53]. This result implies that the type IX collagen molecules might also be designated as a special proteoglycan.

As another kind of cartilage-specific collagen, the type X collagens are assembled by relatively short chains. They can be isolated by selective precipitation from the spent media of cultured hypertrophic chondrocytes. The chains exhibit a M_r of ~59 000 [54]. The experimental data of the isolated protein using this method revealed that there are numerous noncollagenous sequences. In further research, the M_r of the molecules decreases to 45 000 after treating the native molecules with pepsin. In addition, the molecules after treatment by bacterial collagenase generate noncollagenous sequences with a M_r of ~18 000 [55]. The results indicate that there is a short helical zone with a length of 150 nm at one end of the chains of the type X collagen, which accounts for ~25% of the total molecular mass. The rotary shadowing result further confirms the speculation [56]. The structural features of the chains associated with the type X collagen are illustrated in Figure 2.4f.

2.2.3 Molecular Organization and Structure

The molecular species formed by the chains of different collagens are also listed in Table 2.1 [13]. For a specified collagen type in normal tissues, the molecular species are listed from most to least if more than one molecular species is known. The isolation and characterization of designated molecular species have been widely carried on [29].

There are necessary processes for the formation of the molecular species, including hydroxylation of prolyl residues, association of chains, and the propagation of triple helices. These processes are explored in detail in the research on fiber-forming collagens. It should be noted that the results and principles in these works are also appropriate for other molecular species. For the fiber-forming collagens, it is showed

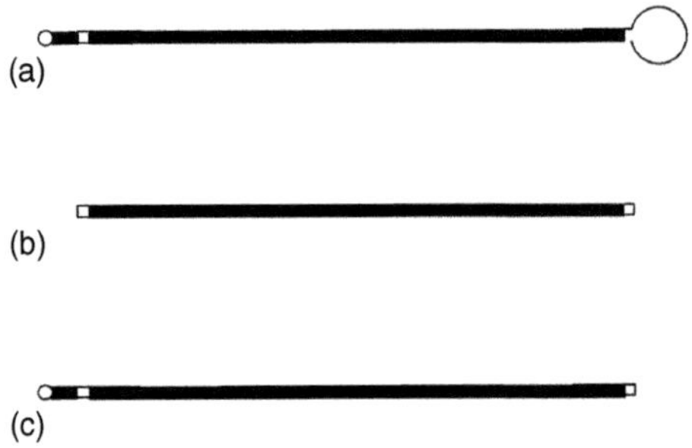

Figure 2.9 Molecular forms of the fiber-forming collagens. (a) A procollagen molecule; (b) a processed collagen molecule after removal of N- and C-terminal propeptides; (c) a pN-molecule after removal of C-terminal propeptides. Source: Reproduced with permission from Nimni [32]; © 1988, CRC Press, Inc.

that molecular assembly starts after the releasing of pro-α chains from ribosomes, including the chain association at the C-terminal nontriplet ends. The chain association is stabilized by the disulfide bonds that form between the chains. The indirect evidence has suggested that a disulfide-isomerase is beneficial to the accurate location of the disulfide linkages [57]. Although it is a common phenomenon to form disulfide bonding between chains in this process, it is necessary for all the collagens. It is shown that some molecular species of the type V collagen can be also formed without the interchain bonding [58, 59]. If there is a normal hydroxyproline complement for the chains, the joined point of the chains acts as the nucleus for the propagation of triple helices to the N terminus. The rate-limiting step for this process is determined by the *cis–trans* isomerization of peptide bonds, including ring nitrogen within proline and hydroxyproline residues [60]. There are many sites for the formation of hydrogen bonds, which are provided by hydroxyproline residues. These hydrogen bonds can provide the forces and play important roles in stabilizing the triple-helical structure. In this mode, the chain assembly and conformation propagation result in asymmetric molecules.

As shown in Figure 2.9a, a procollagen molecule of the fiber-forming collagens has N-terminal and C-terminal propeptides and triple-helical domains [32]. The procollagen molecule is assembled by three chains that have the structural features shown in Figure 2.4a.

It is shown that there is always a globular domain at one end of the triple-helical zone within the procollagen molecule. The triple-helical zone is interrupted by one site for a short span within the N-terminal domain of the molecule. The extracellular processing may lead to the removal of N- and C-terminal propeptide sequences (Figure 2.9b) or C-terminal propeptide sequences of the procollagens (Figure 2.9c) [32]. In specific, the extracellular processing of type I and II procollagens results in the elimination of both N- and C-terminal propeptides at each end of the molecules. For the type III procollagens, their N-terminal propeptides can be retained after processing (Figure 2.10) [40]. In addition, the experimental data also indicated that the extracellular processing of type V procollagen also leads to the removal of the C-terminal propeptides and the retention of N-terminal propeptides, which is similar to type III [61].

Different from the molecular species of fiber-forming collagens, other collagen molecular species have less information about extracellular processing. For the

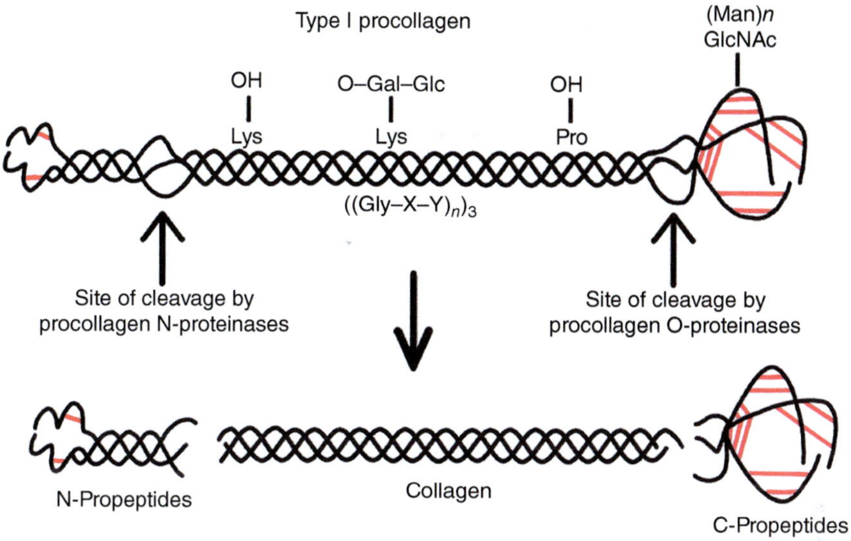

Figure 2.10 Schematic diagram of type I procollagen processing. Source: Reproduced with permission from Kadler et al. [40]; © 2007, Journal of Cell Science.

molecular species of type IV, it seems to be not treated by extracellular processing apart from proteolytic processing, which results in the removal of small sequences [62]. Further research also showed that most of the removed sequences by extracellular processing are derived from the globular domain at one end of the procollagen molecule. Overall, the above results indicate that all molecular species of collagen undergo extracellular processing to some extent. The molecular modification degree may be obviously different.

Table 2.1 also suggests that the chains are not randomly used to assemble molecular species of different collagens. Undoubtedly, the chain segregation is dependent on their different origins in different cells. However, it is obvious that the fiber-forming collagens have common origins [63]. Thus, there should be a mechanism that guides the chain assembly. In this context, it is reported that the nonconserved sequences start in the C-terminal telopeptide and extend into the C-terminal propeptide. These nonconserved sequences might act as recognition sites, which play important roles in the assembly of the chains [64]. This is a convincing proposal that reveals the significance of the C-terminal nontriplet regions to the chain association.

Although the molecules with triple-helical conformation accounted for more than 90% of the total molecules of the fiber-forming collagens, the solution properties of these molecules are closely related to the several flexible loci within the structure [65]. It confirmed the existence of an obvious flexible region surrounding the C-terminal cross-linking site through the rotary shadowing of type I molecules [66]. It should be noted that the flexible region of type I molecules may be due to the absence of imino acids that can provide helix stabilizing. This result implies that this structural feature may contribute to or be beneficial to the flexibility of the collagen molecules. In addition, it is also revealed that the thread-like part of type IV molecules is highly flexible. This result is consistent with the fact that there are numerous sites without helical conformation in these thread-like type IV molecules.

2.2.4 Aggregate Structure

Extracellular aggregates of the molecules are of great importance to the major physiological roles of collagen. Therefore, the construction of the aggregates plays a very important role in the integrity of the organism. Different collagens with different properties have different aggregate structures. Generally, there are four unique aggregation structure modes. As the most common aggregates, fibers are composed of molecules that are arranged in parallel. There are lateral association and axial displacement of these molecules in the fibers. In addition, there are other three aggregate modes in which the molecules are arranged in an antiparallel fashion.

In the initial formation of fiber, the molecules are staggered relative to one another by 268 nm (four D units) to form the linear polymers [67]. The molecule arrangement is displayed in Figure 2.11a [32]. It is a stable association mode since the N- and C-terminal telopeptides of one molecule are next to helical cross-link sites of

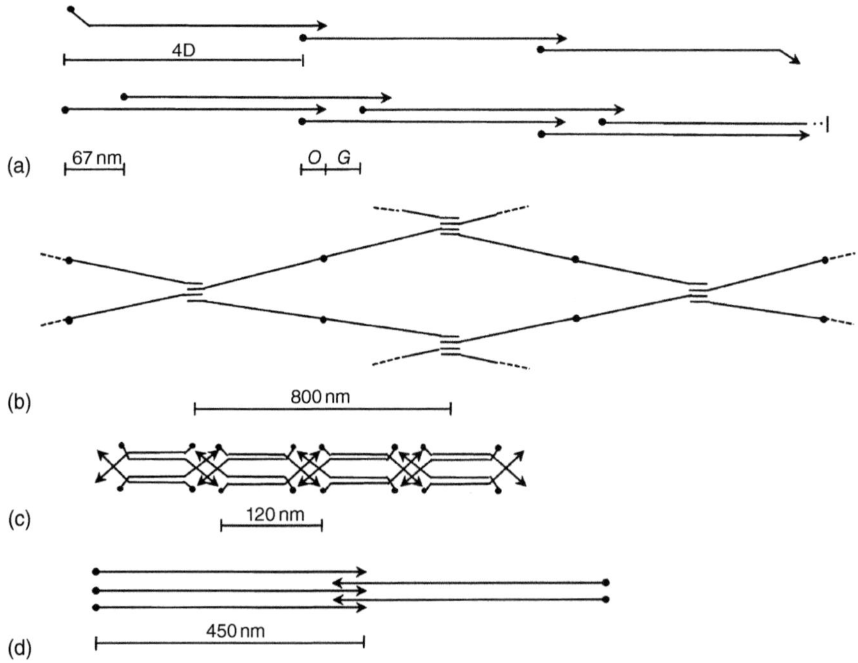

Figure 2.11 Aggregate structures formed by (a) the fiber-forming collagens (types I, II, III, and V); (b) type IV collagen; (c) type VI collagen; (d) type VII collagen. Source: Reproduced with permission from Nimni [32]; © 1988, CRC Press, Inc.

adjacent molecules. For the subsequently added molecules, they are staggered relative to those in the initial polymer by 67 nm (one D). As a result, it creates an overlap (O) region and a gap (G) region within the fibers, causing the periodicity characteristic of the collagen fibers (Figure 2.11a). If the N-terminal propeptides are retained in the molecules, there is no gap region between the nonoverlapping molecules in the same plane. This phenomenon occurs in the fibrils formed by type III molecules. It is confirmed that the fiber-forming molecules are incorporated into fibers by pN-molecules. The pN-domain is removed only from the molecules within the fibers [68].

The deposition, orientation, and diameter of collagen fibers are closely related to the aggregate process [29]. The conversion of procollagen to collagen plays an important role in determining the structure of the collagen fibers. The C-terminal propeptides are too large to make procollagen molecules stabilized in the fibers according to the aforementioned molecular structure mode. It is also reported that the retained N-terminal propeptides within type III molecules can limit the size of fibers by steric hindrance [69]. This mechanism might also be general for other collagen molecules. It is found that the type III collagen is usually detected in the tissues with rich type I collagen [70]. Therefore, the size of fibers assembled by type I molecules might be regulated and controlled when type III molecules with different proportions are mixed with the type I molecules.

The aggregate structures of type IV molecules are obviously different from those of the fiber-forming collagens. Due to the size and conformation, the type IV molecules cannot be incorporated into the aggregate structure like fibers. The rotary shadowing results of the type IV molecules derived from different sources show that protein multimers consist of dimers that are formed through interaction sites within the globular region. In addition, the protein multimers also contain the tetramers that are formed through the monomer association at the helical ends (Figure 2.11b) [12]. These data give the possible aggregate model of type IV collagen. As the basic unit, the tetramers are associated at the globular ends, thus forming the large aggregate structures (Figure 2.12) [71]. This molecule arrangement in the aggregates endows the type IV collagen with a framework containing large openings, which is beneficial to the molecule passage through a basement membrane. As shown in Figure 2.11b, the aggregate structure has a periodicity of ~800 nm, which is the length of two type IV molecules. It should be noted that the aggregate formation is closely dependent on the interaction sites within the terminal regions of the type IV molecules (e.g. C-terminal globular domains). Therefore, the terminal regions of the type IV chains not only determine the chain assembly for the formation of molecules but also provide interaction sites for the formation of aggregates. The above mechanism of aggregate formation for the type IV collagens is totally different from that for the fiber-forming collagens. The C-terminal globular regions should be eliminated before the molecule incorporation into the aggregates of the fiber-forming collagens takes place.

The aggregate structures of type VI molecules have been revealed by rotary shadowing results [43]. It is suggested that the molecules have an aggregate structure of microfibrillar arrays, which is composed of end-to-end tetrameric units. As shown

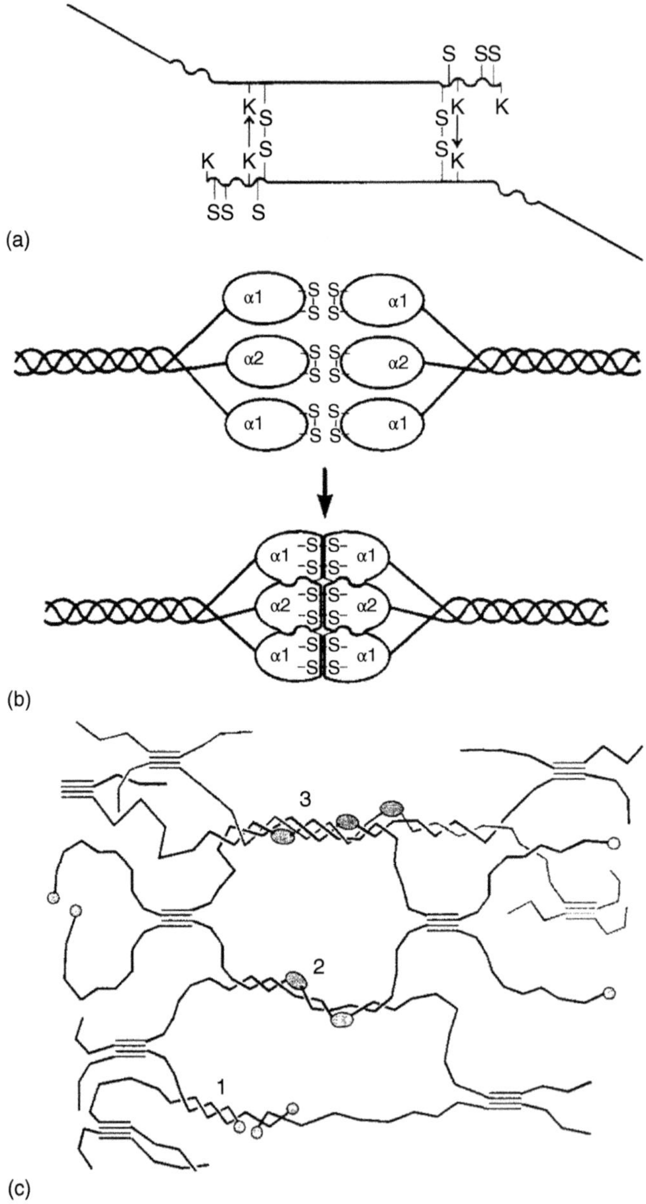

Figure 2.12 Assembly of the collagen IV molecules in the macromolecular network. (a) Assembly via the N-termini. (b) Assembly via the C-termini. (c) Assembly via the triple-helical domains. Source: Reproduced with permission from Kühn [71]; © 1995, Elsevier.

in Figure 2.11c, two lateral monomers intertwin with each other and stagger in an antiparallel way, forming the dimers. The dimers are laterally associated to form the tetrameric units. The microfibril has a periodicity of ~120 nm, which is caused by the alternating regions. One is the thread-like region of an individual molecule. The other is the overlap region of packed globular domains of multiple molecules [72].

The possible aggregation structures of type VII molecules have also been proposed. The molecule precipitation in vitro results in the formation of the aggregates for type VII molecules. The structure features show that the aggregates are formed by laterally associated dimers. The dimeric units are composed of the antiparallel monomers with overlapping of ~60 nm [46]. The aggregate model is displayed in Figure 2.11d. The aggregates of type VII molecules have a periodicity of ~450 nm, which is the length of the helical regions within the molecules.

2.3 Classification, Function, and Distribution of Collagen

Many kinds of collagens have been reported so far [73]. Based on the nature and morphologies of the aggregates, the collagens can be primarily divided into three categories, including fibril-forming collagen, network-forming collagen, and fibril-associated collagen. These collagens are composed of proteins that are closely related but genetically different. All these collagen molecules possess a basic structure of three polypeptide chains containing Gly–X–Y repeats. The polypeptide chains are twisted into triple-helical structures, which are stabilized by intra- and interchain hydrogen bonding. The collagen molecules are then associated with different aggregates to form different collagens. The properties of the collagens are dependent on the raw materials that are used to form them. For example, the isolated collagens derived from fish skins have different physical and chemical properties from those derived from cattle bones [74].

As the most abundant and widespread collagen, fibril-forming collagens accounted for ~90% of the total collagens [17]. Fibril-forming collagens are the primary component of fibrous tissues, which play important roles in transmitting forces, dissipating energy, and preventing mechanical failure [24]. Type I, II, and III collagens are the fibrillar collagens with ropelike structures mentioned above. Among them, type I collagen is the most abundant and widely distributed one of the collagen family [75]. The fibers formed by type I collagens exist in the supporting tissues (e.g. tendon and cornea) with high tensile strength. The fibers derived from type II collagen are primarily in cartilaginous structures, which endow the tissues with shock-absorbing properties [76]. For the type III collagen fibers, they are an important part of distensible tissues (e.g. blood vessels) [15].

Here we focus on the type I collagens, which are the major fibril-forming collagens. Apart from cartilaginous tissues, the type I collagens exist throughout the body [23]. Due to the special structural compositions, the type I collagen has multifunctionalities, including stiffness, elasticity, and toughness. As a triple helix assembled by two $\alpha 1$ chains and one $\alpha 2$ chain, the type I collagen (especially after calcification) endows mechanical properties with tissues, such as torsional stiffness, tensile

Table 2.2 The structures of the different collagen types.

Structure	Collagen type	Tissue distribution
Fibril-forming collagen	Type I	Most connective tissues like skin, bone, dermis, tendon, ligaments, cornea, blood vessels
Fibril-forming collagen	Type II	Hyaline cartilage, intervertebral disk, inner ear, vitreous body, nucleus pulposus
Fibril-forming collagen	Type III	Distensible and developing connective tissues, blood vessels, fetal skin
Network-forming collagen	Type IV	Basement membranes
Fibril-forming collagen	Type V	Tissues containing type I collagen, hyaline cartilage, nervous system
Beaded filament-forming collagen	Type VI	Bone, hyaline cartilage, soft connective tissues; blood vessels, skin, intervertebral disk, peripheral nerves
Network-forming collagen	Type VII	Anchoring fibrils; skin, epidermal joints
Hexagonal network-forming collagen	Type VIII	Many tissues, especially endothelium
Fibril-associated collagen	Type IX	Hyaline cartilage, intervertebral disk, vitreous humor, cornea
Hexagonal network-forming collagen	Type X	Hypertrophic cartilage

Source: Adapted from Refs. [14, 78–80].

strength, and bearing capacity [14]. In addition, the biomechanical scaffold that is used for cell attachment and macromolecule anchorage is primarily realized by the type I collagen. This functionality is special to maintain the structures of tissues [23]. The alignment of the collagen fibers is also of importance to the mechanical properties of the tissues. The numerous reactive groups (e.g. hydroxyl) within the rod-like collagen molecule radially aligned parallel to the fiber, resulting in strong cross-linking between molecules and fibrils themselves. The cross-linked groups can also prevent molecule and fiber from sliding past each other, thus increasing the mechanical strength of the fibers and enhancing the thermal endurance. Moreover, the decreased enzyme accessibility and the reduced fiber hydration enable them to resist enzymatic degradation [20].

Different from fibril-forming collagens, network-forming collagens have a three-dimensional network structure rather than distinct fibrillar structures. These collagens include type IV and VII collagens. The type IV molecules associate with meshwork structures, which are the major components of basement membranes. The membranes therefore can separate cell kinds and divide cell layers from underlying connective tissues [77]. In addition, they can also act as a semipermeable filtration barrier for macromolecules in organs (e.g. kidney and lung) [15]. As the third kind of collagens, fibril-associated collagens include type IX and XII collagens. Their collagens exist on the surface of collagen fibrils and connect the fibrils to other fibers and components in the extracellular matrix. Table 2.2 shows the structures and tissue distributions of the different collagen types [14, 78–80].

References

1 Abraham, J.D., Kumar, N., and Aviva, E. (2011). *Biodegradable Polymers in Clinical Use and Clinical Development*. Canada: Wiley.

2 Highberger, J.H., Gross, J., and Schmitt, F.O. (1950). Electron microscope observations of certain fibrous structures obtained from connective tissue extracts. *Journal of the American Chemical Society* 72: 3321–3322.

3 Schmitt, F.O., Gross, J., and Highberger, J.H. (1953). A new particle type in certain connective tissue extracts. *Proceedings of the National Academy of Sciences of the United States of America* 39: 459–470.

4 Ramachandran, G.N. and Kartha, G. (1954). Structure of collagen. *Nature* 174: 269–270.

5 Rich, A. and Crick, F.H. (1955). The structure of collagen. *Nature* 176: 915–916.

6 Boedtker, H. and Doty, P. (1956). The native and denatured states of soluble collagen. *Journal of the American Chemical Society* 78: 4267–4280.

7 Robert, P.M. (1987). *Biology of Extracellular Matrix: A Series*. New York: Academic Press, Inc.

8 Miller, E.J. and Matukas, V.J. (1969). Chick cartilage collagen: a new type of alpha 1 chain not present in bone or skin of the species. *Proceedings of the National Academy of Sciences of the United States of America* 64: 1264–1268.

9 Miller, E.J. (1971). Isolation and characterization of a collagen from chick cartilage containing three identical alpha chains. *Biochemistry* 10: 1652–1659.

10 Kefalides, N.A. (1971). Isolation of a collagen from basement membranes containing three identical α chains. *Biochemical and Biophysical Research Communications* 45: 226–234.

11 Kuhn, K., Wiedemann, H., Timpl, R. et al. (1981). Macromolecular structure of basement membrane collagens. *FEBS Letters* 125: 123–128.

12 Timpl, R., Wiedemann, H., van Delden, V. et al. (1981). A network model for the organization of type IV collagen molecules in basement membranes. *European Journal of Biochemistry* 120: 203–211.

13 Sorushanova, A., Delgado, L.M., Wu, Z. et al. (2019). The collagen suprafamily: from biosynthesis to advanced biomaterial development. *Advanced Materials* 31: e1801651.

14 Gelse, K., Poschl, E., and Aigner, T. (2003). Collagens – structure, function, and biosynthesis. *Advanced Drug Delivery Reviews* 55: 1531–1546.

15 Domb, A.J. and Jain, J.P. (2011). *Biodegradable Polymers in Clinical Use and Clinical Development*. Hoboken, NJ, Canada: Wiley.

16 Brodsky, B. and Ramshaw, J.A. (1997). The collagen triple-helix structure. *Matrix Biology* 15: 545–554.

17 Kim, S.-K. (2013). *Marine Proteins and Peptides Biological Activities and Applications*. West Sussex: Wiley.

18 Ramachandran, G.N. (1956). Structure of collagen. *Nature* 177: 710–711.

19 Chmielewski, J. (2021). Emerging designs and applications of peptide-based materials. *Peptide Science* 113: e24226.

20 Fratzl, P. *Collagen: Structure and Mechanics.* New York: Springer Science + Business Media, LLC.

21 Brodsky, B. and Ramshaw, J.A. (2017). Bioengineered collagens. *Sub-Cellular Biochemistry* 82: 601–629.

22 Fraser, R.D., MacRae, T.P., and Suzuki, E. (1979). Chain conformation in the collagen molecule. *Journal of Molecular Biology* 129: 463–481.

23 Kadler, K.E., Holmes, D.F., Trotter, J.A., and Chapman, J.A. (1996). Collagen fibril formation. *Biochemical Journal* 316: 1–11.

24 O'Grady, J.E. and Bordon, D.M. (2003). Global regulatory registration requirements for collagen-based combination products: points to consider. *Advanced Drug Delivery Reviews* 55: 1699–1721.

25 Eastoe, J. (1967). *Composition of Collagen and Allied Proteins.* New York: Academic Press, Inc.

26 Ramachandran, G.N. and Reddi, A.H. (1976). *Biochemistry of Collagen.* New York: Springer Science+Business Media.

27 Brinckmann, J., Notbohm, H., and Müller, P.K. (2005). *Collagen: Primer in Structure, Processing and Assembly.* Berlin, Heidelberg: Springer.

28 Golbik, R., Eble, J.A., Ries, A., and Kühn, K. (2000). The spatial orientation of the essential amino acid residues arginine and aspartate within the A1β1 integrin recognition site of collagen IV has been resolved using fluorescence resonance energy transfer. *Journal of Molecular Biology* 297: 501–509.

29 Furuto, D.K. and Miller, E.J. (1983). Different levels of glycosylation contribute to the heterogeneity of alpha 1(II) collagen chains derived from a transplantable rat chondrosarcoma. *Archives of Biochemistry and Biophysics* 226: 604–611.

30 Miller, E.J. (1985). The structure of fibril-forming collagens. *Annals of the New York Academy of Sciences* 460: 1–13.

31 Miller, E.J. and Gay, S. (1987). The collagens: an overview and update. *Methods in Enzymology* 144: 3–41.

32 Nimni, M.E. (1988). *Collagen Volume I Biochemistry.* Boca Raton: CRC Press, Inc.

33 Kohno, K., Martin, G.R., and Yamada, Y. (1984). Isolation and characterization of a cDNA clone for the amino-terminal portion of the pro-alpha 1(II) chain of cartilage collagen. *Journal of Biological Chemistry* 259: 13668–13673.

34 Myers, J.C., Loidl, H.R., Stolle, C.A., and Seyer, J.M. (1985). Partial covalent structure of the human alpha 2 type V collagen chain. *Journal of Biological Chemistry* 260: 5533–5541.

35 Myers, J.C., Loidl, H.R., Seyer, J.M., and Dion, A.S. (1985). Complete primary structure of the human alpha 2 type V procollagen COOH-terminal propeptide. *Journal of Biological Chemistry* 260: 11216–11222.

36 Rahul, D., Meryl, W.H., and Saryu, N.D. (1985). Isolation and partial characterization of a novel basement membrane collagen. *Biochemical and Biophysical Research Communications* 130: 1–8.

37 Sabine, W., Jürgen, E., Hanna, W. et al. (1984). Subunit structure and assembly of the globular domain of basement-membrane collagen type IV. *European Journal of Biochemistry* 139: 401–410.

38 Brinker, J.M., Gudas, L.J., Loidl, H.R. et al. (1985). Restricted homology between human alpha 1 type IV and other procollagen chains. *Proceedings of the National Academy of Sciences of the United States of America* 82: 3649–3653.

39 Glanville, R.W., Qian, R.Q., Siebold, B. et al. (1985). Amino acid sequence of the N-terminal aggregation and cross-linking region (7s domain) of the alpha 1 (IV) chain of human basement membrane collagen. *European Journal of Biochemistry* 152: 213–219.

40 Kadler, K.E., Baldock, C., Bella, J., and Boot-Handford, R.P. (2007). Collagens at a glance. *Journal of Cell Science* 120: 1955–1958.

41 Jander, R., Rauterberg, J., and Glanville, R.W. (1983). Further characterization of the three polypeptide chains of bovine and human short-chain collagen (intima collagen). *European Journal of Biochemistry* 133: 39–46.

42 Ball, S., Bella, J., Kielty, C., and Shuttleworth, A. (2003). Structural basis of type VI collagen dimer formation. *Journal of Biological Chemistry* 278: 15326–15332.

43 Furthmayr, H., Wiedemann, H., Timpl, R. et al. (1983). Electron-microscopical approach to a structural model of intima collagen. *Biochemical Journal* 211: 303–311.

44 Jander, R., Troyer, D., and Rauterberg, J. (1984). A collagen-like glycoprotein of the extracellular matrix is the undegraded form of type VI collagen. *Biochemistry* 23: 3675–3681.

45 Von Der Mark, H., Aumailley, M., Wick, G. et al. (1984). Immunochemistry, genuine size and tissue localization of collagen VI. *European Journal of Biochemistry* 142: 493–502.

46 Bentz, H., Morris, N.P., Murray, L.W. et al. (1983). Isolation and partial characterization of a new human collagen with an extended triple-helical structural domain. *Proceedings of the National Academy of Sciences of the United States of America* 80: 3168–3172.

47 Sage, H., Trueb, B., and Bornstein, P. (1983). Biosynthetic and structural properties of endothelial cell type VIII collagen. *Journal of Biological Chemistry* 258: 13391–13401.

48 Sage, H., Pritzl, P., and Bornstein, P. (1980). A unique, pepsin-sensitive collagen synthesized by aortic endothelial cells in culture. *Biochemistry* 19: 5747–5755.

49 Benya, P.D. (1980). EC collagen: biosynthesis by corneal endothelial cells and separation from type IV without pepsin treatment or denaturation. *Renal Physiology* 3: 30–35.

50 Mayne, R., Van Der Rest, M., Weaver, D.C., and Butler, W.T. (1985). The structure of a small collagenous fragment isolated from chicken hyaline cartilage. *Journal of Cellular Biochemistry* 27: 133–141.

51 Ninomiya, Y. and Olsen, B.R. (1984). Synthesis and characterization of cDNA encoding a cartilage-specific short collagen. *Proceedings of the National Academy of Sciences of the United States of America* 81: 3014–3018.

52 Van Der Rest, M., Mayne, R., Ninomiya, Y. et al. (1985). The structure of type IX collagen. *Journal of Biological Chemistry* 260: 220–225.

53 Vaughan, L., Winterhalter, K.H., and Bruckner, P. (1985). Proteoglycan Lt from chicken embryo sternum identified as type IX collagen. *Journal of Biological Chemistry* 260: 4758–4763.

54 Schmid, T.M. and Linsenmayer, T.F. (1983). A short chain (pro)collagen from aged endochondral chondrocytes. Biochemical characterization. *Journal of Biological Chemistry* 258: 9504–9509.

55 Schmid, T.M. and Linsenmayer, T.F. (1984). Denaturation-renaturation properties of two molecular forms of short-chain cartilage collagen. *Biochemistry* 23: 553–558.

56 Schmid, T.M., Mayne, R., Bruns, R.R., and Linsenmayer, T.F. (1984). Molecular structure of short-chain (SC) cartilage collagen by electron microscopy. *Journal of Ultrastructure Research* 86: 186–191.

57 Myllylä, R., Koivu, J., Pihlajaniemi, T., and Kivirikko, K.I. (1983). Protein disulphide-isomerase activity in various cells synthesizing collagen. *European Journal of Biochemistry* 134: 7–11.

58 Fessler, L.I., Kumamoto, C.A., Meis, M.E., and Fessler, J.H. (1981). Assembly and processing of procollagen V (AB) in chick blood vessels and other tissues. *Journal of Biological Chemistry* 256: 9640–9645.

59 Fessler, L.I., Robinson, W.J., and Fessler, J.H. (1981). Biosynthesis of procollagen [(pro alpha 1 V)$_2$ (pro alpha 2 V)] by chick tendon fibroblasts and procollagen (pro α 1 V)$_3$ by hamster lung cell cultures. *Journal of Biological Chemistry* 256: 9646–9651.

60 Bruckner, P. and Eikenberry, E.F. (1984). Formation of the triple helix of type I procollagen in cellulo. Temperature-dependent kinetics support a model based on cis in equilibrium trans isomerization of peptide bonds. *European Journal of Organic Chemistry* 140: 391–395.

61 Broek, D.L., Madri, J., Eikenberry, E.F., and Brodsky, B. (1985). Characterization of the tissue form of type V collagen from chick bone. *Journal of Biological Chemistry* 260: 555–562.

62 Oberbaumer, I., Wiedemann, H., Timpl, R., and Kuhn, K. (1982). Shape and assembly of type IV procollagen obtained from cell culture. *The EMBO Journal* 1: 805–810.

63 Gay, S., Martin, G.R., Muller, P.K. et al. (1976). Simultaneous synthesis of types I and III collagen by fibroblasts in culture. *Proceedings of the National Academy of Sciences of the United States of America* 73: 4037–4040.

64 Ninomiya, Y., Showalter, A.M., Rest, M.V.d. et al. (1984). Structure of the carboxyl propeptide of chicken type II procollagen determined by DNA and protein sequence analysis. *Biochemistry* 23: 617–624.

65 Silver, F.H. and Birk, D.E. (1984). Molecular structure of collagen in solution: comparison of types I, II, III and V. *International Journal of Biological Macromolecules* 6: 125–132.

66 Hofmann, H., Voss, T., Kühn, K., and Engel, J. (1984). Localization of flexible sites in thread-like molecules from electron micrographs: comparison of interstitial, basement membrane and intima collagens. *Journal of Molecular Biology* 172: 325–343.

67 Silver, F.H., Langley, K.H., and Trelstad, R.L. (1979). Type I collagen fibrillogenesis: initiation via reversible linear and lateral growth steps. *Biopolymers* 18: 2523–2535.

68 Fleischmajer, R., Timpl, R., Tuderman, L. et al. (1981). Ultrastructural identification of extension aminopropeptides of type I and III collagens in human skin. *Proceedings of the National Academy of Sciences of the United States of America* 78: 7360–7364.

69 Fleischmajer, R., Olsen, B.R., Timpl, R. et al. (1983). Collagen fibril formation during embryogenesis. *Proceedings of the National Academy of Sciences of the United States of America* 80: 3354–3358.

70 Henkel, W. and Glanville, R.W. (1982). Covalent crosslinking between molecules of type I and type III collagen. *European Journal of Biochemistry* 122: 205–213.

71 Kühn, K. (1995). Basement membrane (type IV) collagen. *Matrix Biology* 14: 439–445.

72 Bruns, R.R. (1984). Beaded filaments and long-spacing fibrils: relation to type VI collagen. *Journal of Ultrastructure Research* 89: 136–145.

73 Koopmans, G., Hasse, B., and Sinis, N. (2009). Chapter 19: The role of collagen in peripheral nerve repair. *International Review of Neurobiology* 87: 363–379.

74 Sadowska, M., Kołodziejska, I., and Niecikowska, C. (2003). Isolation of collagen from the skins of Baltic cod (*Gadus Morhua*). *Food Chemistry* 81: 257–262.

75 Orgel, J.P., Miller, A., Irving, T.C. et al. (2001). The in situ supermolecular structure of type I collagen. *Structure* 9: 1061–1069.

76 Sung, H.J., Meredith, C., Johnson, C., and Galis, Z.S. (2004). The effect of scaffold degradation rate on three-dimensional cell growth and angiogenesis. *Biomaterials* 25: 5735–5742.

77 Techatanawat, S., Surarit, R., Suddhasthirac, T., and Khovidhunkita, S.O.P. (2011). Type I collagen extracted from rat-tail and bovine achilles tendon for dental application: a comparative study. *Asian Biomedicine* 5: 787–798.

78 Ihanamaki, T., Pelliniemi, L.J., and Vuorio, E. (2004). Collagens and collagen-related matrix components in the human and mouse eye. *Progress in Retinal and Eye Research* 23: 403–434.

79 Fox, M.A. (2008). Novel roles for collagens in wiring the vertebrate nervous system. *Current Opinion in Cell Biology* 20: 508–513.

80 Myllyharju, J. and Kivirikko, K.I. (2004). Collagens, modifying enzymes and their mutations in humans, flies and worms. *Trends in Genetics* 20: 33–43.

3

Synthesis and Applications of Collagen

3.1 Biosynthesis of Collagen

There have been detailed studies on the biosynthesis of collagen. The processes of collagen biosynthesis include intracellular and extracellular steps. All these processes are closely related to the structural and biomechanical properties of the collagen fibrils [1]. The polypeptide precursors of the collagen molecules are generally formed in fibroblasts, bone osteoblasts, and cartilage chondroblasts. They are then secreted into the extracellular matrix. The biosynthesis process starts with the gene transcription within the nucleus and ends with the formation of large fibrils [2–4]. After the procollagen α chains are synthesized on the ribosome, they enter into the rough endoplasmic reticulum. The procollagen α chains then undergo posttranslational modifications and are assembled into procollagen molecules (Figure 3.1) [5]. These steps include hydroxylation of proline and lysine residues, glycosylation, trimerization, disulfide bonding, prolyl *cis–trans* isomerization, and triple-helix folding. The procollagen molecules then pass through the Golgi network and are packaged into secretory vesicles, which further export into the extracellular space. The processing of the procollagens occurs during or shortly after the secretion, which are then assembled into fibrils. The fibrils are finally stabilized by the formation of covalent cross-links (Figure 3.2) [4].

3.1.1 Formation of Pro-α-Chains

As one kind of protein, collagen primarily functions outside of cells. Similar to most proteins that are produced for export, the polypeptide precursors of collagen contain a special amino-acid sequence at the N-terminal ends of the α chains. This structural feature determines that the synthesized polypeptides are destined to come out of the cells. The signal sequence guides the polypeptide chains into the cisternae of the rough endoplasmic reticulum (RER). After the cleavage of the signal sequence in the endoplasmic reticulum, pro-α-chain is produced as a collagen precursor [4].

Collagen-Derived Materials: Synthesis and Applications in Electrochemical Energy Storage and Conversion, First Edition. Feng Wang, Yaqin Huang, and Jin Niu.
© 2022 WILEY-VCH GmbH. Published 2022 by WILEY-VCH GmbH.

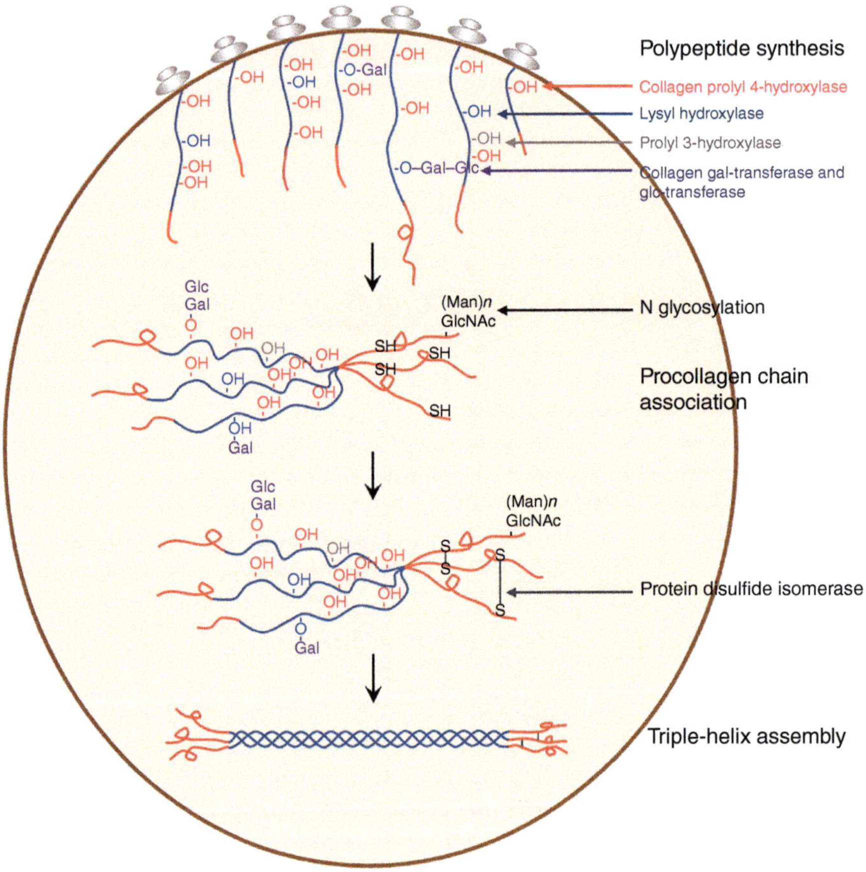

Figure 3.1 Posttranslational modifications and assembly of the procollagen molecule within the endoplasmic reticulum. Source: Reproduced with permission from Brinckmann et al. [5]; © 2005, Springer Nature.

3.1.2 Hydroxylation and Glycosylation

In the lumen of the RER, the produced pro-α-chains are then processed by numerous enzymatic steps. The proline and lysine residues (in the Y position of the Gly–X–Y sequence) are hydroxylated to form hydroxyproline and hydroxylysine residues (Figure 3.2) [4]. The hydroxyproline can stabilize the triple helix of the collagen [6]. The absence of hydroxyproline can result in the unstable structure of the collagen [7, 8]. However, it cannot be secreted from cells at a normal rate [6]. Hydroxylysine plays a very important role in the formation of intermolecular cross-links within collagen [9]. In addition, the glycosylation of the hydroxylysine residues occurs by modifications with glucose or glucosyl-galactose. The glycosylation process is also necessary for a stable tertiary structure and is an alternative way to stabilize the triple-helix structure of the collagen [10].

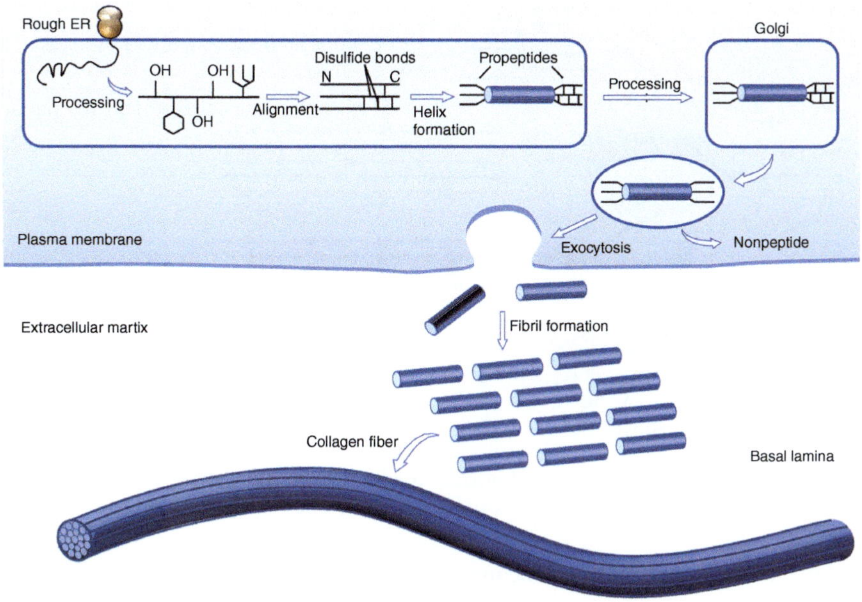

Figure 3.2 Biosynthesis and processing of collagen. Source: Reproduced with permission from Abraham et al. [4]; 2011 John Wiley & Sons, Inc.

3.1.3 Assembly to Triple-Helix Structure

The hydroxylation and glycosylation of the pro-α-chains result in the formation of procollagen. As a precursor of collagen, the procollagen has a triple-helix region at the center and a nonhelical region at the two ends (amino- and carboxyl-terminal extensions called propeptides). The procollagen formation starts when the interchain disulfide bonds form between the carboxyl-terminal extensions of the pro-α-chains, which is beneficial to the alignment of three α chains into helix structure. The procollagen molecules are then transited to the Golgi network and packaged in secretory vesicles. The secretory vesicles fuse with the cell membrane and release the procollagen molecules into the extracellular matrix [4].

3.1.4 Cleavage of Procollagen Molecules

After the release of procollagen molecules, they are cleaved by N- and C-procollagen peptidases, leading to the removal of the terminal propeptides. As a result, triple-helical collagen molecules are released [4].

3.1.5 Formation of Collagen Fibrils

Fibrils are formed through the alignment of the collagen molecules, causing the phenomenon of precipitation. As shown in Figure 3.3 [11], the alignment regularity of collagen molecules is dependent on the ionic strength, pH, or hydrophobicity by additives in the aqueous solution. The formation process of the collagen fibrils is

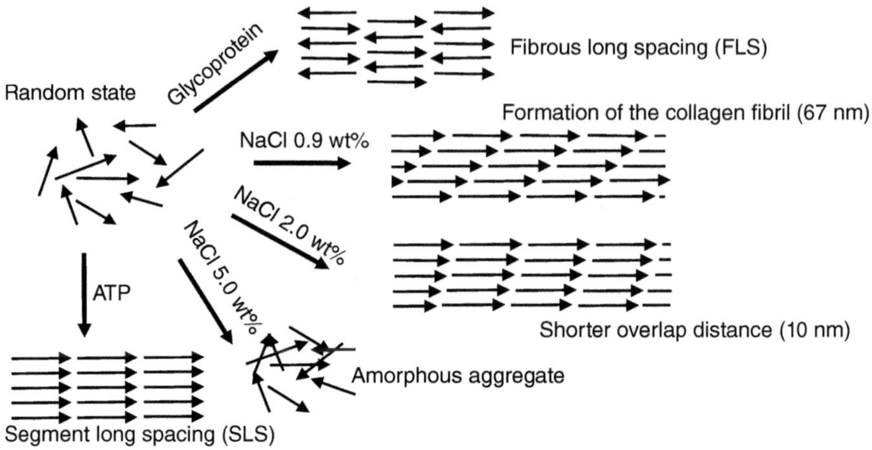

Figure 3.3 The schematic image of fibrillization according to the additives. Source: Reproduced with permission from Ashutosh et al. [11]; © 2012 Scrivener Publishing LLC.

called fibrillogenesis or collagen reconstitution. Fibrillogenesis is an entropy-driven process. The assemblies have a circular cross section because of the solvent molecule loss from the protein molecule surface. Therefore, the final assembles have a minimum ratio of the surface area against volume [12]. The hydrophobic residues within the collagen play important roles in the lateral aggregation of the collagen molecules [13, 14]. Since the fibrillogenesis process occurs in a salt aqueous solution, the formation of salt bridges is considered to be an important factor in fibril formation [15]. In addition, it is also thought that collagen fibrillogenesis is determined by the formation of hydrogen-bonded water clusters that bridge recognition sites on opposing helices [16]. However, the fact is that the neutral pH, physiological ionic strength, and increasing temperature induce the spontaneous assembly of collagen molecules into fibers. Meanwhile, the hydrophobic interactions, salt bridge, and hydrogen-bonded water clusters are also important factors [11].

In the fibril formation process, the orientation of the collagen fibrils is also important. It is reported that the orientation of the collagen molecules or fibrils is determined by applying force. The orientation of the collagen matrix can be realized when a certain driving force enables the collagen fibrils to align. As the most commonly used method for the orientation, the flow chamber enables the collagen molecules to precipitate along the axis of the flow. Lanfer et al. reported that the fibril orientation is affected by shear flow deposition [17]. When the flow rates of the solution increased, the orientation degree of the collagen fibril increased. In addition to orientation, the matrix density increased when the solution concentrations of the collagen molecules increased. The effects of collagen concentrations on fibril density are due to the liquid crystal behaviors of the collagen, which tend to reorganize the fibril structure in high concentration.

The application of uniaxial elongational strain is another way for the alignment of collagen fibrils. The high orientation of the collagen fibrils can be realized by this method, which can be directly applied to ligaments or tendons. Falini et al.

[18] applied uniaxial elongational strain on collagen film and then dehydrated it for 24 hours. It is found that the rearrangement of the collagen molecules occurred when the elongational strain on the film was higher than 12%. In this case, the orientation percent was ~83%, close to the orientation percent of the Achilles tendon. However, this method is only applicable to collagen molecules but not to collagen fibrils [19].

3.1.6 Cross-Linking to Collagen Fibers

After the collagen molecules form fibrils by molecular hydrophobic and electrostatic interactions, they are stabilized in the fibrils by covalent intermolecular cross-links, which are essential for normal physiological function. Specifically, the covalent intermolecular cross-linking endows the fibrillar matrix with tensile strength and viscoelastic characteristics, which are important for their different structural functions. It should be noted that the cross-linking features of the collagen are tissue-specific, which means that the cross-linking patterns are closely associated with their physiological functions. The type, density, and distribution of the cross-links maybe the factors that determine the physiological functions of collagenous connective tissues [17, 20–22].

3.2 Extraction of Collagen

The collagen extraction from animals can be mainly classified into three categories, including acid method, salt method, and enzymatic method. Additionally, there are also hybrid extraction methods (e.g. acid–enzyme dissolution method) [23]. The combination method can increase the extraction efficiency of the collagen. In addition to the primary extraction process, the posttreatment process using advanced equipment and new technologies further increases the extraction efficiency and the purity of the extracted collagen.

3.2.1 Acid Method

The extracted collagen has a triple-helix structure with high thermal stability due to the high content of imino acids. The low-concentration acids can break the salt bonds and Schiff bonds, resulting in the hydrolyzation of collagen fibers without cross-linking or with aldehyde amine cross-linked bonds. Figure 3.4 shows the dialyzed and lyophoilized collagen extracted by acid method [24]. Acetic acid, hydrochloric acid, citric acid, and oxalic acid are commonly used organic acids for collagen extraction. Sinthusamran et al. [25] found that the collagen within the fish swim bladder can be extracted by the acid method. The extracted collagen has a triple-helix structure with high thermal stability due to the high content of amino acids. Sousa et al. [23] extracted the collagen of the Atlantic cod swim bladder by the acid method. They further treated the precipitate with pepsin under acidic conditions; the contents of extracted collagen could be increased from 5.72% (w/w) to 11.14% (w/w). The content of collagen extracted by the acid–enzymatic method is higher than the simplex acid method. However, the latter can unfold the α1 and

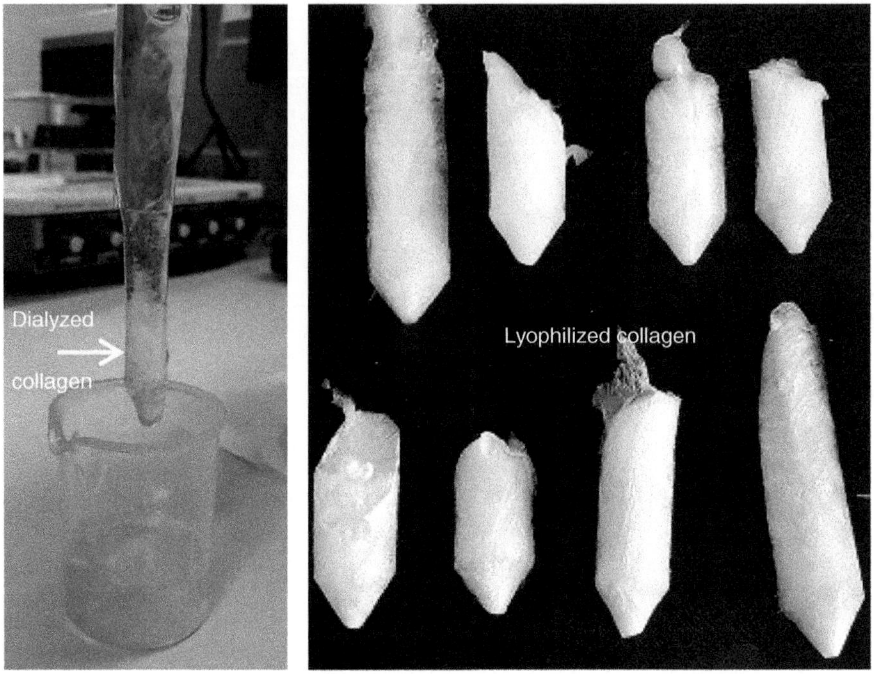

Figure 3.4 (a) Dialyzed and (b) lyophilized collagens extracted by the acid method. Source: Blanco et al. [24]/MDPI/CC BY 4.0.

α2 chains and maintain the initial collagen structure, which is more important for biological application. Blanco et al. [24] obtained the best extraction process of the acid method by optimizing hydrolysis temperature (4 °C), acid concentration (1 mol l^{-1}), and hydrolysis time (34 hours). As a result, the extraction rate could be as high as 61.24% (Figure 3.5).

3.2.2 Salt Method

For the salt extraction method, the efficiency and type of the extracted collagen are different based on the different salt concentrations. For example, sodium chloride can increase the interactions between collagen and water, thus accelerating the dissolution of the collagen. However, the unsuitable salt concentration can destroy the stability of the collagen conformation. In addition, the extraction efficiency of simplex salt is not satisfactory. Therefore, Wen et al. used sodium sulfate with sodium hydroxide to improve the extraction method, therefore increasing the extraction rate of collagen [26].

3.2.3 Enzymatic Method

In the enzymatic method, enzymes are commonly used with organic acids. The enzymes can break the covalent bonds formed between the lysine and hydroxylysine

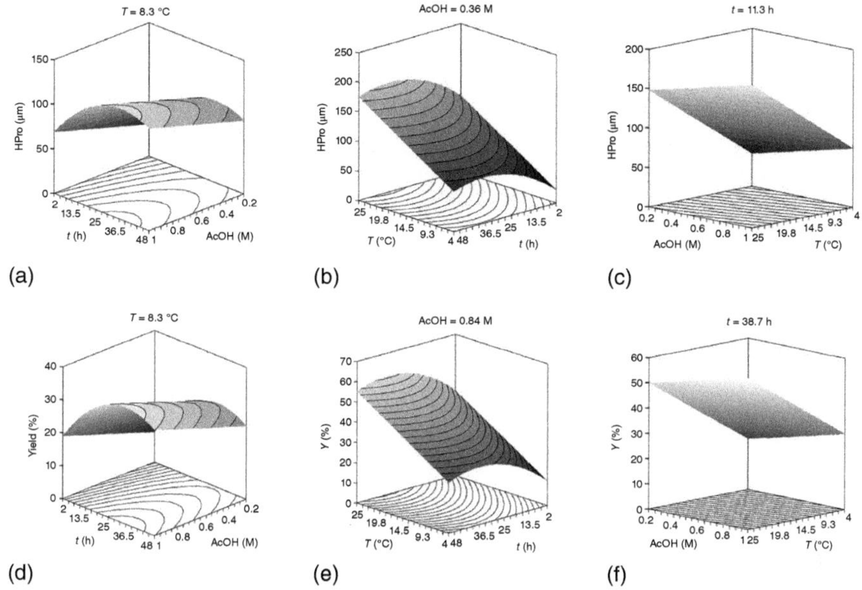

Figure 3.5 Combined effect of acid concentration, hydrolysis time, and hydrolysis temperature on released hydroxyproline (a–c) and extracted collagen yield (d–f) from the skin of *Scyliorhinus canicula*. Source: Blanco et al. [24]/MDPI/Licensed under CC BY 4.0.

residues within the terminal peptide chains of collagen. As a result, the peptide chains are unfolded while the triple-helix structure can be maintained. Generally, the proteases are mainly divided into three categories, which are animal proteases, plant proteases, and microbial proteases. In the research on enzymatic extraction of collagen, pepsin, trypsin, and papain are commonly used. Benjakul et al. [27] used different types of pepsin to extract the collagen. They concluded that the collagens extracted by porcine pepsin and tongol tuna pepsin are different in terms of molecular structures and properties (Figure 3.6). Lin et al. [28] used the connective tissue of abalone viscera, as a raw material, to explore the factors affecting the enzymatic extraction. The experimental results present the best condition for enzymatic extraction of collagen (pH of 2.5, enzyme content of 0.5%, solid–liquid ratio of 1 : 8 [g ml^{-1}], and extraction time of 48 hours).

3.2.4 Posttreatment

After the initial extraction of the collagen, there are also impurities and residual solvents in the samples. Therefore, purification processes are necessary to increase the purity of the extracted collagen, including salting out, dialysis, filtration, chromatography, centrifugation, and electrophoresis. The traditional purification method for collagen is a combination process of salting out, centrifugation, and dialysis. Gaurav Kumar et al. [29] obtained the collagen supernatant by the conventional purification method. They used sodium chloride as salt (2.6 mol l^{-1}). The precipitate was collected by centrifugation and then dialyzed with 0.1 mol l^{-1} acetic acid for 48 hours

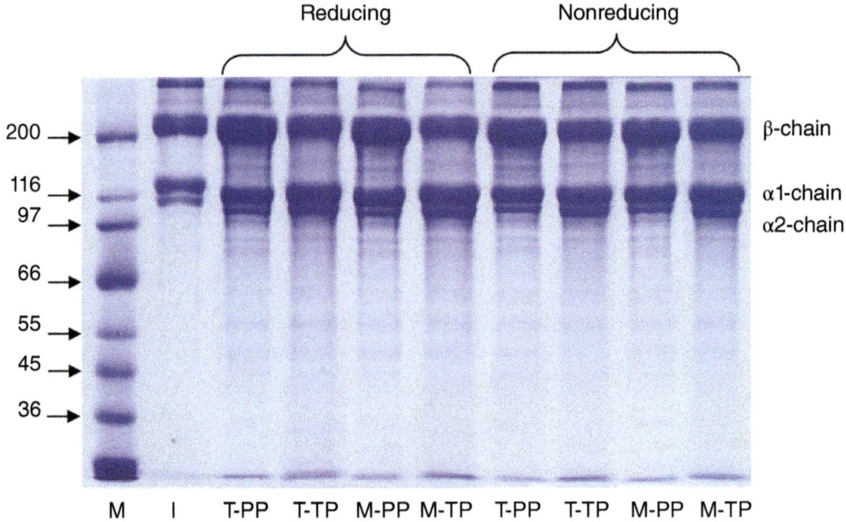

Figure 3.6 Protein patterns of collagens from fish skin extracted with aid of pepsins: T-PP, from the skin of *Priacanthus tayenus* using porcine pepsin; T-TP, from the skin of *Priacanthus tayenus* using tongol tuna pepsin; M-PP, from the skin of *Priacanthus macracanthus* using porcine pepsin; M-TP, from the skin of *Priacanthus macracanthus* using tongol tuna pepsin; M, high-molecular-weight protein markers; I, type I collagen. Source: Benjakul et al. [27]/with permission of © Society of Chemical Industry.

(with a molecular weight of 12–14 kDa). The acetic acid was further replaced with distilled water to obtain a neutral pH value. During salting out, they also used different salt precipitation methods to change the salt concentration, thus extracting different types of collagen [30]. Although the traditional method for collagen purification is commonly used, it is not suitable for large-scale production due to the complicated operation process and low extraction efficiency. Due to the technology development of ultrafiltration membranes, ultrafiltration combined with electrophoresis obviously shortens the time of the purification process. Although this combination method needs certain pressure and related equipment, it is still widely studied and used for collagen extraction. Chen et al. [31] separated and purified the collagen based on the molecular weight of the protein and the characteristics of the membrane. By the hydrophilic ultrafiltration technology, the separation time of collagen can be shortened to ~3 hours and 45 l of collagen solution can be processed at one time, showing a good prospect for industrialization.

3.3 Applications of Collagen

3.3.1 Clinical Applications of Collagen

Due to the excellent biocompatibility and safety, collagen and collagen-based biomaterials have been widely used in biomedical and bioengineering areas. In the very early days, collagen sutures were used by Egyptian surgeons [3]. The main

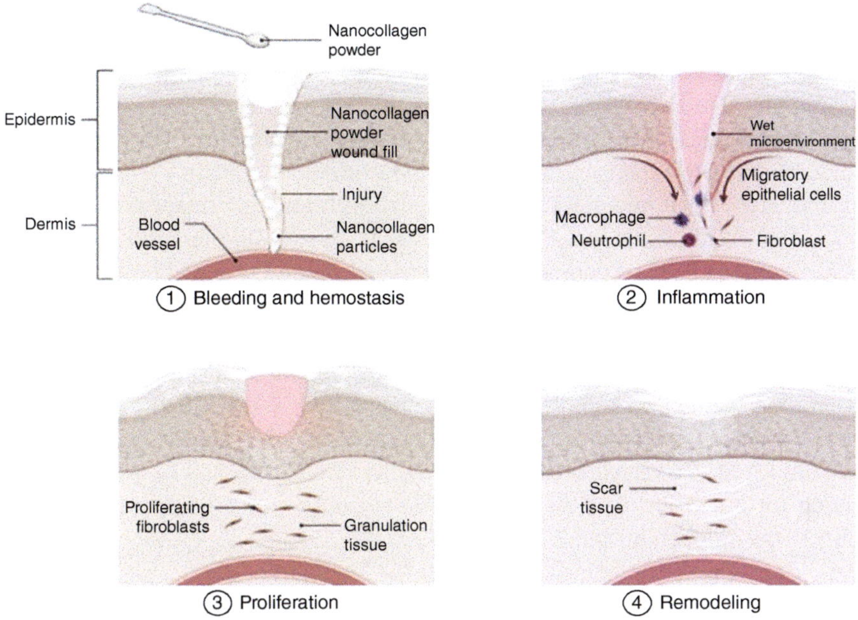

Figure 3.7 An illustrative image demonstrating a proposed skin wound-healing process with the application of nano collagen powders. Source: Lo and Fauzi [37]/MDPI/Licensed under CC BY 4.0.

applications of collagen include shields in ophthalmology, sponges for wounds, and minipellets for protein delivery, as basic matrices for peripheral nerve repair [32]. Collagen can be processed and used in different forms, including powders, rods, tubes, films, sponges, injectable solutions, and dispersion liquids [33, 34].

M. Xeroudaki et al. [35] reported a collagen-based hydrogel that has good transparency and mechanical properties. The collagen-based hydrogel could be used to replace diseased corneal stroma. In in vivo model of human corneal surgery, the hydrogel with porosity not only allowed migration and population by host cells but also maintained their transparency and thickness. It is observed that the host stromal cell was repopulated and the host nerves were regenerated. These are necessary for the regeneration of the corneal stroma. In addition, the hydrogel was also used as a carrier which was loaded with a neurodegenerative drug. It is found that the hydrogel realized the sustained drug delivery in vitro. The prepared hydrogel and its application in therapeutic implantation are promising in the replacement and regeneration of diseased corneal stroma.

Collagen powders can be used in the wound healing of skins because powders can be well filled in the wound fissures. The collagen powders further act as active sites for fibronectin-binding and fibroblasts' growth [36]. Figure 3.7 [37] demonstrates that the nanosized collagen particles can well attach to all the wound crevices. Then, the collagen powders absorb exudate secretion and blood, which creates a wet microenvironment. The microenvironment thus serves as a bio-template in the wound, beneficial to wound healing via cell migration and proliferation.

In addition to collagen hydrogels and collagen powders, the well-aligned collagen fibers are promising for use in the engineering of the tendon tissue, because of their biomimetic compositions and topographic structures. Yang et al. [38] prepared an aligned collagen membrane (CM_a, orientation angle 0° to 15°) from insoluble collagens using counter-rotating extrusion technology. As comparison, a randomly oriented collagen membrane (CM_r, orientation angle −60° to 60°) was also fabricated. It is found that CM_a had a high tensile strength that was comparable to native rat Achilles tendon. Therefore, CM_a is a promising biomimetic scaffold that has similar chemical composition, alignment, and mechanical properties with native tendon tissues. To confirm the practical application of CM_a in engineering of tendon tissue, they investigated the in vitro tenogenic differentiation of rat bone marrow mesenchymal stem cells (rBMSCs) on CM_a and the in vivo tendon regeneration using a rat Achilles tendon defect model. It is verified that CM_a can promote much higher tenogenic differentiation of rBMSCs than CM_r since CM_a could induce an elongated cell shape along the fibers. The in situ tendon repair research further confirmed that CM_a–BMSCs could show a comparable healing quality to the autogenous tendon (Figure 3.8) [38].

3.3.2 Food Applications of Collagen

Collagen can also be a health supplement ingredient since it contains essential amino acids, such as hydroxyproline, lysine, isoleucine, and phenylalanine. Among them, hydroxyproline is beneficial to the transport of calcium from plasma to bone cells. Mizuno et al. [39] reported that bone marrow cells exhibited higher activity when they were cultured in type I collagen. This result revealed that the type I collagen can induce osteoblast differentiation of the bone marrow cells. Lu et al. [40] studied the extracted collagen by the enzymatic method. It is shown that that the extracted collagen has high binding properties with calcium ions since amino and carboxyl functional groups serve as binding sites for the calcium ions [41]. As additives for foods, collagen can not only supplement amino acids but also improve the taste of food. In addition, the extracted collagen can be used to make food itself. Li et al. [42] mixed the collagen (extracted from tilapia scale) with pineapple juice to obtain a nutrient-rich collagen fruity beverage. Yi et al. [41, 43] produced fish jelly products using eel by-products as raw materials through extraction, seasoning, and molding processes, which had rich nutrition and good flavor.

Ahmad et al. [44] evaluated the application potential of collagen/chitosan (CG/CH) and collagen/soy protein isolate (CG/SPI) composite films for food packaging (Figure 3.9). It is showed that the CG/CH film had higher elongation at breakpoint (EAB) and lower tensile strength and elasticity modulus than pure CG film. However, the CG/SPI films exhibited greater tensile strength and elasticity modulus but lower EAB than pure CG film. In addition, the CG/CH film exhibited better water vapor permeability than CG/SPI films. Different from the pure CG film, the transparency values and water solubility of the composite films were obviously decreased. In addition, the ultraviolet (UV) blocking property was obviously increased for the composite films. The Fourier-transform infrared

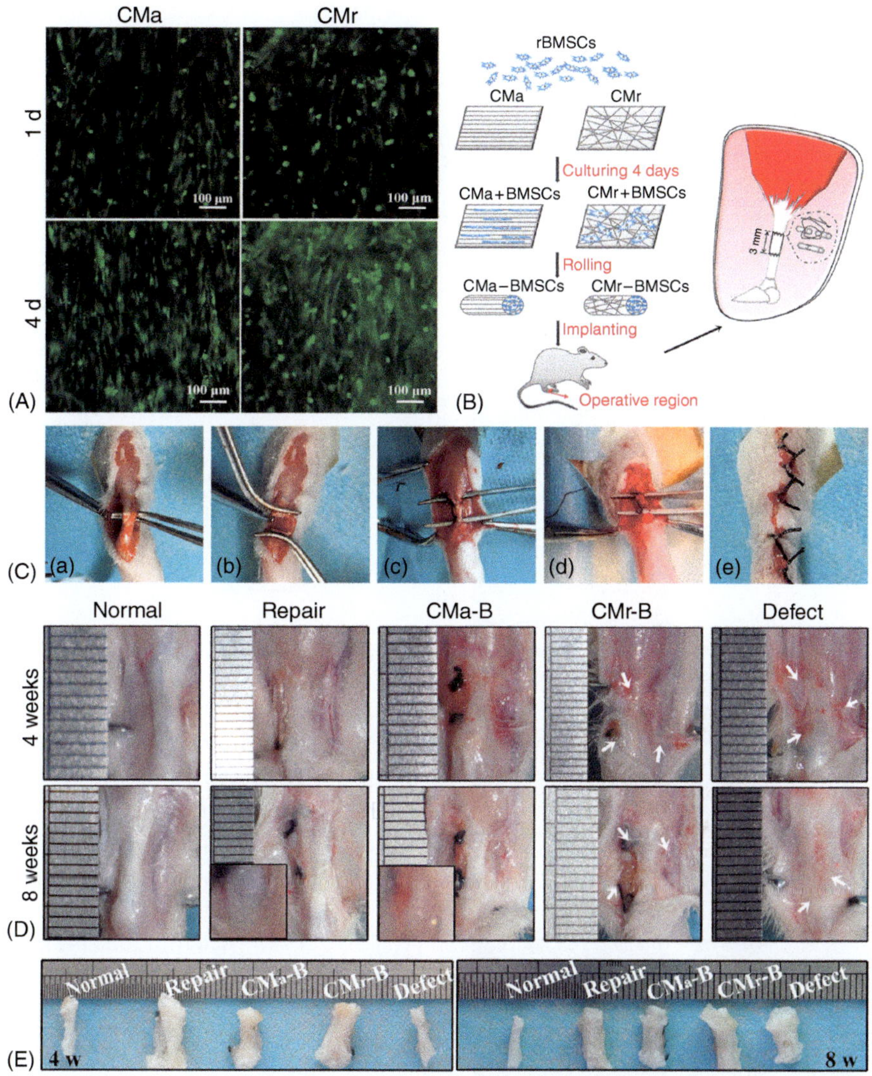

Figure 3.8 Construction of tissue-engineered tendons (A–C) and macroscopic alterations of the regenerated Achilles tendons (D, E). Source: Reproduced with permission from Yang et al. [38]; © 2019, Elsevier.

spectroscopy (FTIR) results showed that the CG/SPI film had rich intermolecular interactions. Thermogravimetric curves demonstrated that the composite films exhibited lower heat susceptibility than the pure CG film. The above performance of the collagen-based films shows great potential for use in biodegradable packaging.

3.3.3 Cosmetics Applications of Collagen

As an important material in the cosmetics industry, collagen is widely used to make products for skin and lip care [45]. The beauty products made from collagen are

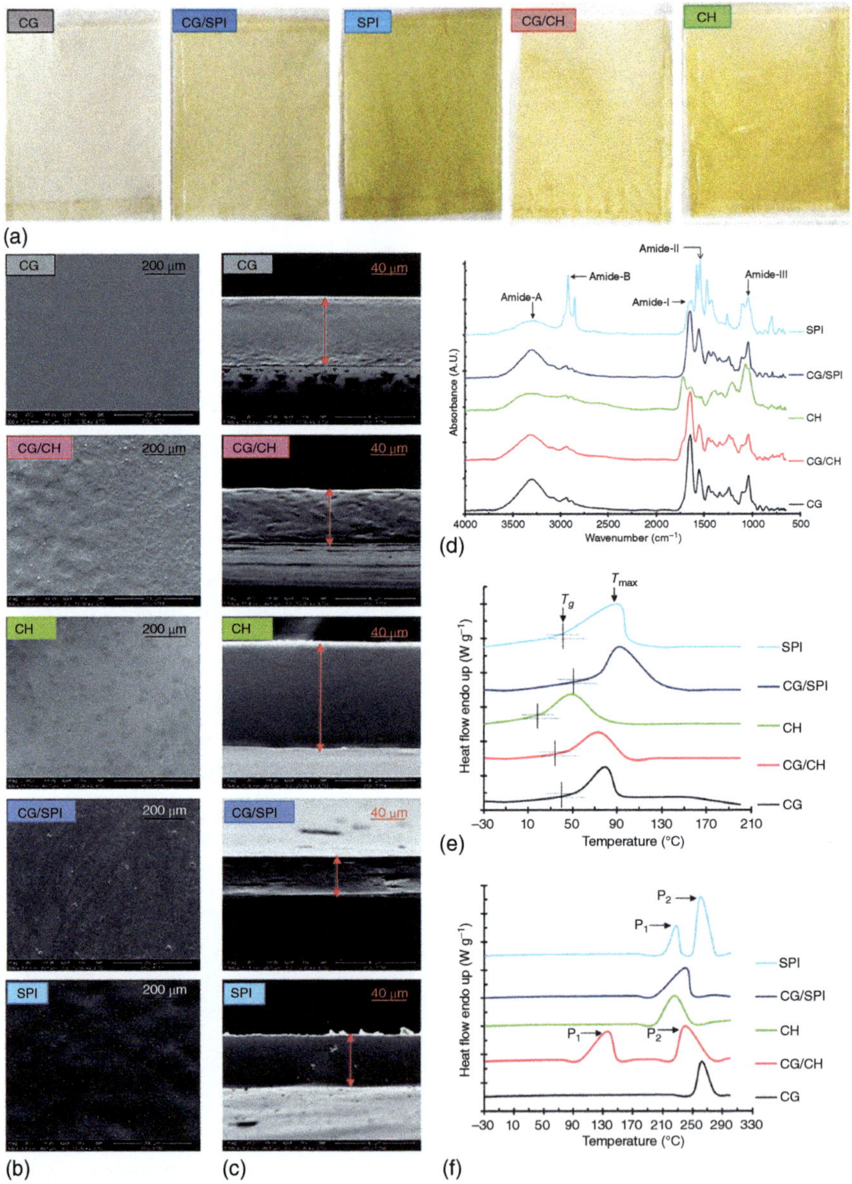

Figure 3.9 (a) Photographs of different films prepared from CG, CH, SPI, CG/CH, and CG/SPI blend; scanning electron microscopy (SEM) micrographs of (b) surface and (c) cryo-fractured cross section of film from CG, CH, SPI, CG/CH, and CG/SPI blend; (d) FTIR spectra of films from CG, CH, SPI, CG/CH, and CG/SPI blend; differential scanning calorimetry (DSC) thermograms of (e) First heating scan and (f) second heating scan of films from CG, CH, SPI, CG/CH, and CG/SPI blend. Source: Reproduced with permission from Ahmad et al. [44]; © 2016, Royal Society of Chemistry.

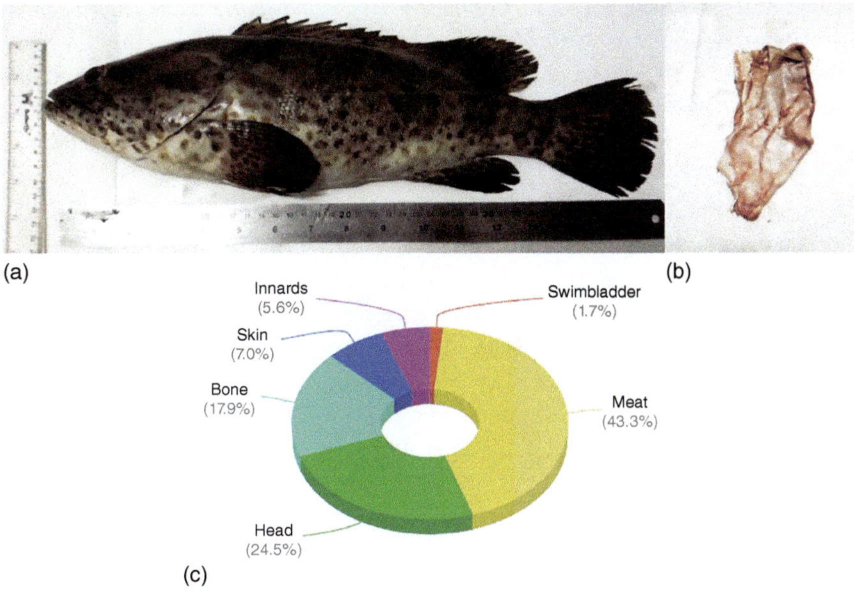

Figure 3.10 Visual appearance of (a) grouper fish and (b) grouper swim bladder. (c) The proportion of the body parts for grouper. Source: Trilaksani et al. [45]/with permission of © IOP Publishing Ltd.

beneficial to skin health, keeping moisture and elasticity and reducing aging signs [46]. In the beauty industry, collagen is often extracted from terrestrial animals, such as cows and pigs [45]. However, these collagens may draw attention to religious beliefs [47]. Moreover, applications of these collagens might also be a risk to public health since they can lead to diseases, such as tapeworm infections and mad cow diseases [45].

Trilaksani et al. [45] used collagen derived from grouper swim bladder as an alternative to terrestrial animal collagen. They investigated and compared the quality of collagen as cosmetic materials. It is shown that collagen derived from grouper swim bladder complied with the standards for cosmetic materials (Figure 3.10). In addition, they also found that the size of the collagen molecules was closely related to their effectiveness. The nano collagen produced positive effects on skin health since it has better penetration abilities to the skin. The above research shows that aquatic collagen is as effective as terrestrial animal collagen, thus avoiding issues in religious beliefs and diseases.

References

1 Fratzl, P. (2008). *Collagen Structure and Mechanics*. Germany: Springer Science+Business Media, LLC.

2 Prockop, D.J. and Kivirikko, K.I. (1995). Collagens: molecular biology, diseases, and potentials for therapy. *Annual Review of Biochemistry* 64: 403–434.

3 Peter, M.R. and Beat, S. (2002). *Connective Tissue and Its Heritable Disorders: Molecular, Genetic, and Medical Aspects*, 2e. New York: Wiley-Liss, Inc.
4 Abraham, J.D., Neeraj, K., and Aviva, E. (2011). *Biodegradable Polymers in Clinical Use and Clinical Development*, 59–89. Hoboken, NJ: Wiley.
5 Brinckmann, J., Notbohm, H., and Müller, P.K. (2005). *Collagen*. Berlin, Heidelberg: Springer.
6 Ramachandran, G.N. and Reddi, A.H. (1976). *Biochemistry of Collagen*. New York: Springer Science+Business Media.
7 Berg, R.A. and Prockop, D.J. (1973). The thermal transition of a non-hydroxylated form of collagen. evidence for a role for hydroxyproline in stabilizing the triple-helix of collagen. *Biochemical and Biophysical Research Communications* 52: 115–120.
8 Jimenez, S., Harsch, M., and Rosenbloom, J. (1973). Hydroxyproline stabilizes the triple helix of chick tendon collagen. *Biochemical and Biophysical Research Communications* 52: 106–114.
9 Eyre, D.R. (1980). Collagen: molecular diversity in the body's protein scaffold. *Science* 207: 1315–1322.
10 Bann, J.G., Peyton, D.H., and Bächinger, H.P. (2000). Sweet is stable: glycosylation stabilizes collagen. *FEBS Letters* 473: 237–240.
11 Ashutosh, T., Murugan, R., Hisatoshi, K., and Anthony, P.F.T. (2012). *Biomedical Materials and Diagnostic Devices*. Salem, MA: Scrivener Publishing LLC.
12 Kadler, K.E., Holmes, D.F., Trotter, J.A., and Chapman, J.A. (1996). Collagen fibril formation. *Biochemical Journal* 316: 1–11.
13 Fratzl, P. (2008). *Collagen: Structure and Mechanics*, 15–47. New York: Springer Science + Business Media, LLC.
14 Parkinson, J., Kadler, K.E., and Brass, A. (1995). Simple physical model of collagen fibrillogenesis based on diffusion limited aggregation. *Journal of Molecular Biology* 247: 823–831.
15 Fei, L. and Perrett, S. (2009). Effect of nanoparticles on protein folding and fibrillogenesis. *International Journal of Molecular Sciences* 10: 646–655.
16 Kuznetsova, N., Chi, S.L., and Leikin, S. (1998). Sugars and polyols inhibit fibrillogenesis of type I collagen by disrupting hydrogen-bonded water bridges between the helices. *Biochemistry* 37: 11888–11895.
17 Lanfer, B., Freudenberg, U., Zimmermann, R. et al. (2008). Aligned fibrillar collagen matrices obtained by shear flow deposition. *Biomaterials* 29: 3888–3895.
18 Falini, G., Fermani, S., Foresti, E. et al. (2004). Films of self-assembled purely helical type I collagen molecules. *Journal of Materials Chemistry* 14: 2297–2302.
19 Henshaw, D.R., Attia, E., Bhargava, M., and Hannafin, J.A. (2006). Canine ACL fibroblast integrin expression and cell alignment in response to cyclic tensile strain in three-dimensional collagen gels. *Journal of Orthopaedic Research* 24: 481–490.
20 Yamauchi, M., Katz, E.P., and Mechanic, G.L. (1986). Intermolecular cross-linking and stereospecific molecular packing in type I collagen fibrils of the periodontal ligament. *Biochemistry* 25: 4907–4913.

21 Yamauchi, M., Banes, A.J., Kuboki, Y., and Mechanic, G.L. (1981). A comparative study of the distribution of the stable crosslink, pyridinoline, in bone collagens from normal, osteoblastoma, and vitamin D-deficient chicks. *Biochemical and Biophysical Research Communications* 102: 59–65.

22 Kuboki, Y. and Mechanic, G.L. (1982). Comparative molecular distribution of cross-links in bone and dentin collagen. Structure–function relationships. *Calcified Tissue International* 34: 306–308.

23 Sousa, R.O., Alves, A.L., Carvalho, D.N. et al. (2020). Acid and enzymatic extraction of collagen from Atlantic cod (*Gadus Morhua*) swim bladders envisaging health-related applications. *Journal of Biomaterials Science, Polymer Edition* 31: 20–37.

24 Blanco, M., Vázquez, J.A., Pérez-Martín, R.I., and Sotelo, C.G. (2019). Collagen extraction optimization from the skin of the small-spotted Catshark (*S. Canicula*) by response surface methodology. *Marine Drugs* 17: 40.

25 Sinthusamran, S., Benjakul, S., and Kishimura, H. (2013). Comparative study on molecular characteristics of acid soluble collagens from skin and swim bladder of Seabass (*Lates Calcarifer*). *Food Chemistry* 138: 2435–2441.

26 Wen, H.F., Chen, L.L., Bai, C.Q. et al. (2016). Comparative study on physical and chemical properties of collagens obtained by different extraction methods from the skin of the Chinese Longsnout Catfish Leiocassis longirostris. *Food Science* 37: 74–81.

27 Benjakul, S., Thiansilakul, Y., Visessanguan, W. et al. (2010). Extraction and characterisation of pepsin-solubilised collagens from the skin of Bigeye snapper (*Priacanthus tayenus* and *Priacanthus macracanthus*). *Journal of the Science of Food and Agriculture* 90: 132–138.

28 Lin, L.H., Zhu, F.Q., Peng, H.J. et al. (2015). Preparation of collagen in abalone organs. *Food Industry* 36: 115–120.

29 Gaurav Kumar, P., Nidheesh, T., Govindaraju, K. et al. (2017). Enzymatic extraction and characterisation of a thermostable collagen from swim bladder of Rohu (*Labeo rohita*). *Journal of the Science of Food and Agriculture* 97: 1451–1458.

30 Cumming, M.H., Hall, B., and Hofman, K. (2019). Isolation and characterisation of major and minor collagens from hyaline cartilage of Hoki (*Macruronus novaezelandiae*). *Marine Drugs* 17: 223.

31 Chen, S.J., Chen, H., Xie, Q. et al. (2016). Separated and purified the collagen based on the molecular weight of the protein and the characteristics of the membrane. *Food Hydrocolloids* 52: 468–477.

32 O'Grady, J.E. and Bordon, D.M. (2003). Global regulatory registration requirements for collagen-based combination products: points to consider. *Advanced Drug Delivery Reviews* 55: 1699–1721.

33 Fujioka, K., Maeda, M., Hojo, T., and Sano, A. (1998). Protein release from collagen matrices. *Advanced Drug Delivery Reviews* 31: 247–266.

34 Lee, C.H., Singla, A., and Lee, Y. (2001). Biomedical applications of collagen. *International Journal of Pharmaceutics* 221: 1–22.

35 Xeroudaki, M., Thangavelu, M., Lennikov, A. et al. (2020). A porous collagen-based hydrogel and implantation method for corneal stromal regeneration and sustained local drug delivery. *Scientific Reports* 10: 16936.
36 Munish, T., Ramneesh, G., Sanjeev, S. et al. (2015). Comparative study of collagen based dressing and standard dressing in diabetic foot ulcer. *Journal of Evoulution of Medical and Dental Sciences* 4: 3614–3621.
37 Lo, S. and Fauzi, M.B. (2021). Current update of collagen nanomaterials – fabrication, characterisation and its applications: a review. *Pharmaceutics* 13: 316.
38 Yang, S., Shi, X., Li, X. et al. (2019). Oriented collagen fiber membranes formed through counter-rotating extrusion and their application in tendon regeneration. *Biomaterials* 207: 61–75.
39 Mizuno, M., Fujisawa, R., and Kuboki, Y. (2000). Type I collagen-induced osteoblastic differentiation of bone-marrow cells mediated by collagen-A2β1 integrin interaction. *Journal of Cellular Physiology* 184: 207–213.
40 Lu, Y., Nie, R., Li, F., and Liu, Z. (2016). Effects of calcium-binding peptide from tilapia scale protein hydrolysates on calcium absorption in caco-2 cells. *Journal of Aquatic Food Product Technology* 25: 1213–1220.
41 Zhou, R. and Chen, S.S. (2021). Research progress on extraction methods of fish skin collagen and its application in food industry. *Acta Agriculturae* 37 (1): 129–135.
42 Li, J.M., Liu, H., Zhang, C.H., and Zhu, Y.L. (2014). The research on the processing techniques and stability of composite beverage of tilapia collagen and fruit juice. *Journal of Anhui Agricultural Science* 42: 2120–2125.
43 Yi, C.P., Zhu, B., and Zhong, C.M. (2014). Preparation of fish jelly from *Elopichthys bambusa* byproduct. *Science and Technology of Food Industry* 35: 193–195.
44 Ahmad, M., Nirmal, N.P., Danish, M. et al. (2016). Characterisation of composite films fabricated from collagen/chitosan and collagen/soy protein isolate for food packaging applications. *RSC Advances* 6: 82191–82204.
45 Trilaksani, W., Adnyane, I.K.M., Riyanto, B., and Safitri, N. (2019). Nano collagen of the grouper swim bladder in compliance with quality standard of cosmetics materials. *IOP Conference Series: Earth and Environmental Science* 404: 012050.
46 Avila Rodríguez, M.I., Rodríguez Barroso, L.G., and Sánchez, M.L. (2018). Collagen: a review on its sources and potential cosmetic applications. *Journal of Cosmetic Dermatology* 17: 20–26.
47 Liu, D., Zhang, X., Li, T. et al. (2015). Extraction and characterization of acid- and pepsin-soluble collagens from the scales, skins and swim-bladders of Grass carp (*Ctenopharyngodon idella*). *Food Bioscience* 9: 68–74.

4 Gelatin

4.1 History of Gelatin

The word gelatin is derived from the Latin word gelatus, which means frozen. It had been confirmed by archeologists that certain forms of crude gelatin had started to be used 8000 years ago. The samples containing collagen glue were obtained from a cave close to the Dead Sea and analyzed by chemists and microbiologists at the Weizmann Institute in Israel. The result definitely indicated that the residents in the New Stone Age were well aware of the adhesive ability of collagen glue and utilized it for various applications. Discoveries in both Thebes (current Luxor) and Deir el-Bahari (in the funerary temple of Queen Hatshepsut) also definitely demonstrated that Egyptians used animal glue at that time. Currently, this kind of glue, namely glutin glue, is still being prepared using the collagen from animal hides. Additionally, a few pyramid texts indicated that the bones were also utilized to produce a variety of soups and broths in Egypt [1].

Therefore, it can be clearly established that gelatin was used to prepare food during the first few centuries CE. In this period, glue was obtained via boiling the bone and hide of animals. Meanwhile, another kind of product was also extracted that became the "original form" of edible gelatin after being cooled and solidified.

Over the centuries, gelatin was an article of luxury, which was utilized for the production of luxurious jellies for dishes placed in aristocratic villas or courts. Up to the late seventeenth century, its physiological and nutritional properties were propagated by scientists.

A kind of pressure cooker, named a "digester," was invented by the French mathematician Denis Papin in 1682, enabling to cook bones until soft [2]. Papin suggested to use the prepared stock for the preparation of soups. In 1681, he advised King Charles II of England that the jellies prepared from bone could be served as a normal food for people.

Gelatin had become a kind of popular foodstuff since the Napoleonic Wars. The French ports were blockaded by the British navy during the wars, which meant a shortage of meat protein for the population. To address this issue, politicians and scientists devoted themselves to find possible substitutes of meat protein and finally found that gelatin was suitable. In 1803, the administrator of the Military Hospital in Paris, Anton Alexis Cadet de Vaux, reported that gelatin prepared from bones could

Collagen-Derived Materials: Synthesis and Applications in Electrochemical Energy Storage and Conversion, First Edition. Feng Wang, Yaqin Huang, and Jin Niu.
© 2022 WILEY-VCH GmbH. Published 2022 by WILEY-VCH GmbH.

be used to produce bouillon. According to these reports, a commission led by the chemist D'Arcel compiled preparation methods for gelatin in 1818. Subsequently, the nutritive value of gelatin as a protein source was systematically researched. The manufacture and application of gelatin were further improved and optimized, which promoted the industrial production of gelatin.

Coignet & Cie., established in 1818 in Lyons, France, was the first company to produce gelatin on a large industrial scale. The company took two steps to improve production: the first one was using "hide split," namely "glue leather," as a raw material and the other one was introducing a technology for the industrial drying of thick gelatin leaves. The production technology was still applied until the mid-twentieth century. The first batch of goods to be sold labeling "powder gelatin" was only gelatin bars or ground thick leaves.

In 1833, a new application of gelatin was developed when the French pharmacist Mothes was authorized a patent to produce gelatin capsules. The "original capsule" was manufactured via soaking a little leather sack full of mercury in gelatin solution with high concentrations. Subsequently, the chilled and dried gelatin film resembling a capsule was stripped from the leather bag. This was the first time that drugs were easier to dose with the assistance of these capsules and to be kept from the influences of humidity, cold, and heat. Moreover, the medicines inside the capsule did not become bitter (Figure 4.1).

The presence of gelatin also changed the history of photography. The Frenchman Louis Jacques Mandé Daguerre (1787–1851) developed the "daguerreotype" process of photographic printing, facilitating the decisive breakthrough in photography. Unfortunately, the process was slightly more complicated and not easy to carry out. Furthermore, producing unique original positives was the only possibility. Thus,

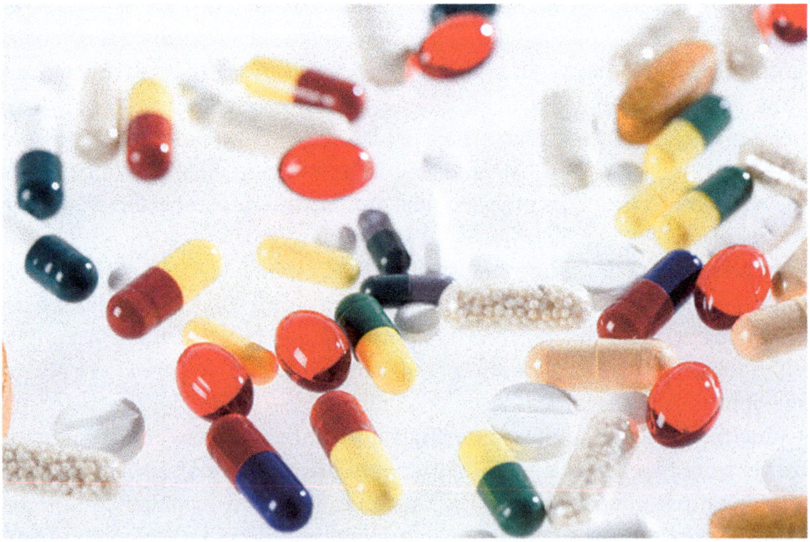

Figure 4.1 Capsule and tablet applications of gelatin. Source: Keenan [3]/with permission of © Elsevier Ltd.

only a few people could enjoy the hobby of photography owing to its expensive cost. After half a century, photography became uncomplicated with the help of gelatin, which was significant for its widespread popularity. Since 1880, the ready-made dry photoplate coated with gelatin emulsion has been available to make it possible for the production of negatives and positives. Nevertheless, photography became widely popular only when the famed "Kodak Number 1" camera was promoted by George Eastman in 1888.

Another significant step in making gelatin widely used was the appearance of family size packages of gelatin granules in America due to the hard work of Charles Knox and in Germany, in the stage of global economic depression, through DGF (Deutsche Gelatin Fabriken AG). This single event prompted gelatin to become a basic and common element for a lot of foodstuffs and contributed to its use around the world.

In 1845, Peter Cooper in America got the first patent for a gelatin dessert, which was another primary event to enable gelatin to be more widely known. In 1897, a new product, called JELL-O®, was prepared via adding fruit flavors. Currently, the JELL-O are annually sold for approximately 300 million boxes in America. The company Rowntree in Great Britain developed a different method to offer gelatin desserts to the citizens. Beginning with an analogous product to JELL-O, they began selling concentrated cubes, named Jelly Tablets, in 1932. The cubes included various fruit flavors as well. Though the product had achieved a big success in Great Britain, it had never been sold massively in other countries.

From then on, the gelatin producers and food industry have been consecutively creating applications of gelatin. More multifunctional properties of gelatin were discovered, and then gelatin was made into new products to users. Many products were actually on the lips of everybody: marshmallows had been marketed in America since the beginning of the twentieth century and were very prevalent in the 1950s, the famed gummy bears were developed in 1930, and fruit yogurts adding gelatin, which had no whey filmy layer on the surface, were first launched into the market by the 1970s (Figure 4.2).

In the late 1970s, hydrolyzed gelatin became a popular topic on the talk shows on television in America. Thus, the "liquid protein" of decreasing calorie beat the record in demand. Lage Sundstroem in Sweden invented the low-fat margarines and sandwich using gelatin, importantly bringing in the worldwide success. These products started to be manufactured on an industrial scale by 1984. Gelatin was served as the emulsifying agent and stabilizing agent to satisfy the demands of spreading ability and consistence by users.

Currently, gelatin is widely applied as a modern product. Although inconspicuous at the first sight, it is an essential ingredient in our daily lives. For instance, match heads include gelatin and digital holiday snapshots can be printed on ink-jet photographic paper of high quality or on traditional photographic paper in the photo lab with coated gelatin. Additionally, gelatin has been to be shown an effective cleansing agent. It can be used to clean up and refurbish constructions polluted by asbestos. It is applied to restore historical files as well. Healthy joints and bones can be supported with the help of gelatin. Although gelatin has a large number of applications, its

Figure 4.2 Food applications of gelatin. Source: Keenan [3]/with permission of © Elsevier Ltd.

many multifunctional properties have not been studied in detail. As a matter of fact, the gelatin with a highly diverse history has proven that its future has just started.

The demand for gelatin has augmented substantially around the world in the last few decades. Recent reports indicate that the annual worldwide output of gelatin amounts to around 326 000 tons of which pig-skin-derived gelatin accounts for the highest (46%) proportion followed by bovine hides (29.4%), bones (23.1%), and other sources (1.5%). The primary production region of gelatin is in western Europe with 40% production; the remaining is in North America 20%, Latin America 17%, eastern Europe 2%, and elsewhere 21% [4].

Although gelatin has such extensive applications, most consumers hold pessimistic and anxious attitudes toward its usage [5]. This is largely caused by religious sentiments (Judaism and Islam do not consume pork-related products while Hinduism forbids the consumption of cow-related products) as well as the enhanced and more rigorous insistence on vegetarianism all over the world [6]. Additionally, there is a constantly increasing attention among researchers about whether collagen and gelatin derived from animal tissue are able to transmit pathogenic vectors, such as prions [7]. However, studies carried out by a variety of authorities have demonstrated that the manufacturing process of gelatin provides a powerful barrier to defend possible bovine spongiform encephalopathy (BSE) prions.

In March 2003, the scientific steering committee of the European Union affirmed that the risk related with bovine bone gelatin was almost zero. On 18 January 2006, the European Food Safety Authority (EFSA) made a statement proclaiming that the remaining BSE risk in gelatin that stemmed from bone is considered as being extremely low [8].

A great amount of gelatin applied to pharmaceutical and food industries originates from either cow hide and bone or pigskin. Intensive research has been conducted to search for an alternative source for gelatin derived from mammalian sources owing to the outbreak of zoonotic problems, such as BSE, and religious sentiments. Moreover, the demand and cost have increased due to the fierce competition between manufacturers for the procurement of pigskin or other mammalian species. However, only a small number of alternatives are available to date, and hence it has been impossible to eliminate mammal-derived gelatin. Researchers from academia and industry are consecutively discovering an alternative to either gelatin or mammalian gelatin, such as gelatin from fish and poultry by-products. Although there is low commercial production currently, poultry skin and bones are expected to produce gelatin in the near future. Fish gelatin is also a better alternative to mammalian gelatin, particularly with properties such as a lower melting point (leading to faster dissolution). Unfortunately, the output of fish gelatin only accounts for 1% of the annual world gelatin production [9] (Figure 4.3).

4.2 Structures and Compositions

4.2.1 Molecular Structures

The presence of gelatin shows the multifunctionality of nature. Protein is the essential constituent of gelatin. Gelatin is mainly composed of protein, the content of which accounts for between 85% and 92%. Additionally, gelatin contains a small quantity of mineral salts and traces of moisture. Gelatin is prepared by the partial hydrolysis of native collagen, which is the primary structural protein in the connective tissue of animal skin and bone. The fundamental unit of collagen is the tropocollagen rod, which is a triple-helical protein structure. Collagen and gelatin are composed of different amounts of 18 amino acids, while gelatin includes relatively more nonpolar amino acids (>80%), for example, proline, alanine, and glycine [10–12]. Figure 4.4 illustrates the molecular structure of gelatin and Figure 4.5 shows the amino-acid composition of gelatin.

The proline content is especially significant due to the fact that it tends to facilitate the formation of the polyproline II helix defining the tropocollagen trimer. The primary structure of gelatin is closely similar to that of collagen, a triple-helical structure that can have two different compositions. The triple helix of type I collagen from skin and bone is made up from two α1 (I) and one α2 (I) chains, while type III collagen from skin (not present in bone) is composed of three α1 (III) chains [14]. Collagen consists of three α chains intertwined in this way to offer a three-dimensional structure, which provides a perfect conformation for interchain hydrogen bonding [15]. The major distinction between these three α-chain types is the composition of amino acid. The triple helix has a length of about 300 nm and the chain is about 105 kDa in molecular weight. It is stabilized

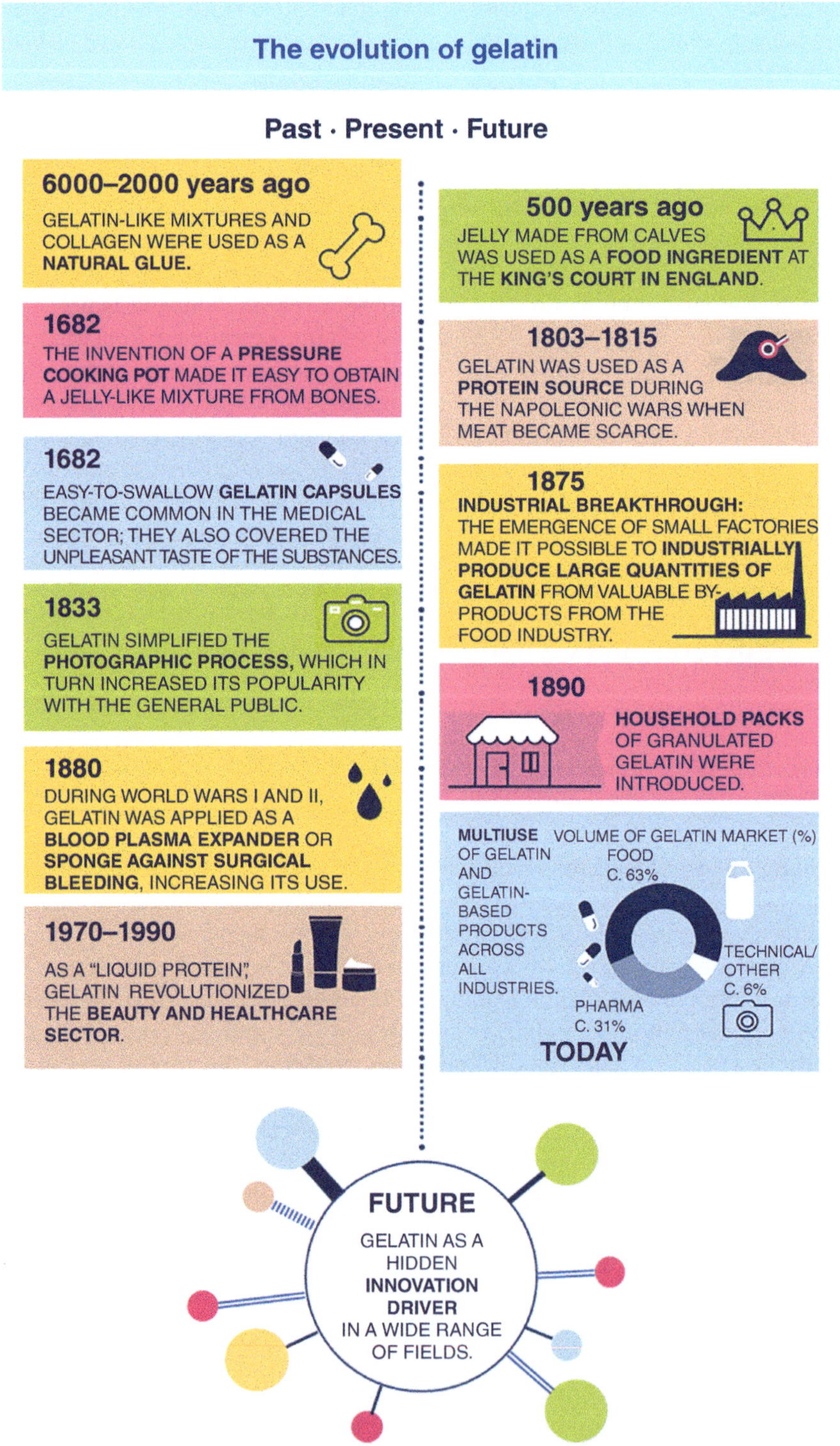

Figure 4.3 The evolution of gelatin. Source: From GME, Gelatine – Yesterday, today and tomorrow, The Evolution of Gelatine, 2021. https://www.gelatine.org/en/gelatine/history.html.

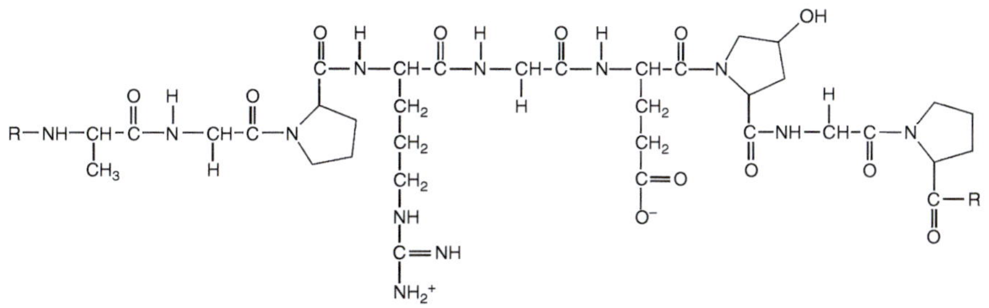

Figure 4.4 Molecular structure of gelatin. Source: Reproduced with permission from Kim [13]; © 2013, John Wiley & Sons.

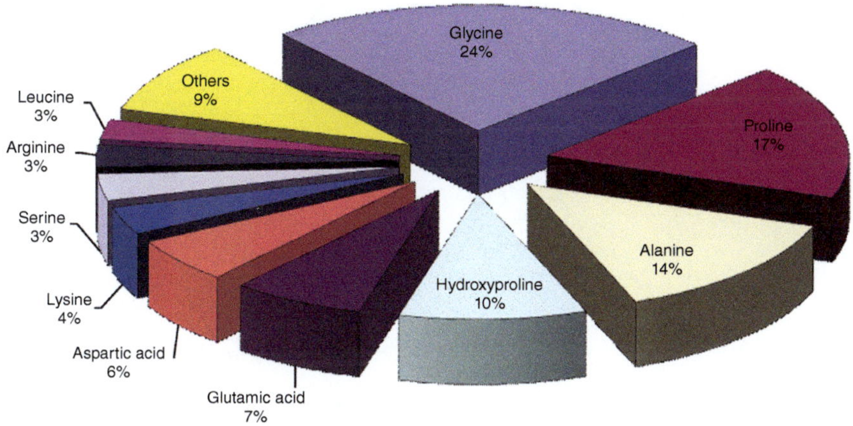

Figure 4.5 Amino-acid composition of gelatin. Glycine is the most abundant; the essential amino acid, tryptophan, is not present in gelatin. Source: Reproduced with permission from Keenan [3]; © 2012, Elsevier Ltd.

via the interchain hydrogen bonds. Owing to the fracture of hydrogen bonds leading to loss of triple-helix conformation, collagen denaturation results in the separation of the rods and total or partial separation of the chains, followed by the polymer in the form of coil. Industrial gelatins are mixtures of different polypeptide chains, which are composed of α chains (one polymer chain), β (α-chain dimer), and γ (α-chain trimer, higher molecular weight) ingredients with a molar mass of approximately 90, 180, and 300 × 103 g mol^{-1}, respectively [16]. Gelatin with more α chains would exhibit the property of higher gel strength. With increasing numbers of α chains, the existence of oligomers becomes more complicated and harder to read [17]. Therefore, it is necessary to separate these molecular weight fractions. Polyacrylamide gel electrophoresis (PAGE) is employed to get highly accurate molecular weight spectra of both commercial and laboratory gelatins to implement quantitative separation. Slight differences of structure result from the source of raw material as well as from the use of pretreatment and extraction procedures. Table 4.1 shows the component of an alkaline hide sample and an acid pigskin sample, both with very analogous Bloom and viscosity values [18].

Table 4.1 Component of gelatin [18].

Source	<A	A	α	β peptide	β	γ	γ-X	1-4	Q	Bloom (g)	Viscosity (mPa s)
Limed hide	28	4	30	6	13	3	8	4	3	250	5.7
Acid pigskin	35	4	10	12	11	6	11	7	4	250	5.5

Source: Johnston-Banks [18]/Springer Nature.

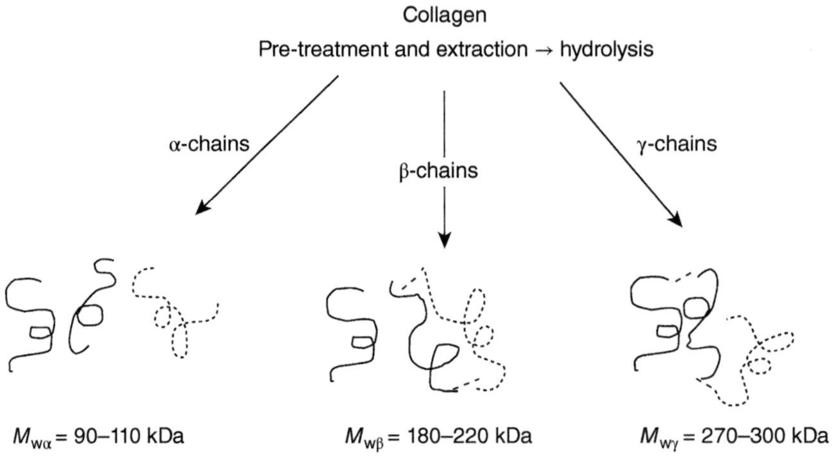

Figure 4.6 Gelatin is not a monodisperse protein but rather consists of a mixture of different chain types with varying molecular weights resulting in polydispersity. Source: Reproduced with permission from Haug and Draget [19]; © 2011, Elsevier.

Table 4.1 indicates that the contents of α structure and β peptides are the major differences between the samples. Generally, the Bloom value is proportional to the total amount of the α and β fractions, as well as their larger peptides, but the viscosity is proportional to the content of higher-molecular-weight fractions. The triple-helical structure of gelatin does not conform with any of the common structural categories, for example, the α helix, β sheet, or β turn discovered in the majority of globular proteins (Figures 4.6 and 4.7).

4.2.2 Amino Acids

During the manufacturing process of gelatin, the collagen molecule unfolds into a mixture of chains. Among these chains, the amino-acid composition is extremely unique. The three α chains involved in a collagen molecule may have a slight difference in the amino-acid compositions. Normally, the amino-acid sequence is Gly-XY, where X is proline and Y is hydroxyproline. This indicates that the proportion of glycine is approximately a third of all the remainder in gelatin and collagen (Figure 4.8). Amino acids with sulfur are not actually existent, and the

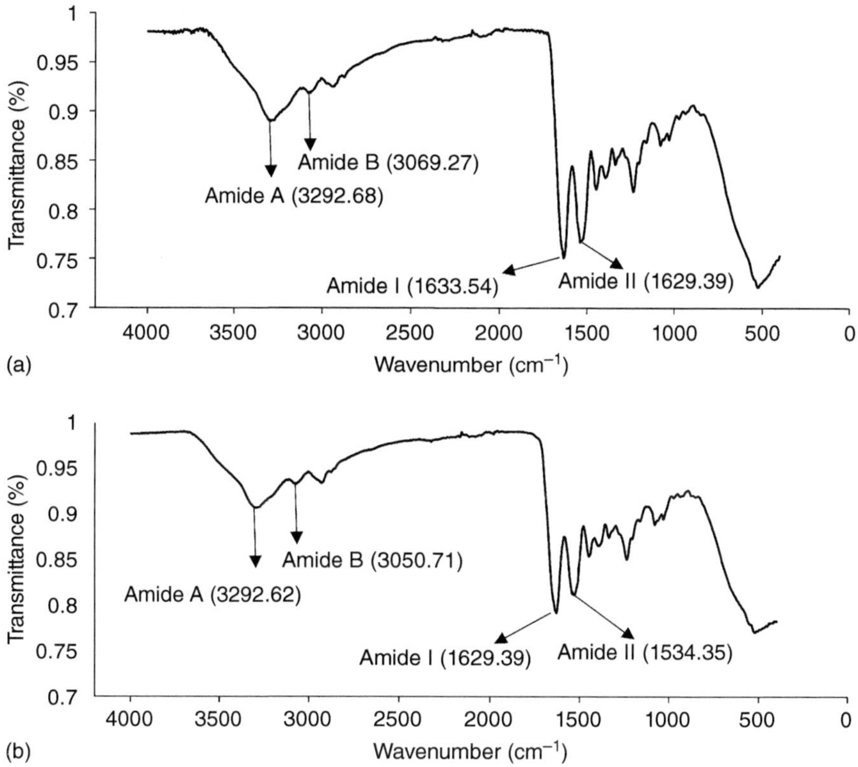

Figure 4.7 The Fourier transform infrared spectroscopy (FT-IR) spectra of fish by-products gelatin (a) and commercial halal bovine gelatin (b). Source: Reproduced with permission from Mirzapour-Kouhdasht et al. [20]; © 2020, Elsevier.

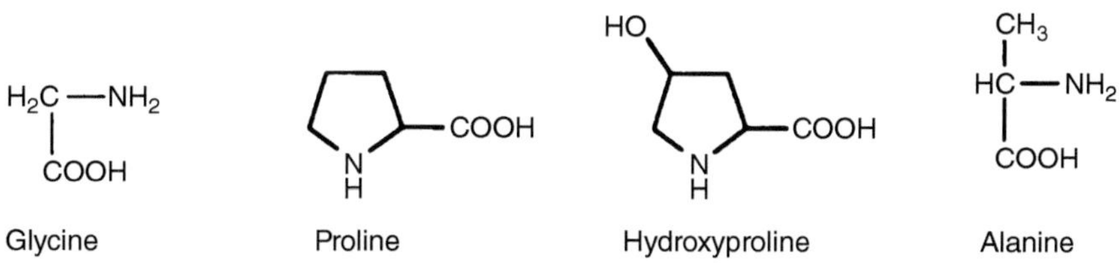

Figure 4.8 Chemical structures of the most important amino acids of collagen and gelatin. Source: Reproduced with permission from Nijenhuis [15]; © 1997, Springer-Verlag.

cross-links between the chains themselves do not involve these types of residues. Table 4.2 lists the amino-acid compositions of various gelatin and collagen.

Collagen is not totally dissolved in water, while a few fractions can be dissolved in salt or dilute acid solution. However, the solubility of collagen declines with the acceleration of aging processes in the majority of mammal animals, which is owing to an increase in the number of intra- and intermolecular covalent linkages generated in the source tissue after aging. Only if subjected to the use of strict processing,

Table 4.2 Amino-acid composition of collagen and four different gelatin-amino-acid residues per 1000 residues [19].

Amino acid	Type I collagen (bovine)[a)]	Type A gelatin[a)]	Type B gelatin[a)]	Cold-water fish gelatin[b)]	Warm-water fish gelatin[c)]
Alanine	114	112	117	112	123
Arginine	51	49	48	49	47
Asparagine	16	16	46	48	48
Aspartic acid	29	29			
Glutamine	48	48	72	72	69
Glutamic acid	25	25			
Glycine	332	330	335	347	347
Histidine	4	4	4	11	6
4-Hydroxyproline	104	91	93	60	79
Hydroxylysine	5	6	4	5	8
Isoleucine	11	10	11	11	8
Leucine	24	24	24	21	23
Lysine	28	27	28	28	25
Methionine	6	4	4	3	9
Phenylalanine	13	14	14	13	13
Proline	115	132	124	96	119
Serine	35	35	33	63	35
Threonine	17	18	18	24	24
Tyrosine	4	3	1	9	2
Valine	22	26	22	18	15

a) Babel [21].
b) Product information from Norland Products Inc-Cod, Pollack and Haddock.
c) Sarabin et al. [22].
Source: Haug and Draget [19]/Elsevier.

the steady intra- and intermolecular cross-links produce dissolved gelatin, subsequently resulting in the weight distribution of a polydisperse molecule. Amino-acid compositions of type A gelatin are almost similar to its parent collagen. As a result, the average isoelectric point of type A gelatin closely resembles that of collagen ranging from 7 to 9.4. However, type A bovine hide gelatin is one exception from this, which has a lower isoelectric point with the range of 5.7–7.4 as a result of specific pretreatments. The reason why type B gelatin lacks numerous the nonionizable glutamine and asparagine residues is that these amino acids are transformed into their carboxyl forms via alkali deamidation, and the gelatin becomes more acidic. Therefore, the average isoelectric points of gelatin by alkali processing are lower, ranging from 4.8 to 5.5.

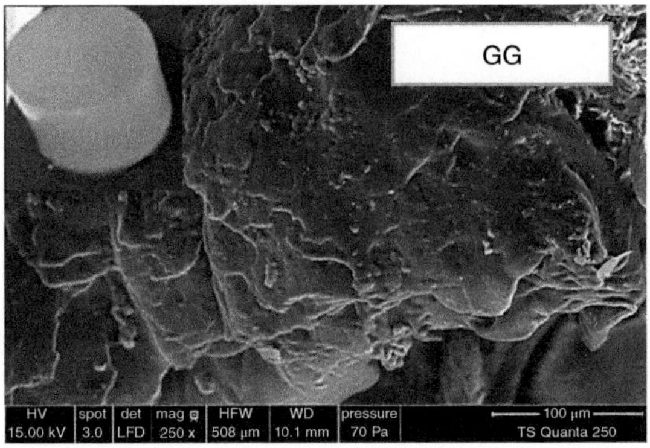

Figure 4.9 Microstructures of European eel gelatin gel. Source: Taktak et al. [23]/with permission of Elsevier Ltd.

Table 4.2 implies a significant difference between gelatin extracted from various sources. The total content of proline and hydroxyproline in collagen and gelatin differs quite dramatically between species. Different from residues (156%) of gelatin from cold-water fish, there are 217–223% residues for gelatin and collagen obtained from warm-blooded mammal animals (e.g. pig and cattle). For example, compared with gelatin from a cold-water fish species, gelatin from tilapia, a warm-water fish, shows a larger content of proline and hydroxyproline, and the sum of these imino acids is close to 200 [22].

The α chains of gelatin and collagen contain polar and nonpolar regions. The nonpolar regions are composed of the tripeptide Gly-Pro-R, where R is a nonpolar amino acid, dominated by hydroxyproline. These nonpolar regions are distributed with polar regions, which are slightly short of both proline and hydroxyproline. The unique properties of gelatin are endowed by the existence and distribution of the charged, polar and nonpolar amino acids. Owing to the existence of charged amino acids, gelatin is easily soluble in water under appropriate conditions and forms colloidal solutions, indicating that gelatin by definition is a hydrocolloid. As a result of its chemical properties, gelatin is a multifunctional hydrocolloid with significant surface power (Figures 4.9 and 4.10).

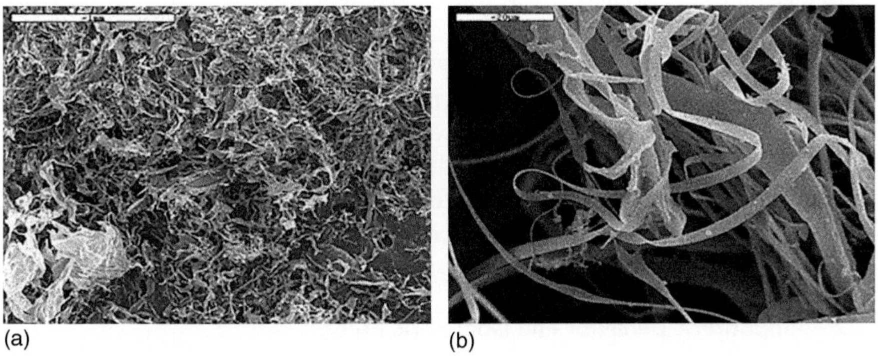

Figure 4.10 Images of scanning electron microscopy (SEM) of giant squid skin gelatin: (a) 150× magnification; (b) 1500× magnification. Source: Uriarte-Montoya et al. [24] / With Permission of © Elsevier Ltd.

4.3 Properties and Characterization

The physicochemical characteristics of gelatin are primarily dependent on the amino-acid sequence of molecules, the molecular mass distribution, the resulting spatial structure, and the environmental conditions (e.g. ionic strength, pH, and reaction with other compositions).

The functional characteristics of gelatin can be classified into two categories. The characteristics of the first category are related to gelling, such as melting and setting temperature, gel strength, viscosity, and gelling time. The second category is associated with the surface behavior of the gelatin. These characteristics include the adhesive characteristics, dissolution behavior, the formation and stabilization of emulsions and foams, and so on.

4.3.1 Viscosity and Adhesion

It is likely that the adhesive property of gelatin is known and used for the longest period of time (about 8000 years) among its surface activities. Cohesion results from the interaction between the molecules in a substance, while adhesion results from the interaction between the molecules of the different compositions in a system.

Gelatin solutions in high concentration totally enable to cover the surface profiles of the particles to be affixed to each other, which results in the formation of adhesion forces. Nevertheless, their cohesion is quite small in this condition. When the gelatin solution has been evenly distributed on the surfaces to be connected, it begins gelling during cooling and maintaining its shape. This step greatly enhances the cohesive power of the binding layer, hence generating the ideal adhesive effect. Therefore, the adhesion of gelatin is directly determined by the viscosity of the used solution rather than its gelling ability. If there is more liquid to be used for dissolution, a medium-Bloom gelatin is a better choice.

The manufacturer normally provides viscosity data for a gelatin, as this property is paramount in many applications and since the viscosity to a greater extent reflects the average molecular weight of a gelatin sample than the Bloom strength. The calibrated pipette is a major equipment to measure the viscosity for standard gelatin. It would record the flow time for 100 ml 6.67% (w/w) gelatin solution at $60.0 \pm 0.1\,°C$, and the viscosity values are described in mPas. Apart from the calibrated pipette, other measuring techniques are used as well. For example, the viscosity of gelatin hydrolysates is determined for 10% or 20% solutions at 25 or 30 °C. For some applications, rotational viscosimeters, rheometers, or other special-purpose devices may also be used at various temperatures and concentrations.

4.3.2 Amphoteric Behavior and Isoelectric Point

The isoelectric point (IEP), namely the electric zero, is quite significant in the effects of surface activities for gelatin. It is also named as the isoionic point according to the method of determining this parameter. In theory, there are a few differences between

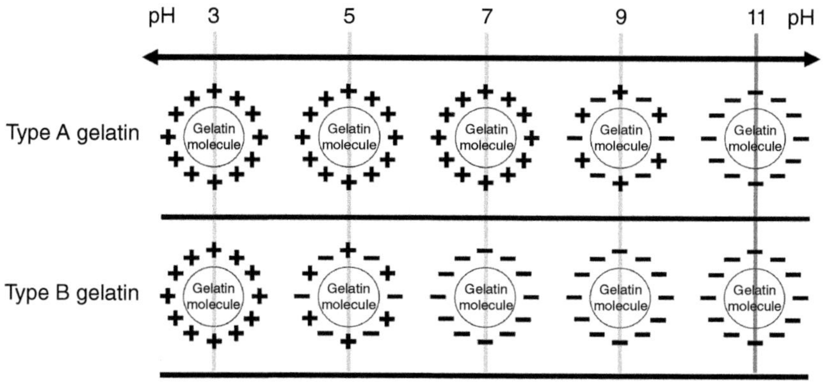

Figure 4.11 Charge distribution pattern of type A and B gelatins in aqueous solutions of different pH. Source: Reproduced with permission from Schrieber and Gareis [8]; © 2007, John Wiley & Sons.

the isoelectric point and the isoionic point. However, they can be considered as being identical in practice.

Figure 4.11 indicates that gelatin is neutral in charge if the pH of environment around the gelatin corresponds to the IEP. The gelatin becomes negatively charged with the increase in pH. Otherwise, the gelatin becomes positively charged. Thus, in gelatin solutions with a pH of about 5.0–9.0, alkaline-conditioned gelatin and acid-conditioned gelatin are negatively and positively charged, respectively. Over pH 9.0, all types of gelatins are negatively charged, while below pH 5.0, all are positive.

At its IEP, a protein molecule has a so-called random coil structure due to its neutral charge. However, the charge on the molecule of gelatin will also be changed when the pH of the matrix deviates from the IEP. After that, the charges are released and the molecule unfolds. This structural variation has positive effects on the surface activity of gelatin.

According to the pH of the solution to be whipped or emulsified in combination with the state of charge of its constituents, both type A and type B gelatin have been shown to be considerably applicable. Under some circumstances, gelatin mixtures are designedly employed to combine the unique properties of both categories.

For further manufacturing, it is paramount to know that turbidity and loss of firmness can appear if the pH of a particular product is exactly in accord with the IEP under the circumstance of low concentrations of gelatin. The IEP influences the compatibility of gelatin with other hydrocolloids as well. Therefore, the pH of the end product is a determining criterion in selecting the best suitable gelatin type besides the other constituents.

4.3.3 Surface Activity

The reason why gelatin possesses surface activities is that the gelatin side chains are similar to those of all proteins with electriferous groups and that hydrophobic or hydrophilic amino acids exist in some parts of the collagen sequence (Table 4.3).

Table 4.3 Hydrophilic and hydrophobic amino acids in gelatin [8].

Amino acid	Hydrophilic	Hydrophobic	Information
Alanine		+	Aliphatic, nonpolar side chain
Arginine	+		Basic
Asparagine	+		Weak polar side chain
Aspartic acid	+		Acid
Glutamine	+		Weak polar side chain
Glutamic acid	+		Acid
Glycine		+	Aliphatic, nonpolar side chain
Histidine	+		Basic
Hydroxyproline		+	Weak polar side chain
Hydroxylysine	+		Basic
Isoleucine		+	Nonpolar side chain
Leucine		+	Nonpolar side chain
Lysine	+		Basic
Methionine		+	Aliphatic, weak polar side chain
Phenylalanine		+	Aliphatic, nonpolar side chain
Proline		+	Nonpolar side chain
Serine	+		Weak polar side chain
Threonine	+		Weak polar side chain
Tyrosine	+		Aromatic, weak polar side chain
Valine		+	Nonpolar side chain

Source: Schrieber and Gareis [8]/John Wiley & Sons.

Since either hydrophilic or hydrophobic parts tend to move to the surfaces of aqueous solutions, the surface tension is reduced and the needed identically charged film is formed around the components of the dispersed phase, which is able to be further enhanced by the formation of gel. Thus, the gel firmness and the distribution of charge (IEP) of the used gelatin are two significant criteria for the selection of a suitable type of gelatin. The higher gel firmness leads to the larger Bloom value and a firmer gel-like protective sheath around the air bubbles or oil droplets under the same concentration and temperature.

In the meantime, several characteristics of gelatin can protect and stabilize the formed surfaces. These multifunctional characteristics of gelatin are used to prepare and stabilize foams and emulsions.

Emulsifiers are soluble substances that are capable of bringing together other substances which are normally separated from each other. If such small amounts of emulsifiers are added into a dispersion of an oily phase in an aqueous phase, their molecules incline to gather at the interface between the two phases. After that, the emulsifier covers the single-oil droplets with a quite thin film. These films repel each

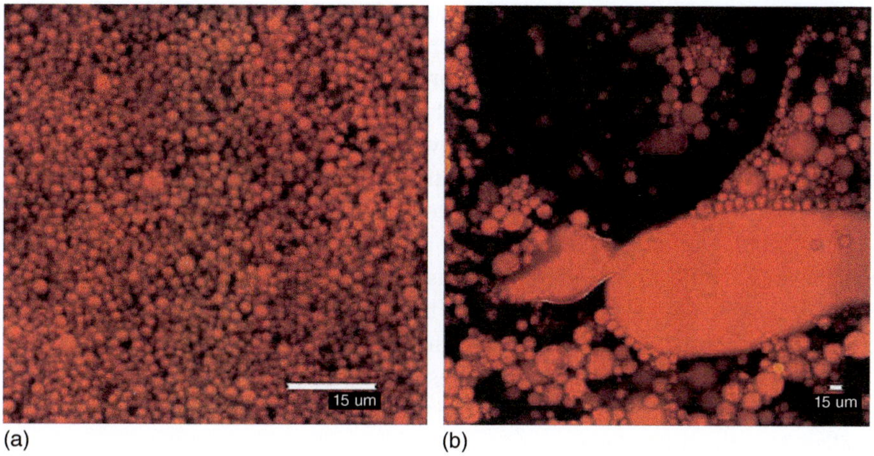

Figure 4.12 Stable emulsion with gelatin (a) and "broken" emulsion without gelatin (b). Source: Schrieber and Gareis [8]/with permission of © John Wiley & Sons.

other if they have the same electrical charge, hence essentially preventing the larger drops from forming. However, the fine emulsion film is generally destroyed by high mechanical stress and high temperatures, leading to destruction of the emulsion. The "breakdown" of emulsion is able to be inhibited via supplying it with extra stability (Figure 4.12).

The used stabilizer is supposed to enhance the viscosity of the emulsion and form a specifically steady protective sheath near the droplets or inhibit the droplets from being electrically neutralized. This last-named characteristic is referred to as the "protective colloid effect" of the stabilizer as well.

Type A gelatin has a slightly high isoelectric point ($pI \geq 7.0$), which is more appropriate for preparing positively charged oil-in-water emulsions over a wider range of pH values than conventional protein emulsifiers (e.g. soy, casein, or whey proteins) [25]. Early researches by Kim, Jeon, Lee, and Lee reported that cod-bone gelatin possessed emulsifying properties resembling those of a commercial emulsifier (e.g. Tween-80) [26]. Just like with gelling properties, the emulsion ability of fish gelatin is usually lower than that of mammal-derived gelatin. At the same temperature and concentration, the higher the gel firmness is, the firmer the gel-like protective sheath will be around the oil droplets [8]. Thus, gel firmness is a significant criterion for the selection of a suitable gelatin type besides the charge distribution. For instance, the emulsion activity index of pig-skin gelatin was higher than that of tuna-fin gelatin at the same protein concentration [27]. For both types of gelatin, the emulsion ability enhanced with increasing protein concentration from 2% to 5%, with high protein concentrations promoting more protein adsorption at the interface. Nevertheless, gelatin with a relatively low concentration (0.05%) could prepare oil-in-water emulsions, such as the gelatin from the skin of bigeye snapper fish [28]. The aforementioned authors ascribed the higher value of emulsion ability recorded via increasing gelatin concentrations from 0.1% to 0.2% to a higher degree of polypeptide unfolding during the shearing involved in the emulsifying process.

Apart from protein concentration, the molecular weight was probably a determining factor in affecting the capability of gelatin to form and stabilize oil-in-water emulsions. In this regard, fish-gelatin emulsion with low-molecular weight (~55 kDa) included more large droplets and showed more oil destabilization than that with high-molecular weight (~120 kDa) [29]. Owing to the relatively low surface activity of fish gelatin compared to globular proteins (e.g. β-lactoglobulin), there were a few large droplets existing in the emulsions after homogenization.

The molecular structure determines the capability to form and stabilize foams of the substance in question. Generally, it has to have surface activity, for only in this way is it likely to lead a product to foam by decreasing the surface tension at the interface between liquid and air. Increasing the viscosity of the continuous phase is essential for preventing the rapid collapsing of such a foam. Under an ideal condition, the foaming agent will enable the entire product to be stiffened to totally stabilize the structure of foam.

Gelatin is capable of lowering the surface tension at the interface between liquid and air by improving the viscosity of the aqueous phase, causing gelatin to exhibit fit-foaming characteristics even without gelling [8]. Foaming characteristics largely depend on the features of the raw material. For adsorption at the interface between air and water, molecules should include hydrophobic regions. Once protein unfolds, these hydrophobic regions become more exposed to facilitate the formation and stabilization of foam [30]. In this respect, gelatin from farmed giant catfish was discovered to have a higher foam ability than that from calfskin, which was possibly attributed to the higher content of hydrophobic amino-acid residues in the gelatin from farmed giant catfish. Furthermore, the nearly four-times higher viscosity of the gelatin from farmed giant catfish skin was one of the major reasons for its greater foam stability as well [31].

Due to the multifunctional characteristics, gelatin possesses to a high degree the necessary functions in the three application areas of an emulsifier, a stabilizer, or a whipping agent. Nevertheless, the emulsifying ability of gelatin is commonly weaker than that of other surface-active substances (e.g. globular proteins and gum arabic). Thus, gelatin usually creates relatively droplet sizes during homogenization when used in isolation. Additionally, to enhance the effectiveness as an emulsifier, it is necessary for gelatin to be either hydrophobically modified via attaching nonpolar side groups or used combining with anionic surfactants.

4.3.4 Gel Formation and Strength

Gelatin is classified as a physical gel, namely the interactions or bonds between the chains that physically make up the material in nature (van der Waals' interactions and hydrogen bonds with $E \approx 2\,\text{kcal mol}^{-1}$). Some physical gels (e.g. alginate) are thermally irreversible. However, gelatin is able to form thermoreversible gels due to the relatively weak bonding energy in gelatin. On cooling, these sols transform to gels, and on warming they return back to sols. Besides gelatin, pectin, agarose, carrageen and casein (a milk protein that forms spherical aggregates, approximately 20–300 nm in diameter) can reversibly form gels as well [32] (Figure 4.13).

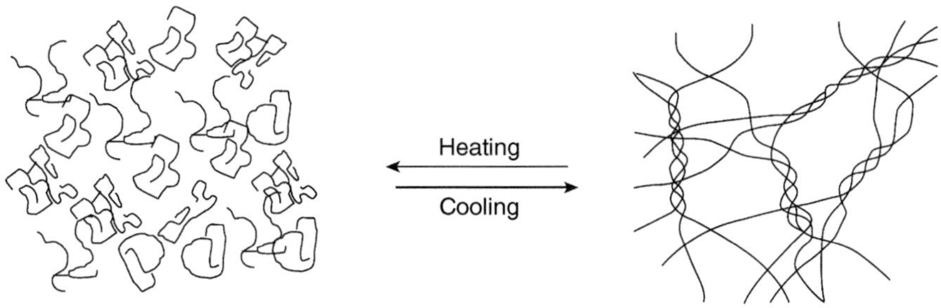

Figure 4.13 The thermoreversible gelling process for gelatin. Source: Reproduced with permission from Haug and Draget [19]; © 2011, Elsevier.

Gel strength is the main physical characteristic of gelatin gels. These are controlled by molecular weight, in combination with complicated interactions depending on the amino-acid composition and the ratio of α/β chains existing in the gelatin [33]. According to Schrieber and Garies [8], the gel strength is primarily determined by the proportion of fractions possessing a molecular weight of about $100\,000\,g\,mol^{-1}$. Additionally, the gel strength correlates strongly with the α-chain content in gelatin. Gelatin, including more α chains, would exhibit greater gel strength. However, the gel strength would be decreased if there is a high ratio of peptides, the molecular weight of which is lower or higher than the α chains [34]. Bloom values are used to express the gel strengths of commercial gelatin. These values are intrinsically used to measure the rigidity of a gelatin gel formed and tested in standard conditions. In 1925, O.T. Bloom invented the Bloom gelometer (US Patent No. 1540979), which is used to test the rigidity of gelatin gel via measuring the requested weight (force in grams) to compress the surface of a gelatin gel by 4 mm utilizing a flat-bottomed, cylindrical plunger of 12.7 mm in diameter. This tested method has been normalized by the American Organization of Analytical Chemists (AOAC), Gelatin Manufacturers Institute of America, and Gelatin Manufacturers of Europe (Standardized Methods for the Testing of Edible Gelatin). It is paramount that the production of the gelatin solution and gel, and the measurement of the Bloom strength are performed accurately based on the standardized methods to ensure reliable and repeatable results.

According to definition, the Bloom strength refers to gelatin gels matured only under the temperature of 10 °C. A Bloom value is unable to be confirmed when the gelatin solution does not gel at 10 °C for over 16–18 hours. Currently, the Bloom values are generally confirmed via employing a texture analyzer from Lloyds, a Stevens LFRA, or Stable Micro Systems. The used test plunger should meet the criterion of AOAC. Although with the same dimensions as the old BSI plunger (not used since 1998), the AOAC plunger has a sharp edge providing a higher surface area and a little larger Bloom value. The aforementioned standard methods also describe the measuring parameters of the instruments, which are supposed to be rigorously obeyed. It is significant that the test-sequence standard should include gelatins with known Bloom values to explain any other impacts on the test procedure, for example, preparation errors and instrumentation failures.

Table 4.4 Gel strength and melting point of various fish gelatins [36].

Type of gelatin	Bloom value (g)	Gelling temperature (°C)	Melting point (°C)
Sole fish gelatin	350	18–19	19.4
Tilapia fish gelatin	273	Not reported	25.4
Grass fish gelatin	267	19.5	26.8
Tuna fish gelatin	426	18.7	24.3
Catfish gelatin	243–256	15–18	23–27
Porcine gelatin	220–240	22–25	31–35
Bovine gelatin	210–240	20–24	30–33.8

Source: Domb et al. [36]/John Wiley & Sons.

The gelling strength of commercial gelatins ranges from 50 to 300, but the most optimal range of gelatins with Bloom values is between 250 and 260 [35]. Table 4.4 presents the Bloom values, gelling temperatures, and melting points of gelatins derived from bovine, porcine, and fish collagen as reported in the literature [4]. Compared with the bovine or porcine gelatin (both have Bloom values of 200–240), fish gelatin characteristically has a lower Bloom value with the range of 0–270 (measured under the conditions of the standard Bloom test). The various gelatins show the wide range of Bloom values, which result from the different sources of gelatins and are also related to the amino-acid composition and extraction conditions that affect the physical characteristics of gelatins, such as gelling point and gelling strength. For example, the use of sulfuric acid with high concentration, citric acid, and sodium hydroxide has been reported to cause low Bloom values, showing that the gel-forming ability of the gelatin is sensitive to acid and alkali hydrolysis due to the fact that both influence the cross-linking degree in the collagen [37]. Gelatin with a high Bloom value typically possesses higher melting and gelling points, shorter gelling times, and lighter color, as well as is more neutral in odor and taste in the final product. The higher gelling capability indicates that fewer gelatins are needed to result in the ideal gel firmness in the final product as well.

Gelatin concentration shows a good correlation with the Bloom value, which can be illustrated mathematically. In a pure aqueous solution, the gelatin concentration times and the square root of the Bloom value are constant for gelatin prepared by any of the following way:

$$C_1 \times B_1^{1/2} = C_2 \times B_2^{1/2}$$

or

$$C_2 = C_1(B_1/B_2)^{1/2}$$

Based on this formula, a manufacturer of products including gelatin can decrease the specific gelatin concentration from 8% to 7.7% by changing the Bloom value (such as from 250 to 280 Bloom). However, the same degree of gel firmness can still be obtained as these calculated values only give a preliminary rough indication. It is

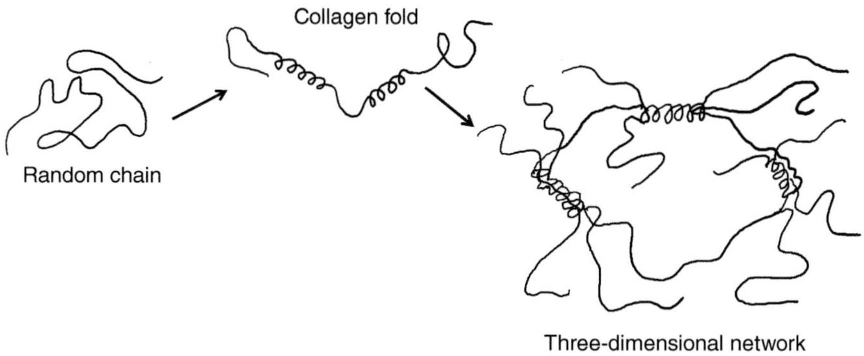

Figure 4.14 Model of gel formation (from sol to gel upon cooling). Source: Reproduced with permission from Schrieber and Gareis [8]; © 2007, John Wiley & Sons.

necessary for sensory and other analytical tests because all the other compositions of the formulation can affect the gelatin behavior and thus the dependence degree of gel firmness on the concentration. Alternatively, a harder, less rubber-like texture can be created by utilizing larger Bloom gelatin with constant concentration in the formulation. Such products are known as having "a short bite" by the food industry.

The molecular-weight distribution of gelatin determines all the hydrodynamic characteristics. The temperature at which the system is observed is the only distinction between gel formation and viscosity. This is easy to explain, as the mobility of the molecules in the liquid is dependent on the viscosity, while the gel firmness mainly depicts the structure formed via intermolecular hydrogen bonds (Figure 4.14). The mobile molecules aggregate to small clusters when a gelatin solution is cooled, followed by consecutively growing and finally forming a gel. Under this circumstance, hydrophobic and electrostatic interactions make some impact as well. Therefore, there is a consecutive transition between these two characteristics of gelatin.

However, there is no linear relationship between gel formation and viscosity. The standard viscosity and Bloom value, prescribed by the gelatin producer, are single-point measurements that are unable to be applied to come to conclusions in regard to the kinetics of forming gel, mainly due to the fact that the various molecular weight fractions also work. Therefore, the gel strength is primarily determined by the function of fractions with a molecular weight of about 100 000 g mol^{-1}, while the viscosity is mainly a proportion of those with the molecular weight in the range from 200 000 to over 400 000 g mol^{-1}. Thus, according to the method of production, gelatins with the same Bloom value can have considerably different viscosities.

4.3.5 Rheology of Gelatin Solution

To confirm the viscoelastic behavior of a special gelatin type during the gel process, it is necessary to use an oscillating rheometer. This makes the elastic and viscosity properties to be tested synchronously as a function of temperature and time. In an ideal case, the elastic and viscosity properties are defined as storage modulus G' and the so-called viscosity or loss modulus G″, respectively. According to the Bloom measurement conditions and using the storage modulus G′ within 17 hours, gel firmness can be evaluated using the Bloom gelometer.

During the process of gelation, there is a geometric point where the storage modulus G′ (elastic part) is in equilibrium with the loss modulus G″ (viscous part). This crossover point of G′ and G″ on the cooling curve represents the temperature of the sol–gel transition, namely the gelation point. In contrast, this crossover point is defined as the gel–sol transition temperature or gel melting point on the heating curve.

The cooling curve is never consistent with the heating curve. It presents an obvious hysteresis degree, and the melting process is at a higher energy level. Utilizing an empirical rule of thumb, the difference in temperature between the melting and the setting points in a pure aqueous gelatin solution of 5–25% is about 5 °C (Figure 4.15).

To summarize, the rheological behavior of aqueous gelatin solutions can be expressed as below:

A dilute gelatin solution is a Newtonian fluid exhibiting purely viscous behavior. However, cooling leads to variations in structure due to the formation of hydrophobic and hydrophilic interactions between the gelatin molecules. The system can be referred to as a solution in this initial phase. After further cooling, the structural formation process proceeds with the decrease of temperature until reaching the phase-transition point. At this point, some much more stable interactions are formed, and gelling subsequently begins. A permanent gel is formed if the temperature continues to decrease. This gel is essentially viscoelastic. The system possesses the property of a solid. This phase transition is referred to as a sol–gel transition. However, from the perspective of technical applications, the importance of the rheological model is limited. In industrial practice, phase transitions of gelatin/water systems are determined to a large extent by the temperature, the gelatin type, the ratio of water to gelatin, and other factors. Generally, transitions between sol and gel are assessed in a temperature range from approximately 5–60 °C, and the gelatin concentration changes between 0.5% and 50% (determined by the quality of the gelatin). However, in practical applications, the melting and setting properties of the final product, including gelatin, are largely affected by the property and the quantity of other compositions (Figure 4.16).

Mechanical and thermal "history" is another example about deviations in the setting property of complicated formulations. In the period of processing solution containing gelatin, if the cooling speed is too rapid, the rheological parameters such as melting and setting temperatures are incomparable with the well-defined results in the laboratory. Due to the fast-increased viscosity, the system hardly has time to achieve equilibrium (characterized by the best three-dimensional structure) via the formation of a steady network with the help of hydrogen bonds. Therefore, the tested gel strength is lower compared with a generally desired value of the used gelatin grade for fast cooling. Owing to the inferior alignment of the gelatin molecules caused by fast cooling, there are deviations from 10% to 15% in gel strength. Similarly, during the process of dissolving gelatin, the rheological performance of solutions produced at temperatures below 50 °C differs from that

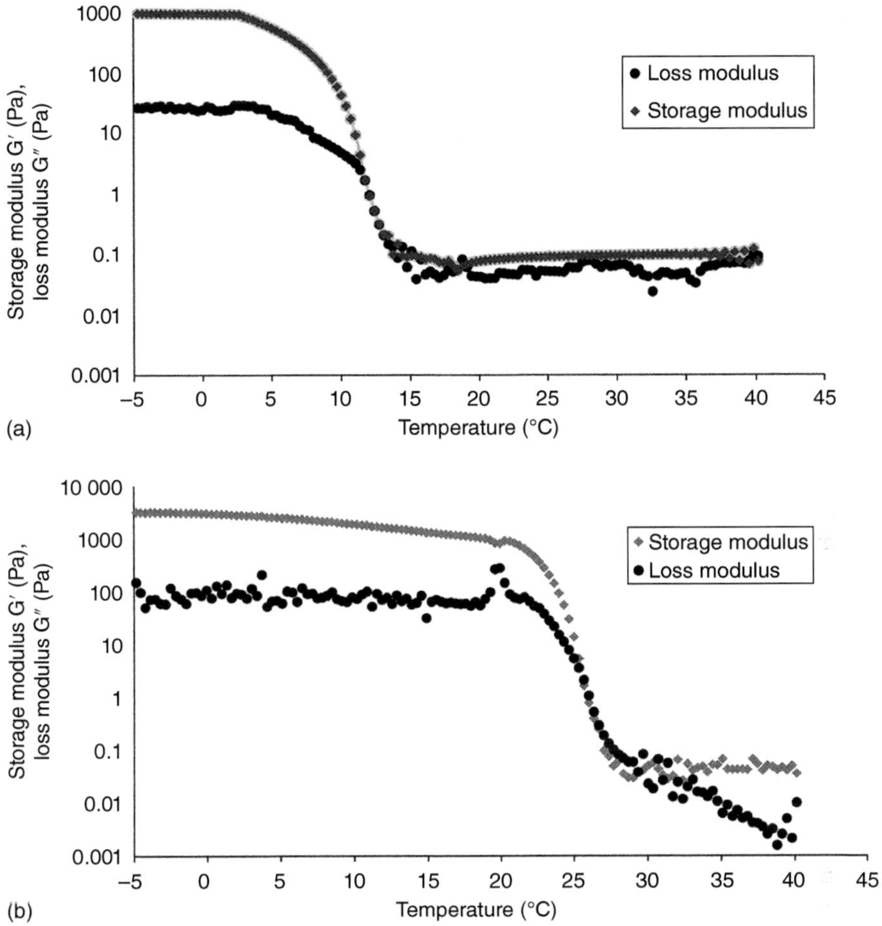

Figure 4.15 The temperature sweep test of commercial gelatin. Cooling (a) and heating (b) processes. The gelling and melting points are the crossover of G′ and G″ during cooling and heating, respectively. Source: Reproduced with permission from Mirzapour-Kouhdasht et al. [20]; © 2020, Elsevier.

of solutions prepared under higher temperatures, showing a type of prestructure under lower temperatures (Section 3.1.2).

Moreover, every extra ingredient changes the behavior of the overall system. In a pure gelatin/water system, transition between sol and gel occurs in a quite narrow temperature range. The complicated viscosity increases quickly in several degrees Celsius. For rheology, the gelatin in a high-solids system, including a small amount of water, cannot or just quite slowly form and stabilize the three-dimensional network feature of transition between sol and gel after cooling. In this case, gel formation occurs within a wide temperature range and is also intensively restrained by the extra ingredients. Meantime, the setting point is increased due to the decreased amount of water. The influence of all these factors on actual systems is only capable

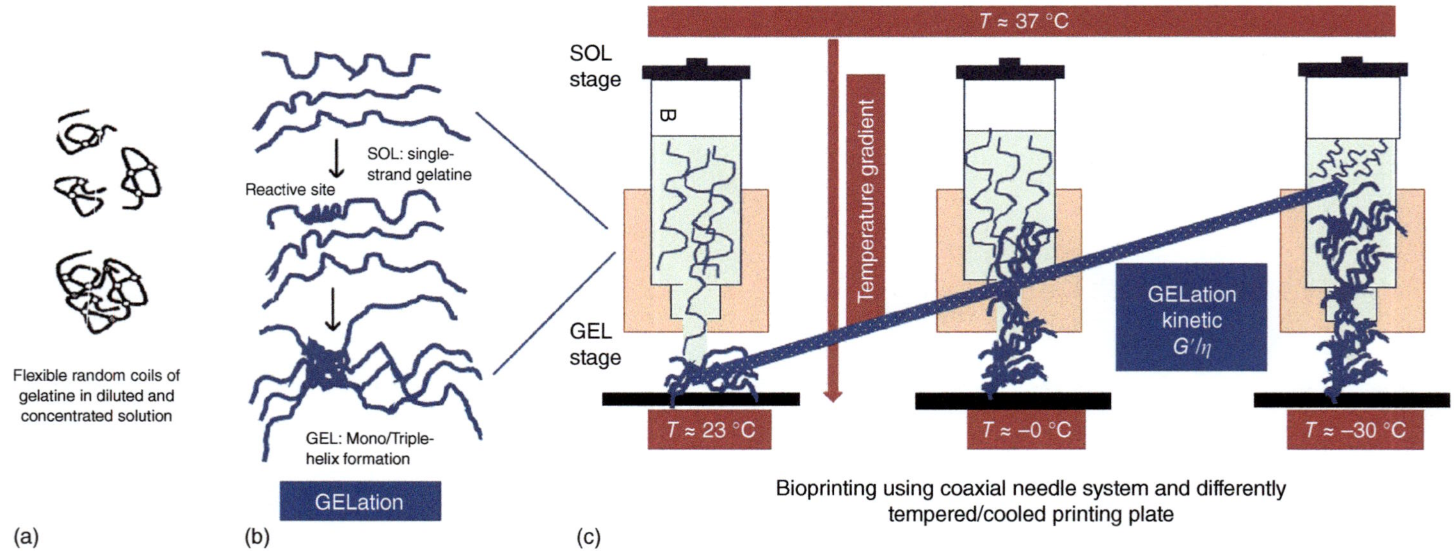

Figure 4.16 Schematic presentation of (a) flexible random coils of gelatin in dilute and concentrated solutions (indicating intra- and intermolecular interactions), (b) The SOL–GEL transition from mono-helix to multi- or triple-helix structures with physically cross-linked junction zones, which strengthen the GEL structure, and (c) By getting oriented in the direction of flow during printing, and are affected by the cooling rate. Source: Reproduced with permission from Kokol et al. [38]; © 2021, Elsevier.

of being evaluated via real experiments, which are able to be implemented successfully and rapidly utilizing modern analytical methods and comprehensive background information of gelatin.

Some methods applied to the rheological characteristics (such as gel strength and viscoelasticity) and chemical characteristics (such as molecular-weight distribution and amino-acid composition) of gelatin are given below.

4.3.5.1 Amino-Acid Analysis

For amino-acid analysis, dry gelatin is typically dissolved in distilled water and subsequently hydrolyzed in vacuum-sealed glass tubes at 108 °C for 18 hours in the presence of 5.7 N HCl, including 0.1% phenol and norleucine as the internal standard. After hydrolysis, the sample is vacuum dried, followed dissolved in a buffer, and analyzed for amino-acid composition [39].

4.3.5.2 Determination of Total Protein Content

Though protein determination could be implemented using the Kjeldahl method, it is not used for regular determination owing to time and sample consumption. Due to simple, rapid, and relatively more precise characteristics, both Biuret and Lowry assays are normally used for total protein determination [40], while the difference between the Biuret and Lowry assays is that the color response of gelatin is determined at 540 and 650 nm, respectively. Bovine serum albumin is served as the reference protein for both assays. The concentration of bovine serum albumin is determined via using the absorbance at 320 nm applying as a background-scattering correction.

4.3.5.3 Electrophoretic Analysis

Gelatin is dissolved in distilled water at 60 °C. After that, a two- or threefold concentration loading buffer, including β-mercaptoethanol, is added. Subsequently, samples are heat denaturated for 5 minutes at 90 °C and analyzed by sodium dodecyl sulfate–polyacrylamide gel electrophoresis (SDS–PAGE) using 4% stacking gels and 5% resolving gels. Coomassie Brilliant Blue R-250 is used to stain protein bands. The protein bands stained on the gel are scanned. Fetal calfskin is served as the marker for α, β, and γ components' mobility [39].

4.3.5.4 Viscoelastic Properties

The rheometer is applied to test the dynamic viscoelastic characteristics of gelatin. Cold distilled water and dry gelatin are mixed for 60 minutes and subsequently put in a water bath held at 45 °C. The mixture solution of gelatin is produced by stirring for 20 minutes. The viscosity test of the sample is carried out at a scan rate of 1 °C min^{-1}, a frequency of 1 Hz, an oscillating applied stress of 3.0 Pa, and a gap of 0.15 mm. The gelation solution is cooled from 40 to 5 °C, held at 5 °C for 10 minutes, and subsequently heated to 40 °C; the viscoelastic characteristics of gelatin slices maturating in low temperature are investigated. The gel slices are heated and cooled on the rheometer. The viscosity modulus (G_v), elastic modulus (G_0), and phase-angle (rad) values are expressed as a function of temperature [34].

4.3.5.5 Gel Strength

The gel strength is measured by dissolving dry gelatin in distilled water (6.6% w/v) at 60 °C. The solution of gelatin matures at 10 °C for 16–18 hours in a refrigerator. Subsequently, the gel-strength measurement is implemented on a texture analyzer with a 5-kN load cell, which is equipped with a flat-faced, cylindrical Teflon plunger of 1.27 cm in diameter. The maximum force (in g) is determined when the plunger has penetrated into the gelatin gel of 4 mm in depth [34].

References

1 Wilson, H. (1988). *Egyptian Food and Drink*. Aylesbury, UK: Shire Publication Ltd.
2 Tode, S. (2003). *The GELITA Story*. Hamburg: Verlag Hanseatischer Merkur.
3 Keenan, T.R. (2012). *Polymer Science: A Comprehensive Reference*. Amsterdam: Elsevier Ltd.
4 Karim, A.A. and Bhat, R. (2009). Fish gelatin: properties, challenges and prospectus as an alternative to mammalian gelatins. *Food hydrocolloids* 23: 563–576.
5 Asher, D.M. (1999). The transmissible spongiform encephalopathy agents: concerns and responses of United States regulatory agencies in maintaining the safety of biologics. *Developments in Biological Standardization* 100: 103–118.
6 Karim, A.A. and Bhat, R. (2008). Gelatin alternatives for the food industry: recent developments, challenges, and prospects. *Trends in Food Science & Technology* 19: 644–656.
7 Wilesmith, J.W., Rayan, J.B., and Atkinson, M.J. (1991). Bovine spongiform encephalopathy: epidemiological studies on the origin. *The Veterinary Record* 128: 199–203.
8 Schrieber, R. and Gareis, H. (2007). *Gelatine Handbook*. New York: Wiley.
9 Arnesen, J.A. and Gildberg, A. (2006). Extraction of muscle proteins and gelatine from cod head. *Process Biochemistry* 41: 697–700.
10 Kim, S.K. and Mendis, E. (2006). Bioactive compounds from marine processing byproducts: a review. *Food Research International* 39: 383–393.
11 Mendis, E., Rajapakse, N., Byun, H.G., and Kim, S.K. (2005). Investigation of jumbo squid (*Dosidicus gigas*) skin gelatin peptides for their in vitro antioxidant effects. *Life Sciences* 77: 2166–2178.
12 Mendis, E., Rajapakse, N., and Kim, S.K. (2005). Antioxidant properties of a radical-scavenging peptide purified from enzymatically prepared fish skin gelatin hydrolysate. *Journal of Agricultural and Food Chemistry* 53: 581–587.
13 Kim, S. (2013). *Marine Proteins and Peptides: Biological Activities and Applications*. New York: Wiley.
14 Gómez-Estaca, J., Montero, P., Fernández-Martín, F. et al. (2009). Physical and chemical properties of tuna-skin and bovine-hide gelatin films with added aqueous oregano and rosemary extracts. *Food Hydrocolloids* 23: 1334–1341.

15 Nijenhuis, K.T. (1997). *Thermorevesible Networks: Viscoelastic Properties and Structure of Gels*. Berlin, Heidelberg: Springer-Verlag.

16 Rbii, K., Surel, O., Brambati, N. et al. (2011). Study of gelatin renaturation in aqueous solution by AFlFFF–MALS: influence of a thermal pre-treatment applied on gelatin. *Food Hydrocolloids* 25: 511–514.

17 Buice, R.G. Jr., Gold, T.B., Lodder, R.A., and Digenis, G.A. (1995). Determination of moisture in intact gelatin capsules by near-infrared spectroscopy. *Pharmaceutical Research* 12: 161–163.

18 Johnston-Banks, F.A. (1990). Gelatine. In: *Food Gels* (ed. P. Harris). Dordrecht: Springer.

19 Haug, I.J. and Draget, K.I. (2011). *Handbook of Food Proteins*. Amsterdam: Elsevier Ltd.

20 Mirzapour-Kouhdasht, A., Moosavi-Nasab, M., Krishnaswamy, K., and Khalesi, M. (2020). Optimization of gelatin production from Barred mackerel by-products: characterization and hydrolysis using native and commercial proteases. *Food Hydrocolloids* 108, 105970.

21 Babel, W. (1996). Gelatin-ein vielseitiges biopolymer. *Chemie in unserer Zeit* 30 (2): 86–95.

22 Sarabia, A.I., Gómez-Guillén, M.C., and Montero, P. (2000). The effect of added salts on the viscoelastic properties of fish skin gelatin. *Food Chemistry* 70: 71–76.

23 Taktak, W., Hamdi, M., Chentir, I. et al. (2021). Development of emulsion gelatin gels for food application: physicochemical, rheological, structural and thermal characterization. *International Journal of Biological Macromolecules* 182: 1–10.

24 Uriarte-Montoya, M.H., Santacruz-Ortega, H., Cinco-Moroyoqui, F.J. et al. (2011). Giant squid skin gelatin: chemical composition and biophysical characterization. *Food Research International* 44: 3243–3249.

25 Dickinson, E. and Lopez, G. (2001). Comparison of the emulsifying properties of fish gelatin and commercial milk proteins. *Journal of Food Science* 66: 118–123.

26 Kim, S.K., Jeon, Y.J., Lee, B.J., and Lee, C.K. (1996). Purification and characterization of the gelatin from the bone of cod, Gadus macrocephalus. *Korean Society of Life Science* 6: 14–26.

27 Aewsiri, T., Benjakul, S., Visessanguan, W., and Tanaka, M. (2008). Chemical compositions and functional properties of gelatin from pre-cooked tuna fin. *International Journal of Food Science and Technology* 43: 685–693.

28 Binsi, P.K., Shamasundar, B.A., Dileep, A.O. et al. (2009). Rheological and functional properties of gelatin from the skin of bigeye snapper (*Priacanthus hamrur*) fish: influence of gelatin on the gel-forming ability of fish mince. *Food Hydrocolloids* 23: 132–145.

29 Surh, J., Decker, E.A., and McClements, D.J. (2006). Properties and stability of oil-in-water emulsions stabilized by fish gelatin. *Food Hydrocolloids* 20: 596–606.

30 Townsend, A.A. and Nakai, S. (1983). Relationships between hydrophobicity and foaming characteristics of food proteins. *Journal of Food Science* 48: 588–594.

31 Jongjareonrak, A., Rawdkuen, S., Chaijan, M. et al. (2010). Chemical compositions and characterisation of skin gelatin from farmed giant catfish (*Pangasianodon gigas*). *LWT - Food Science and Technology* 43: 161–165.

32 Papon, P., Leblond, J., and Meijer, P.H. (2002). *The Physics of Phase Transitions*. Berlin: Springer.

33 Cho, S.M., Kwak, K.S., Park, D.C. et al. (2004). Processing optimization and functional properties of gelation from shark (*Isurus oxyrinchus*) cartilage. *Food Hydrocolloids* 18: 573–579.

34 Liu, H., Li, D., and Guo, S. (2008). Rheological properties of channel cat fish (*Ictalurus punctaus*) gelatine from fish skins preserved by different methods. *LWT - Food Science and Technology* 41: 414–419.

35 Holzer, D. (1996). Gelatin production. US Patent, 5, 484,888, filed 3 August 1994.

36 Domb, A.J., Kumar, N., and Ezra, A. (2011). *Biodegradable Polymers in Clinical Use and Clinical Development*. New York: Wiley.

37 Gudmundsson, M. and Hafsteinsson, H. (1997). Gelatin from cod skins as affected by chemical treatments. *Journal of Food Science* 62: 37–47.

38 Kokol, V., Pottathara, Y.B., Mihelčič, M., and Perše, L.S. (2021). Rheological properties of gelatine hydrogels affected by flow- and horizontally-induced cooling rates during 3D cryo-printing. *Colloids and Surfaces A: Physicochemical and Engineering Aspects* 616: 126356.

39 Gómez-Guillén, M.C., Turnay, J., Fernández-Díaz, M.D. et al. (2002). Structural and physical properties of gelatin extracted from different marine species: a comparative study. *Food Hydrocolloids* 16: 25–34.

40 Zhou, P. and Regenstein, J.M. (2006). Determination of total protein content in gelatin solutions with Lowery or Biuret assay. *Journal of Food Science* 71: C474–C479.

5

Synthesis of Gelatin

5.1 Raw Materials

For the production of gelatin, the raw material could be any tissue containing collagen. For reasons of availability and achievable product quality, the main raw material used is derived from the collagen contained in skins, bones, and hides of mammalian sources (e.g. bovine and porcine). Additionally, there are a small number of raw materials being derived from fish and poultry, which are mainly produced to meet the demands of special religious user communities. Currently, they only occupy niche markets. In despite of their essentially differences, these raw materials all have the same characteristics: they are derived from animals and have been permitted for human consumption after being subjected to strict controls. The production process of gelatin includes cleaning of the source tissues followed by pretreatment, extraction of gelatin, filtration/purification/sterilization, concentration, drying, and finally milling.

Bones from mammalian sources are the major sources of collagen for the preparation of gelatin. During the preparation process, the bones are first washed to remove impurities. The crushed and defatted bone chips are exposed to the acidic environment (generally 4–7% hydrochloric acid) for at least two days after washing, crushing, and rewashing. Throughout this period, the calcium carbonate and calcium phosphate, both of which are combined with the bone material and supply it with its firmness, are transformed into their soluble forms and are then isolated by sedimentation. This demineralization process is referred to as maceration, which is carried out using a process of countercurrent. The freshly crushed and defatted bone chips are first subjected to a salt-rich acid treatment, which is prepared ahead of time. In the last process stage, the essentially demineralized bone is subjected to fresh acid treatment. This countercurrent program obviously improves the integral efficiency of the maceration process and inhibits undesired warming and foaming. Following maceration, the rest of the proteinaceous structural framework for the bone is the practically raw material or "ossein," which is processed directly or dried and stored for later use. After the above preliminary treatment, the raw material is pretreated with either acid or alkaline followed by gelatin extractions based on the source of the collagen and the requested quality of the ultimate gelatin.

Collagen-Derived Materials: Synthesis and Applications in Electrochemical Energy Storage and Conversion, First Edition. Feng Wang, Yaqin Huang, and Jin Niu.
© 2022 WILEY-VCH GmbH. Published 2022 by WILEY-VCH GmbH.

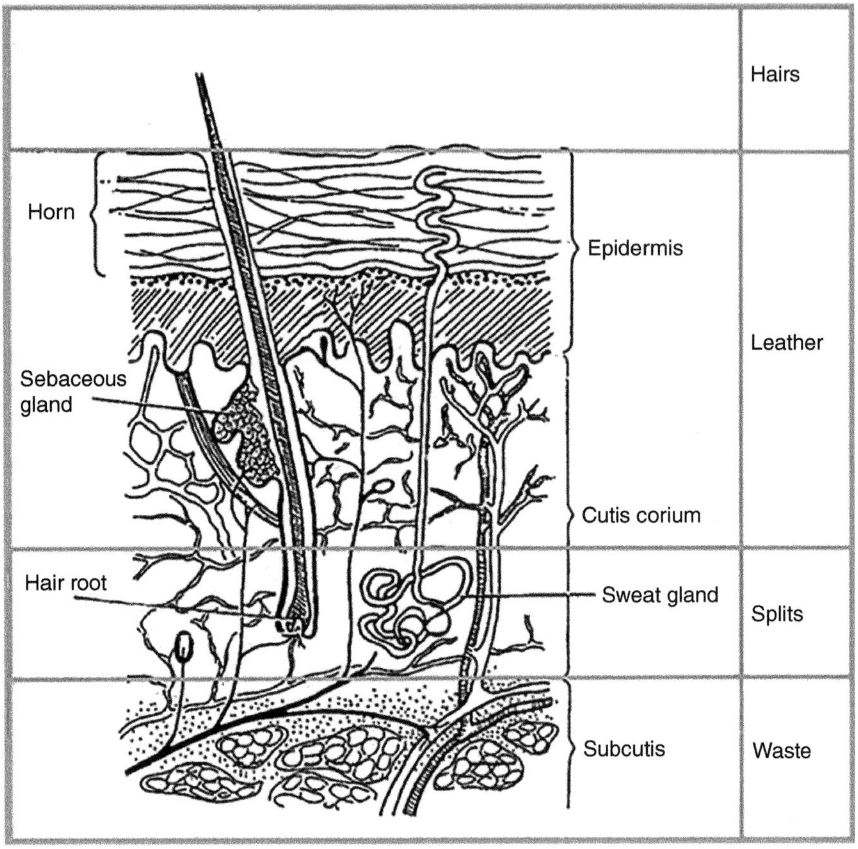

Figure 5.1 The central layer of the cattle hide ("split") is an excellent source of collagen for the manufacture. Source: Reproduced with permission from Schrieber and Gareis [1]; © 2007, John Wiley & Sons.

Fresh cattle hide is the second main source of collagen for the production of gelatin. The climate where the cattle are raised is the determining factor for the thickness of the hide. Warmer climate makes the hide become thinner. The outermost layer in this kind of hide is comprised of less collagen and is almost specially applied to produce leather. The meat side containing fatty tissue is professionally removed. The central layer is actual pure collagen, which is a raw material of high quality for the gelatin production (Figure 5.1).

During the process of producing gelatin using fresh cattle hide, the hide is first subjected to alkaline pretreatment, the process of which swells the hide and makes it depilated. Subsequently, horizontal cutters are employed to cut the hide into three separate layers. This process was known as "split" by the industry. According to the age of the animal, the separate layers through the splitting process can reach the maximum size of 4 m^2. The collagen layer is cut into pieces about the size of hand using cutting machines followed by the immediate treatment with acid or alkali.

Reducing the size of the hides is the foundation of great conditioning. The acid and alkali treatments are beneficial to accelerating extraction of collagen. In principle, hide pieces without the splitting process can be processed as well only if they are subjected to the depilating step with alkali or sulfide prior to being processed. However, their low content of collagen greatly decreases the production and hence reduces the productivity and operating efficiency.

Pigskin was first used to product gelatin in the 1930s, which is currently the most significant raw material for gelatin around the world. Every pig slaughtered can produce processable skin of approximately 3–4 kg. The pigskin is first isolated from the fatty layer during meat processing and is subsequently transported into the gelatin factory by cooling or freezing if it is not requested to process into luncheon meat or sausages. The cooling process restrains microbial cells from degrading and oxidizing of any remaining fat in the skin during the transportation process. The fat tissue of pigskin also comprises collagenous protein, which can be used to produce gelatin after the fat has been melted into the "greaves" form. Nevertheless, a special process using special equipment is requested due to the turbidity resulted by emulsified fat. In countries such as China that have a large number of pigskin leather factories, the pigskin is cut into separate layers according to the above description for cattle hide followed by further processing to gelatin.

Due to its cross-linked nature, the collagen in the connective tissue of animal dissolves quite slowly, even in boiling water. Therefore, it is necessary for it to undergo mild chemical treatment to destroy these cross-links before extraction. The animal age determines the degree and type of cross-linking. Thus, it is imperative for the gelatin manufacturer to accurately adjust the parameters of process conditioning and the following extraction conditions based on the processed raw material to enable get gelatin with the needed characteristics (Table 5.1).

Europe and North America are the main producing areas of acid pigskin gelatin, and South America mainly produces bovine hide for gelatin production. Figure 5.2 displays the distribution of various raw materials of producing gelatin in Europe

Table 5.1 Raw material conditioning [1].

	Raw material conditioning	
Raw material type	**Acid**	**Alkali**
Bones	x	X
Cattle hide splits	x	X
Pigskin splits	x	X
Pigskin	x	
Fish skin	x	
Poultry skin	x	
Poultry feet	x	

Source: Schrieber and Gareis [1]/John Wiley & Sons.

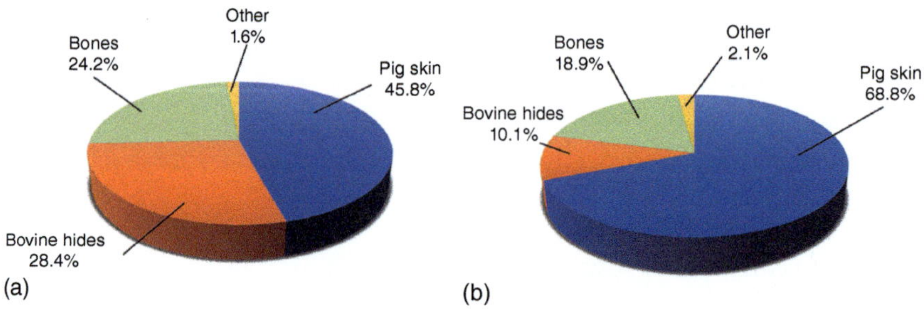

Figure 5.2 The raw material consumption for gelatin production in 2006, worldwide (a) and in Europe (b). A higher percentage of pigskin gelatin was produced in Europe compared to the rest of the world (from www.gelatin.org). Source: Reproduced with permission from Haug and Draget [2]; © 2011, Elsevier.

and the whole world in 2006. Compared with worldwide, the gelatin produced in Europe is mainly derived from pigskin. This is probably due to the outbreak of bovine spongiform encephalopathy (BSE) in Europe in the 1990s. Moreover, there are a few Muslim countries in Europe and consequently the use of gelatin from pig sources is not strictly restricted by religion or ethnicity. However, gelatin from bovine sources may be the priority selection in other areas of the world, such as Asia and Africa, due to such dietary restrictions.

The "other" in the two charts of Figure 5.2 contains poultry and fish sources, and their proportion is only about 2%. Figure 5.3 shows the poultry gelatin as an alternative for mammalian gelatin. Compared with mammalian gelatin, cold-water fish gelatin possesses suboptimal physical characteristics, which limit their demand and application. Gelatin from warm-water fish have more similar physical characteristics to gelatin from mammalian and consequently can directly take the place of

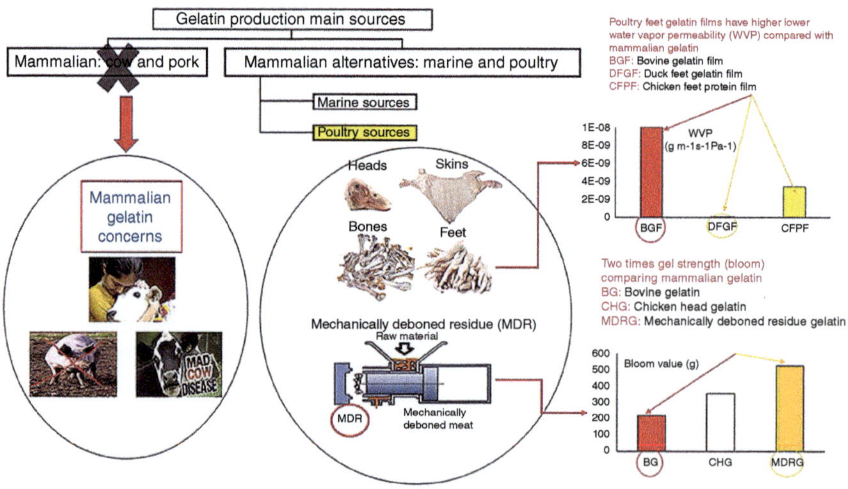

Figure 5.3 Poultry gelatin: a sustainable alternative for mammalian gelatin. Source: Abedinia et al. [3]/with permission of © Elsevier Ltd.

gelatin from mammalian in numerous products. Unfortunately, the raw materials for producing gelatin from warm-water fish are in short supply, which restricts the gelatin yield.

Gelatin manufacturers around the world primarily include four big associations, and the web pages of the different associations provide the information about gelatin suppliers:

- Gelatin Manufacturers of Europe (GME): www.gelatine.org
- Gelatin Manufacturers Institute of America (GMIA): www.gelatin-gmia.com
- Gelatin Manufacturers Association of Asia Pacific (GMAP): www.gmap-gelatin.com
- South American Gelatin Manufacturers Association (SAGMA): www.sagma-gelatina.org

5.2 Extraction

The processes of manufacturing gelatin include three major steps: (i) pretreating the raw material, (ii) extracting the gelatin, and (iii) purifying and drying gelatin. Figure 5.4a shows a typical industrial process for gelatin manufacture using natural sources (bovine and porcine by-products) as raw materials [4]. As comparison with conventional method, enzymatic method has higher extraction efficiency and yield, showing good prospect for practical application (Figure 5.4b). In the industry, type A gelatin was regarded to derive from porcine bones and skins or beef hides and type B gelatin was normally found in beef-based raw materials, occasionally from pig bones as well. Sodium carbonate or caustic lime are general acids and alkaline agents applied to extract minerals and bacteria from these raw materials. At the first step of gelatin manufacture, type A gelatin and type B gelatin were produced by acidic pretreatment and alkaline hydrolysis, respectively. Acid pretreatment is most appropriate for the collagen with less covalent cross-links from fish or pigskins, whereas alkaline pretreatments are fitter for the more complicated collagen from bovine hides. Some marine-derived gelatin, such as fish gelatin, is usually produced by using weak-acid pretreatment of the raw material and mild temperature conditions in the process of extraction [5]. Acid and alkaline pretreatments are the most suitable and the most widely applied among all pretreatment methods [6].

During the extraction process, the general parameters and process, especially using the raw materials from pig and bovine, include five steps as follows:

(i) **Defatting**: immersing pretreated raw materials after washing in hot water followed by roasting them for a minimum of half an hour at 100 °C to make pretreated raw materials reduce fat around 2%,

(ii) **Pretreatment**: immersing in a large tub of acid (generally a 4% HCl, $pH < 1.5$) or alkali (K_2CO_3 or Na_2CO_3, $pH > 7$) for approximately five days,

(iii) **Extraction**: loading into extractors and boiling in distilled water; subsequently, the consequent liquid gelatin is sterilized via rapid heating at around 140 °C in 4 seconds,

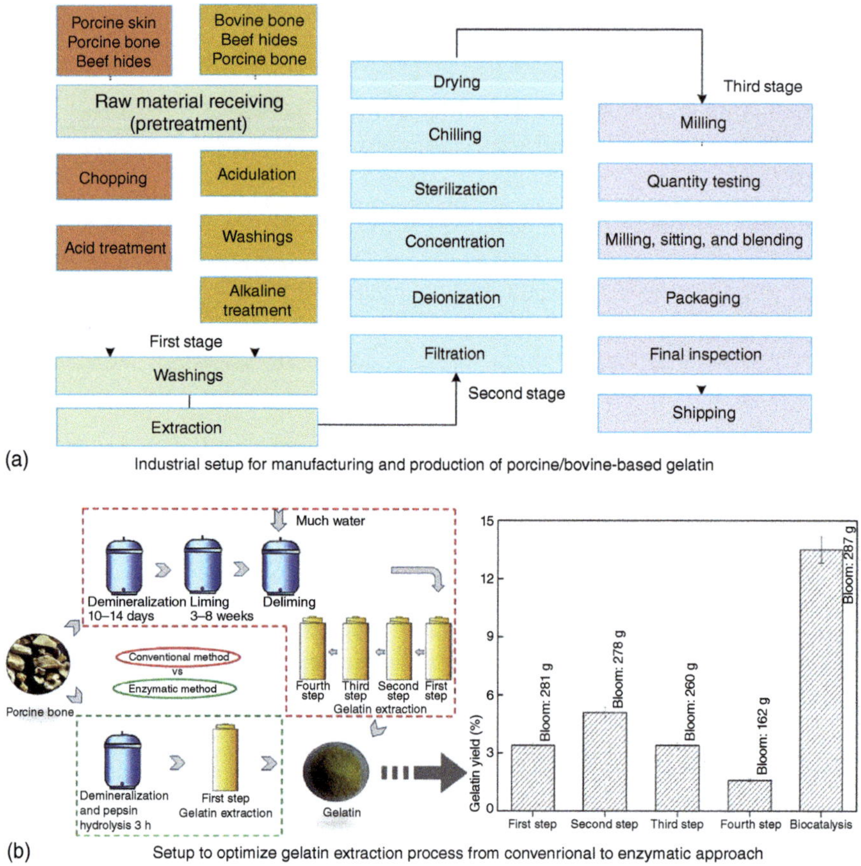

Figure 5.4 (a) A typical setup of gelatin extraction/production in the industry. (b) The extracted gelatin yields and bloom from porcine bone using the proposed bio-friendly extraction process (one-step biocatalysis: the green color line). Source: Alipal et al. [4]; © 2021, Elsevier Ltd.

(iv) **Evaporating**: the liquid is piped through filters to separate out residual tissue, skin, or bone. The filtered liquid is piped into evaporators to separate the liquid from the solid gelatin,

(v) **Grinding**: the solid gelatin is pressed into sheets and is then ground to a fine powder (flavorings, colorings, or sweeteners could be added into the powder).

At the initial stage of gelatin extraction, the pretreated animal skins, fish, or poultry by-products are in a state of mixing between insoluble native collagen and immature liquid gelatin [7]. To break the mixture into an optimal production of gelatin, the pretreated materials are further boiled in water at over 45 °C. The chemical agents (normally HCl or NaOH) applied on collagen during pretreatment sunder the non-covalent bonds to disorganize the protein structure, leading to adequate swelling and solubilization. At the same time, the hydrogen and covalent bonds are broken by the extra heat treatment to destabilize the triple helix, resulting in the formation of helix-to-coil transition and transformation to dissolved gelatin.

5.2.1 Alkali-Treated Method

Type B gelatin is the end product using the alkali treatment. During the alkaline process, both the ossein produced from bone and the chopped split material are subjected to the treatment with alkali in vats or pits with the volume of 125 m^3, with or without stirring. Among a series of alkaline agents utilized for this treatment, saturated lime water ($Ca(OH)_2$, pH 12.0) is normally the most used curing liquid. The used concentrations and temperatures have a great influence on the conditioning process; for instance, the process lasts from several days to four months using a 1% NaOH solution at 20 °C with supersaturated lime milk. Additionally, stirring accelerates the conditioning process.

The quality of the gelatin related to viscosity and Bloom may be a result of the relationship between temperature, concentration of NaOH, and conditioning duration. Generally, the stronger the conditioning is, the higher the viscosity is.

Even if the process using lime milk seems to be slightly inefficient at the first sight, it possesses many advantages in technology: though the solution is supersaturated, a constant degree with relatively mild alkalinity is still retained due to the low solubility of the $Ca(OH)_2$. During the treatment process that takes a few months, the substances without protein (e.g. mucopolysaccharides and sulfur-containing compounds) and proteins without collagen, particularly globulin and albumin, found in the raw material, are effectively dissolved out, which causes the raw material to be reliably purified. Meanwhile, the slightly mild $Ca(OH)_2$ is beneficial to balance out some diversities in the raw materials. This applies to the thickness of the hide pieces, the particle size of the bone chips, and the age of the animals. Thus, unwanted losses of production can be avoided. If alkali treatment is excessive, the collagen will dissolve in cold water. Thus, collagen will be soluble in the aqueous phase when cleaning the raw materials, which contributes to lower productions.

In the process of alkaline treatment, the solution is continuously enriched with soluble protein without collagen and other substances. As a result, it is frequently varied over months from a daily to a weekly basis. After that, the treated material is washed to remove the alkali and is neutralized by adding acid. A majority of the neutral salts prepared in this process are basically removed by repeated cleanings.

5.2.2 Acid-Treatment Method

The treatment method with acid is primarily applied for the production of gelatin derived from fish skin, pigskin, and bones. Figure 5.5 shows the microstructure of bovine bone gelatin with different pretreatments. During the acid-treatment process, the following steps are carried out to enable the material to get ready for the primary step: reducing size, cleaning, removing hairs, and defatting. After pretreatment, the material is infused into cold dilute mineral acid, such as in 2–4% dilute HCl or H_2SO_4 solution, which enables it to be acidified to around pH 4. Then, the resulted material is heated gradually from 50 °C to boiling. This process needs 10–40 hours. Phosphoric and organic acids are appropriate for this processing stage as well. However, they have a high cost and a negative impact on the odor and taste of the end product.

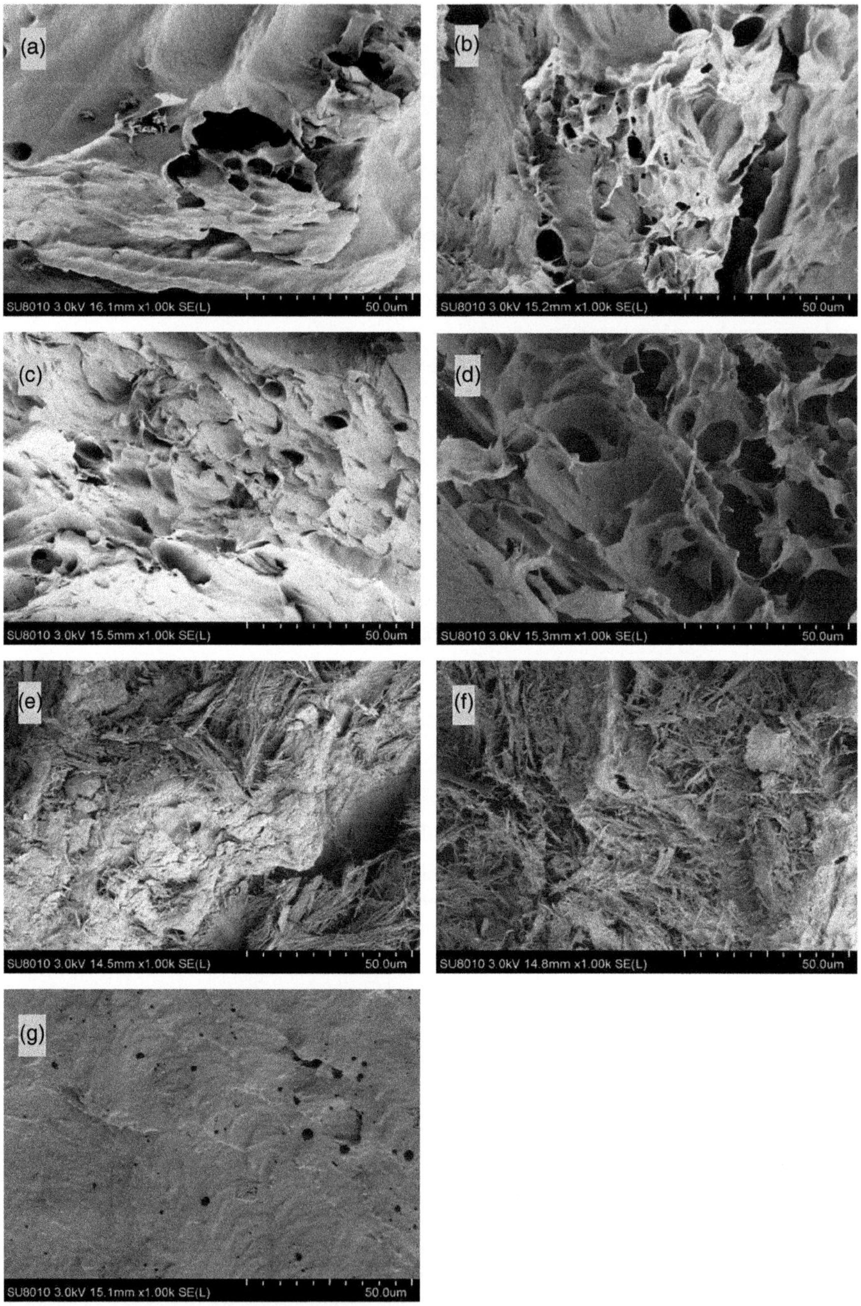

Figure 5.5 Microstructure of bovine bone collagen after different pretreatments. Magnification: 1000×. (a): hydrochloric acid pretreatment (26 hours), (b): hydrochloric acid pretreatment (21 hours) and pepsin hydrolysis (5 hours), (c): acetic acid pretreatment (26 hours), (d): acetic acid pretreatment (21 hours) and pepsin hydrolysis (5 hours), (e): citric acid pretreatment (26 hours), (f): citric acid pretreatment (21 hours) and pepsin hydrolysis (5 hours), (g): un-treated bovine bone collagen. Source: Cao et al. [8]/with permission of © Elsevier Ltd.

Table 5.2 Raman peaks position of gelatin extracted from bovine bone powder using different pretreatments [8].

	Peak wavenumber (cm^{-1})						
Bond	**G-P**	**H-1**	**A-1**	**C-1**	**H-2**	**A-2**	**C-2**
Amide-III	1245.01	1244.84	1245.01	1246.04	1245.36	1245.18	1247.23
δ_{C-H}	1455.67	1455.67	1452.57	1455.67	1455.67	1455.67	1452.57
Amide-I	1666.61	1666.61	1669.72	1669.72	1669.72	1672.82	1672.82
ν_{C-H}	2944.65	2938.44	2941.55	2941.54	2944.65	2941.55	2944.65

G-P denotes commercial gelatin from pigskin. H-1, A-1, C-1, H-2, A-2, and C-2 denote the gelatins extracted with different pretreatments: H-1:hydrochloric acid pretreatment (26 hours); H-2: hydrochloric acid pretreatment (21 hours) and pepsin hydrolysis (5 hours); A-1: acetic acid pretreatment (26 hours); A-2: acetic acid pretreatment (21 hours) and pepsin hydrolysis (5 hours); C-1: citric acid pretreatment (26 hours); C-2: citric acid pretreatment (21 hours) and pepsin hydrolysis (5 hours).
Source: Cao et al. [8]/Elsevier.

In the Raman spectrum of the gelatin samples, there were no important variations in the location of δ_{C-H} and ν_{C-H}, showing that pepsin and acid do not have an important influence on δ_{C-H} and ν_{C-H} (Table 5.2). The amide-I band located near the 1666-cm^{-1} spectrum is defined as the peptidic bond C=O stretching of the Gly–X–Y tripeptide sequence (Figure 5.6a) [9]. Table 5.2 shows that the prominent characteristic of the amide-I band shifted to the right in C-1 (1669.72 cm^{-1}), H-2 (1669.72 cm^{-1}), A-2, and C-2 (1672.82 cm^{-1}), indicating that HCl facilitates to form more β-sheet and random-coil structures compared with acetic acid and citric acid in the swelling process of bovine bone collagen. Furthermore, the formation of more β-sheet and random-coil structures was induced with the structural destruction of collagen by pepsin, which leads to a greater effect in dissolving the collagen (Figure 5.6b,c).

After that, the desaturated collagen or gelatin must be defatted and filtered to a high clarity. Due to the mechanical stirring used, this process separates off a large amount of fat, which floats on the surface and is thus easy to be removed. The fat content of pigskin, carefully isolated from the bacon fat, is about 30%; by contrast, the protein content is only around 20%.

When using acid to regulate bovine hide splits, the manufacturer usually makes the acid react over 48–72 hours. However, it is worth noting that extraction has to be performed in an acid medium too after acid conditioning. Subsequent to acid treatment, the gelatin producer increases the pH to around 2–4 by the addition of alkali. After that, the numbers of the salts formed are removed over 24 hours utilizing repeated variations of water. The pH and the extraction temperatures/time are used to adjust the ratio of gel firmness to viscosity during extraction. Unfortunately, the two parameters have the effect on the extraction speed. Thus, it is necessary for the

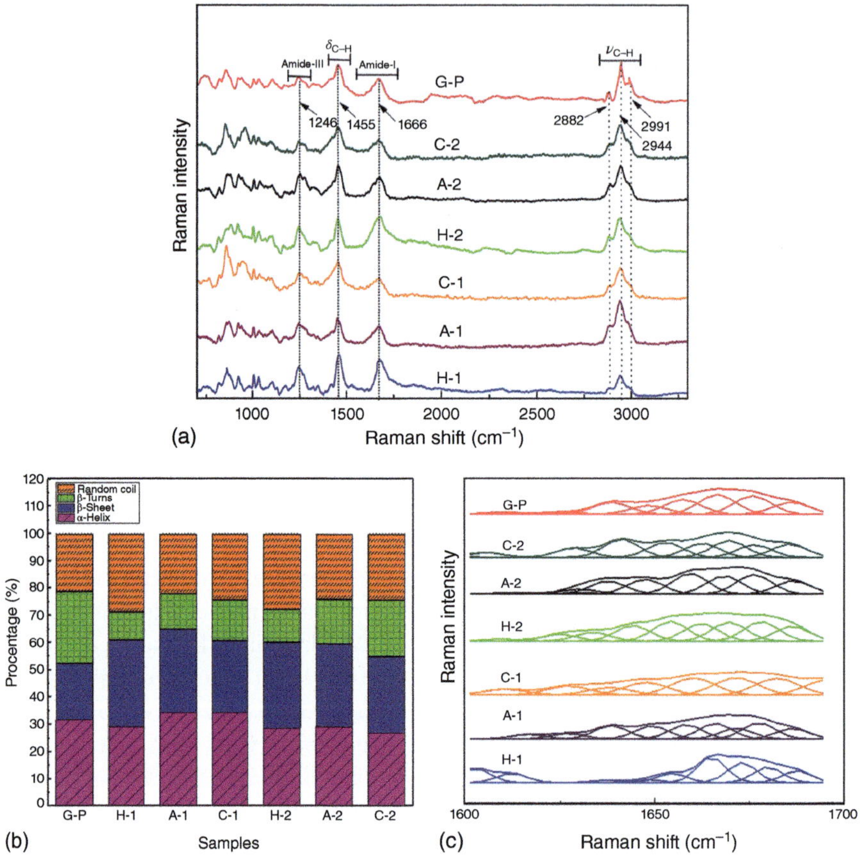

Figure 5.6 (a): Raman spectrum between 700 and 3300 cm^{-1} of gelatin extracted from bovine bone power using different pretreatments. (b): Secondary structure fractions of gelatin samples extracted from bovine bone power using different pretreatments estimated from the amide-I band (1600–1700 cm^{-1}). (c): Fitting diagram after deconvolution of the amide-I band (1600–1700 cm^{-1}) of gelatin extracted from bovine bone power using different pretreatments. Source: Reproduced with permission from Cao et al. [8]; © 2020, Elsevier.

manufacturer to discover the best ratio of the ideal fast extraction to the undesirable chemical/thermal hydrolysis of the gelatin that has been melted out. In the end, the gelatin is concentrated to an appropriately high concentration in virtue of either membrane ultrafiltration or vacuum evaporation. The resulted material is dried by passing dehumidified air followed by being grounded and mixed.

5.2.3 Salt-Fraction Method

The method of salt fraction is divided into two extraction processes, namely acid salt method and saline alkali method. During the process of producing gelatin using the acid salt method, the raw material is first soaked in the mixed salt solution (e.g. Na_2SO_4, NaCl) and acid (e.g. HCl, H_2SO_4, H_3PO_4). If the acid concentration is too large, the backbone of the collagen molecule is easy to be broken and dissolved. The way of adding salt in acid solution is conducive to stabilizing the helical structure of collagen molecules, preventing excessive expansion of raw materials as well as lowering the loss that resulted from excessive dissolution of collagen. Compared with the acid method, the acid salt method is much better in the aspect of the yield, gel strength, and the viscosity of gelatin. However, it is difficult for the acid salt method to realize industrial production due to its uncontrollability.

The saline alkali method is the method of gelatin extraction by immersing raw material using NaOH and Na_2SO_4 to replace milk of lime. During the process, the raw material is first immersed in the mixed solution of 10–14% Na_2SO_4 and 2–6% NaOH. Subsequently, the raw material is washed to remove the alkali and salt and then neutralized with weak acid (1–1.5% acetic acid or formic acid) to decrease its pH to 5.0–5.5. In the immersing process, NaOH in the mixed solution enables the collagen fibers to sufficiently expand, which is beneficial to promoting the speed of producing gel. Nevertheless, because of the strong alkalinity of NaOH, the collagen fibers will be converted from maximum expansion to dissolution when the concentration of NaOH is too high, which causes the decomposition of raw material. However, Na_2SO_4 plays an important role in the dehydration and salting out for protein, which enables the collagen fibers to shrink and stops them from achieving the maximum expansion. The salt alkali method is appropriate to produce gelatin from certain raw materials (e.g. the dry skin), which has difficulty in treatment with lime milk. Immersing wet or dried cow skin and dried pigskin using the saline alkali method can not only boost the production of gelatin, but also enhance the gel strength and viscosity of gelatin. During the production process of gelatin, the saline alkali method may cause environmental pollution due to the usage of plenty of strong alkaline solutions with high concentration. Thus, it is necessary to use waste acid to recycle or neutralize alkaline solution.

Giménez et al. researched the influence of pre-extraction salt treatments (KCl, NaCl, $MgCl_2$, and $MgSO_4$) on extraction yield, the content of mineral, molecular weight distribution, gel strength, and the aggregation phenomena of gelatin from Dover sole (Solea vulgaris) skins [10]. Compared with unsalted gelatin, skin washing with KCl or NaCl generated distinguishing variations in molecular weight distribution and hence in gel strength. Washing with Mg^{2+} salt caused larger retention of Mg^{2+} in the recovered gelatin, which resulted in worse quality. Salt treatment had an important effect on the gel strength. The largest gel strength for gelatin was obtained using KCl (176.83 ± 11.57) and NaCl (181.74 ± 10.19). By comparison, a soft gel was achieved using $MgCl_2$ (82.84 ± 3.80) or unsalted treatment (78.24 ± 5.42).

5.2.4 Enzymatic Method

Collagen is difficult to be broken by a common proteolytic enzyme but can be hydrolyzed by collagenase from an animal or a microorganism. Subsequent to automatic denaturation, the broken fractions can be hydrolyzed by the general proteolytic enzyme. It is believed that the stable helical structure of natural collagen may be the determining factor for preventing it from being hydrolyzed by the proteolytic enzyme. Nevertheless, later research has found that the general proteolytic enzyme has almost no impact on the procollagen with helical structure but can act on the nonhelical structure of the peptide region to prevent the intramolecular and intermolecular cross-linking. In the presence of a denaturation reagent or above denaturation temperature, the helical structure of collagen was lost and collagen was further hydrolyzed by the enzyme. This indicates that certain fragments of collagen molecules are easy to be broken by enzymes; in other words, the nonhelical structure of collagen can be attracted by other proteolytic enzymes and further degraded besides collagenase.

As a multistage degradation product of collagen, gelatin could be prepared by two kinds of enzymatic method. The first one is using enzyme solution to treat the collagen and stirring it in acid to obtain collagen solution. Subsequently, the precipitated fiber is obtained by isoelectric point precipitation. Gelatin is prepared after separating the precipitate and then heating. The second one is using enzyme solution treatment instead of the liming process. After some time, gelatin is extracted by treating the skin or bone material and then heating it. The prepared gelatin is named as E-type gelatin. There are some advantages of using the enzymatic method, including short processing time and low waste generation [11, 12].

The type of enzyme plays an important role for the gelatin extraction. Based on their sources, the enzymes applied to the enzymatic method contain plant enzymes (such as bromelin, papain, and ficin) and non-plant enzymes (such as pepsin, trypsin, and chymotrypsin). Certain intrinsic enzymes in the skin can be activated at the high temperature of extraction, which results in the long chains of gelatin to be broken into short chains [13].

As pretreatment reagents during the extraction of gelatin, pepsin, bromelain, papain, neutrase, and crude proteases prepared from the viscera of different fish can improve the yield of gelatin but degrade the α and β chains of the obtained gelatin, which leads to reduce the quality of gelatin, especially the gel strength. The sustained action of added proteolytic enzymes results in the degradation, which demonstrates that a long-chain gelatin with high quality can be prepared by searching for novel enzymes that enable to cleave long chains of gelatin only at a few sites. After being activated at high temperature during the extraction process, certain thermally stable skin endogenous proteases lead to degradation of gelatin. Soybean trypsin inhibitor and protease inhibitor are added to prevent gelatin from degrading, but this causes reduced gelatin yield. It is necessary to study the effects of the activation of endogenous proteases during gelatin extraction on the gelatin quality. Additionally, if the source of gelatin raw materials contains the endogenous proteases, it is paramount to investigate the action of the endogenous proteases inhibitor during gelatin extraction and its following influence on the quality of gelatin. Mutant enzymes and ultrasound can be used to increase the gelatin yield without reducing the gelatin quality.

5.2.5 Heat–Pressure Method

The extraction of gelatin from bones usually uses the heat–pressure process. Compared with acid and alkali processes, this process takes shorter time and needs no chemical agents.

Generally, bone materials are chopped and cleaned with water after defatting. Subsequently, the chopped bone materials are placed into a high-pressure tank with a ratio for bone materials to water of 1–3 at 104 °C and a pressure of 1.2 atmospheres for 5–6 hours. All the solid material is separated from the above mixtures by filtering. After that, the solid material is degreased by phase separation. Finally, the gelatin is obtained by concentration, drying milling, and blending. The gelatin prepared with this method is limited in terms of quality and application.

In recent years, a new process, named "pressure hydrolysis," has been introduced by some manufacturers for the gelatin manufacture from raw bone. This process that can function without maceration was invented by modifying the traditional process for the manufacture of bone glue. During this process, the bone chips are treated in batches in autoclaves with an aqueous medium at around 140 °C for over 20 minutes. Consequently, the collagenous protein is dissolved step by step. The resulted gelatin has a strong odor, taste, and color but a quite poor gelling ability. The applicability of gelatin is greatly limited by these properties.

Due to the process conditions of pressure hydrolysis, the extraction process follows a basic rule: the particle with smaller size leads to a shorter process. Bone chips obtained by fine grounding are transformed into gelatin with outstanding gelling ability and good sensory characteristics in minutes and are very appropriate for consecutive extraction processing.

5.2.6 Microwave- and Ultrasonication-Assisted Extraction Methods

The two methods of microwave- and ultrasonication-assisted extraction have also been used for producing gelatin. Microwaves are electromagnetic waves with a frequency range from 300 MHz to 300 GHz [14]. Compared with gelatin extracted from rabbit skin by conventional hot-water extraction, the gelatin extracted with the assistance of short-term (5–30 minutes) microwave (2450 MHz, 500 W) showed greater gel strength and lower yield [15]. However, the yield would be increased and gel strength decreased if microwave-assisted times were extended to 60–90 minutes [15]. The research by Kim et al. showed that the viscosities of gelatin from duck skin with microwave-assisted extraction (2450 MHz or 200 W for 10 minutes) were greater than those of gelatin by hot-water extraction [16]. The ultrasonication-assisted extraction method is used to produce gelatin with improved characteristics. Gelatin extracted from fish scale with ultrasonication assistance (200 W, 40 kHz, 1 hour) had higher gel strength, rheological behavior, and emulsifying characteristics than that with the hot-water method [17]. Furthermore, ultrasonication (750 W, 20 kHz) modes and times were determining factors for the gelatin characteristics. For example, gelatin extracted by ultrasonication with 10 minutes on-time and 50 minutes off-time showed great gel [18]. Compared with the microwave-assistance method (2450 MHz, 350 W, 1–5 minutes), the ultrasonication-assistance method (50–150 W, 5–45 minutes) increased viscosity, gel strength, gelling point, and melting point of gelatin from common carp by-products, e.g. scales and fins [19]. Additionally, the viscosity and gel strength of gelatins reduced when increasing ultrasonication powers and times [19].

5.2.7 Posttreatment

Subsequent to the process of extraction, the obtained gelatin solution must be treated with a rigorous purification process, which is very varied (Figure 5.7). During this process, manufacturers usually combine methods with apparatus based on experience. Herein, only a small fraction of the possible process stages is introduced.

5.2.7.1 Filtration and Clarification

First, the gelatin solution is divided into three phases using consecutive high-performance separators. The insoluble solids and fats are separated out from the aqueous gelatin solution. This kind of technical grade fat is subjected to be repurified and mainly applied for further processing or saponification by the chemical industry. The fat from pigskin is applied in the animal feed industry in many countries. During the processing of pigskin, the fat yield is higher than that of the gelatin. Therefore, fat is a valuable and significant by-product. However, fat is more of an undesirable by-product in the hide split and ossein due to its much smaller proportion.

After being subjected to separation or other pre-clarification steps, such as fine sieving, the gelatin solution still falls short of the requirement of clarity. Thus, it is necessary to be treated with a number of other filtration steps.

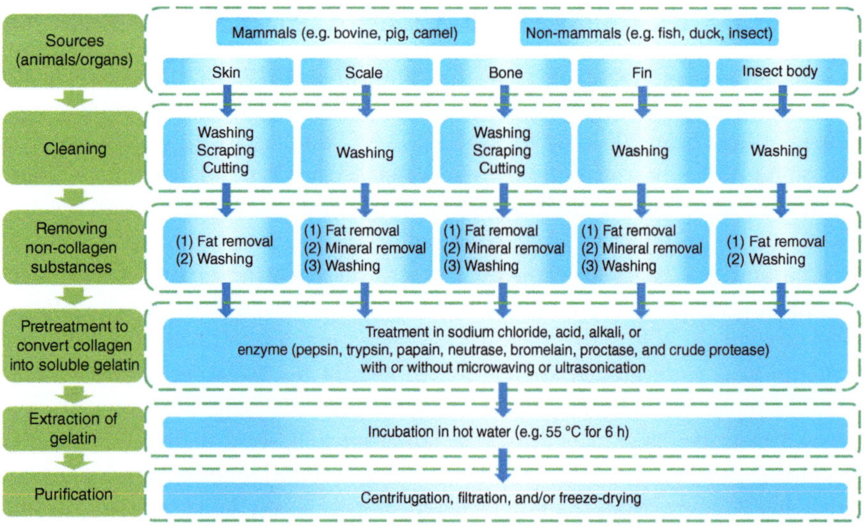

Figure 5.7 Common gelatin extraction process from various gelatin sources. Source: Reproduced with permission from Zhang et al. [20]; © 2020, Elsevier.

Generally, centrifugal pre-coated alluvial filtration is first applied during this purification process. Substances induced by turbidity are removed by perlite or diatomaceous earth. Filtration using pressed cellulose cakes is also another method to be widely employed. In the subsequent multistage fine filtration, the perlite layers or diatomaceous earth and the cellulose filters are employed until a certain pressure difference is observed. After that, they are discarded except cellulose filters.

For gelatin with some low-molecular-weight types, modern membrane microfiltration techniques would take place of centrifuges and filters. The membranes enable majority of linear gelatin molecules with pore sizes of around 1 μm to pass through but hold back globular fat particles or solids.

5.2.7.2 Deionization

Owing to neutralization and conditioning, the gelatin solution still consists of quite a number of mineral salts although many washing steps are used. The ash content in this gelatin accounts for about 2–3%, which is usually higher than the specified value (under 2%) of many food and pharmacopeia regulations. In fact, most consumers in the pharmaceutical industry request an ash content less than 1% for some applications. Gelatin without salt is practically requested for particular photographic layers. The contents of the certain cations within the gelatin should be only a few parts per million.

The soluble salts have an important effect on most physical properties of gelatin. These properties and the sensory parameters become worse with the increase of salt content. However, only some salts are sometimes undesired. For instance, if gelatin containing high sulfate content is dissolved in drinking water with amounts of calcium, the obtained dried product may become turbidity induced by calcium sulfate or other undissolved salts. For the purpose of avoiding such undesirable impacts, ion exchangers can be used to pass through the dilute gelatin solution. Ultrafiltration is also appropriate for desalting gelatin. However, the salts are filtered apart from the influence on concentration of the dilute gelatin solution. Nanofiltration with a pore size of about 5 nm is another possible technique. Unfortunately, it is just applicable in the case where the mineral salts are to be partially removed due to the low throughput.

5.2.7.3 Concentration

There is over 95% water in the purified and demineralized gelatin solution. This water must be almost totally removed. From the perspective of microbiology, only dried gelatin containing a water content of 10–12% possesses a limitless shelf life. Additionally, gelatin solutions without dilution cannot be easily stored or transported.

Normally, concentration occurs in a multistep vacuum system with thin-layer, circular, or platelike evaporators. In such a system, the water is slowly evaporated in a one-stage process at approximately 52 °C or in a multistage process at 50–100 °C, followed by being concentrated with energy recovery. Subsequently, it is reprocessed and recycled.

Membrane filtration, e.g. nanofiltration, is attractive due to its thermal and energy merits and hence is widely applied in practical applications. It is possible to highly separate water and mineral salts using the membrane with pore sizes of around 50 nm. However, gelatin with small molecules is also capable of passing through the membrane, which in turn reduces the yield of gelatin. Thus, it is necessary for the permeate to be reprocessed. The type of gelatin to be concentrated determines the pore size of the used membrane.

The evaporator is used to adjust the final concentration. The size of the evaporate is extremely smaller than it would be without ultrafiltration. Concentrations of 50% can be successfully realized based on the gelatin type. Generally, the gelatin producer hopes to get highly condensed solutions after this step to save the requirement of energy for a later drying step. This is approximately 10 times that is needed to remove water using a multiple-effect evaporator. Due to the heat used during the process of evaporation, tiny amounts of albumin or globulin in the solution could be degenerated and hence precipitated, which needs an extra filtration process to guarantee a glittering and translucent gelatin solution.

5.2.7.4 Final Sterilization

After a highly concentration and filtration process, gelatin solutions are finally subjected to be sterilized. This process utilizes two methods of direct vaporization sterilization and indirect sterilization by platelike thermal transducers (Figure 5.8). The two methods are highly safe in microbiology. Use of the platelike thermal transducers is capable of recovering heat energy up to 95%, while energy cannot be recovered using direct steam. Nevertheless, the direct vaporization sterilization process is more moderate on the gelatin product.

5.2.7.5 Drying Process for Granulated Gelatin

During this process, the gelatin solution sterilized is cooled via a scraper thermal transducer. The highly condensed solution gels are squeezed out in the shape of noodles and then moved down a conveyor belt of drying machine. In the drying stage,

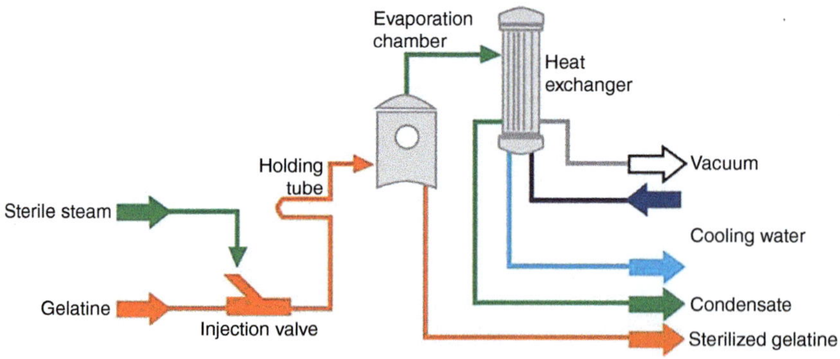

Figure 5.8 Sterilization by direct steam injection is the most common technology in the gelatin industry. It has widely replaced indirect sterilization by plate heat exchangers. Source: Reproduced with permission from Schrieber and Gareis [1]; © 2007, John Wiley & Sons.

Table 5.3 Typical particle sizes of gelatin blends in the market [1].

Particle size to (mm)	Corresponds to (mesh)	Degree of grinding
0.1	140	Powder form
0.5	35	Fine mesh
0.8	20	Standard mesh
2.5 or 3.5	6–8	Coarse mesh
10	3/8	Very coarse mesh

Source: Schrieber and Gareis [1]/John Wiley & Sons.

the remaining water is slowly removed utilizing filtered, dry, and microbiologically clean air.

Gelatin is incapable of being dried via direct warm air due to a low melting point. Consequently, the relative humidity and temperature of the drying air are originally 10–15% and 30 °C, respectively. The temperature of air is normally risen depending on the drying degree of the gelatin to enhance the capacity to absorb water. When the process finishes, the temperature of air is up to around 60 °C, and air has a high humidity. The gelatin gets away from the drying machine in the shape of noodles with around 10% of a water content. Nevertheless, prior to intermediate storage, the noodle-like gelatin is squashed and grinded to considerably coarse granules.

5.2.7.6 Normalization and Packaging of Granulated Gelatin

Finally, the obtained gelatin with uniform batches can be up to 10 tons or more, which varies depending on the used raw materials, the employed producing steps, and other related treatment processes. For particular applications, certain factors such as the properties of the raw materials (such as the species and age of animals) need to be considered. Nevertheless, in order to guarantee that gelatin with a homogeneous and consistent quality is supplied for later treatment by the user, it is necessary for gelatin to be normalized before delivery.

This normalized procedure is carried out by mingling the different batches according to the chemical, physical, and possibly photographic information of each batch obtained during analysis. Thus, it is possible to obtain totally homogeneous blends with any needed size.

Some normalization process involves sieving and grinding as well due to which the coarse particles subjected to drying are usually not the best for later treatment by the user. However, the size of particles in normalized gelatin is not always homogeneous. The normalization process only guarantees that all particles are smaller than the specified restrictions. The sizes of particles for various gelatins in the market are listed in Table 5.3.

The gelatin particles are packed in paper bags containing a weight of 10–15 g for selling retail to families. Small enterprises obtain gelatin products packed in cardboard boxes and plastic bags or containers with weights of 1, 5, or 10 kg. For industrial users, weights of 25 or 50 kg are packed in bags and weights of 500–1000 kg are packed in big bags (Super Sacks). Some large clients can also receive their deliveries in batches transported by silo car followed by being preserved in their own silos until needed to be further processed.

5.3 Modification

5.3.1 Physical Modification

Modification of the physical characteristics for polymers means altering the structure of the polymeric molecule and the supramolecule in its processing. Structure-modified technology could be related to alterations of the phase state, conformational alterations of polymer macromolecules, and orientational alterations at the levels of molecule and supramolecule.

Physical modification of gelatin is principally divided into two categories, namely orientation of gelatin macromolecules and renaturation of the collagen-like helix structure.

5.3.1.1 Orientational Methods of Modification

The orientational method is used to prepare gelatin fibers or other materials with high strength in the stretching direction by orientational stretching. Bergmann and Jacobi were perhaps the first to research the stretch strength of gelatin. They declared that gelatin possessed a tensile strength of 9.3 $kg\,mm^{-2}$ along the stretch axis and 4 $kg\,mm^{-2}$ along the direction perpendicular to that axis. Anokhin reported much stronger gelatin fibers with a range of 10–20 $kg\,mm^{-2}$ obtained by casting fibers from 60% gelatin solutions and using fourfold stretching. This method was investigated to develop the technology for producing fibers for special applications, such as medicine. Unfortunately, these gelatin fibers are unable to be applied in practice unlike other commercially supplied protein fibers, such as casein and collagen [21].

5.3.1.2 Conformational Methods of Modification

Controlling the conformational transformation processes is regarded as a considerably convenient method for manufacturing polymers. Gelatin processing is a good example of manufacturing polymers with specific macromolecular conformations. Gelatin molecules are formed into coils in water at elevated temperatures and then undergo transformation to form collagen-like helix during the casting process at decreased temperatures. Therefore, such a technique of modification just involves chilling solution and storing obtained gel for a while for the purpose that the molecules achieve the highest helicity. However, this technique is incapable of meeting the large-scale production requirements of gelatin materials. The latest polymer industry needs intentional renaturation of gelatin without involving the chilling step (gel formation), which is not yet resolved in spite of large theoretical and practical significance.

Moreover, other physical methods, e.g. radiation, pressure, and ultraviolet (UV) irradiation, are also able to be applied to modify gelatin. Owing to the formation of covalent cross-link bonds, molecular aggregation induced by radiation causes enhanced thermostability of gelatin. Galant reported that radiation (^{60}Co gamma rays, 4–10 kGy) has two types of impacts on the collagen-like structure stability [22]. The impact of ionizing radiation can result in an important improvement in the retention and stabilization of gelatin gel structures, which is discovered during irradiation of gelatin gels. Another impact causes radiation damage of gelatin molecules, leading to weakening of helix structure folding and forming collagen-like gels. Galant also indicated that the radiation cross-linking process was augmented when irradiating gelatin solution [22]. Benbettaïeb et al. suggested that the thermostability of the gelatin films could be increased with the assistance of irradiation (electron beam accelerator, 60 kGy) through the elevated glass transition temperatures and degradation temperatures [23]. They showed that intermolecular cross-linking of gelatin might be induced by the formation of free radicals during irradiation.

Devi et al. found that the melting temperature (T_m) of 5% w/w gelatin rose by 2–3 °C after pressurization (600 MPa) for 15 minutes [24]. A theory about the rise in T_m of gelatin gels via high-pressure treatment has been raised, showing that the needed critical length of the helix shortens with the increase of pressure [25]. Walkenström and Hermansson showed that gelatin gels treated by high pressure had relatively higher melting temperature and larger G′ as well as similar microstructures with the heat-treated gels [26]. Thus, they drew a conclusion that the gelatin gels could be stabilized under high pressure (600 MPa, 20 minutes) due to a special hydration structure formed by the cross-linking connections.

In recent years, UV irradiation has been studied due to its physical, nonthermal, cost-effective, and eco-friendly advantages [27]. UV irradiation is able to generate free radicals at the aromatic residues of collagen and gelatin from amino acids, for instance, phenylalanine and tyrosine [28]. These radicals bond with each other, contributing to cross-linking. Bhat and Karim reported that dry-fish gelatin particles exposed to UV irradiation (254 nm) for 30 and 60 minutes possessed a considerably improved gel strength and an obviously decreased viscosity and an unaltered melting point [27]. Additionally, the positive enthalpy (ΔH) also increased markedly after UV irradiation, which was perhaps due to the cross-linking of fish gelatin caused by UV. They suggested that the considerably decreased viscosity showed chain fragmentation and the importantly increased gel strength indicated a higher cross-linking despite the fact that the reduced chain length might be beneficial to enhance gel strength. Otoni et al. investigated alterations in the molecular weight of gelatin from fish skin using refractive index and gel electrophoresis [28]. Their research showed that cross-linking of gelatin chains after UV-B treatment generated a stronger gel strength and enhanced viscosity with the increase of UV doses (>14.9 J cm^{-2}). Nevertheless, UV irradiation had no significant influence on the melting behavior and maturation of gelatin from fish skin. In comparison, gelatin after exposure to gamma irradiation (up to 10 kGy) had a reduced solution viscosity and gel strength. This was probably due to the fact that more activated cross-linking sites with gelatin by gamma irradiation led to reduced gel strength of gelatin hydrogels [29].

5.3.2 Chemical Modification

Chemical modification is quite an effective method to alter the conformations, structures, surficial net charges, and hydrophobic characteristics of protein. Chemical modification of protein is normally implemented by the cross-linking of chemical reagents with the reactive groups in the protein polypeptide chain. The reactive groups in the gelatin chain contain amino (–NH_2), hydroxyl (–OH), carboxyl (–COOH), and sulfhydryl (–SH) groups [30, 31]. However, a majority of modifications are performed through hydroxylysine, lysine, and the amino groups of the N-terminal amino acid. These can be modified in the range of neutrality to alkalescence, which does not powerfully hydrolyze and degrade gelatin. The modified reaction generally utilizes the sulfonyl chlorides, chlorides and anhydrides of carboxylic acids, epoxides, and isocyanates.

Table 5.4 lists some examples of chemical modifications, indicating that chemical modification can obviously change the characteristic of a specific gelatin. For instance, a kind of gelatin with an enhanced swelling property in water is produced via modifying the lysine group with succinic acid anhydride.

Modified reagents with a polyfunctional group (e.g. polyisocyanates and polyepoxides) can react with a few pendant groups of the gelatin, resulting in the gelatin cross-linking (Figure 5.9). The cross-linking degree is determined by many various parameters, such as the pH, the temperature and duration of reaction, the type of gelatin, as well as the used cross-linking reagent. Some typical cross-linking reagents for gelatin include metallic cross-linking reagents (Al^{3+}, Cr^{3+}, Fe^{3+}, Ce^{3+}, La^{3+}, Zr^{4+}) and organic cross-linking reagents (epoxides, aldehydes, carbodiimides, isocyanates, ketones). According to the cross-linking degree, the obtained gelatin shows the increased melting point, decreased swellability, and enhanced resistance to mechanical stress. These characteristics enable gelatin to have the ability of resisting boiling water.

Table 5.4 Overview of possible chemical modifications [1].

Functional group	Modification/reagent	Reagent example
Amino group –NH_2	Organic acid halides/acid anhydrides	Succinic acid anhydride Phthalic acid anhydride
	Aryl-substituted reagents	Phenylisothiocyanates
	Activated organic halogen compounds	Benzyl bromide Biphenylhalomethane
	Carbamoylation	Cyanates
	Reaction with double bonds	Maleimides
	Epoxides	Glycidyl ether
	Guanidylation	*O*-methylisourea
Carboxyl group –COOH	Methylation of carboxyl groups	Methanolic sulfuric acid
	Reaction with carbodiimides	Carbodiimides
Hydroxyl group –OH	Acetylation	Acetic acid anhydride

Source: Schrieber and Gareis [1]/John Wiley & Sons.

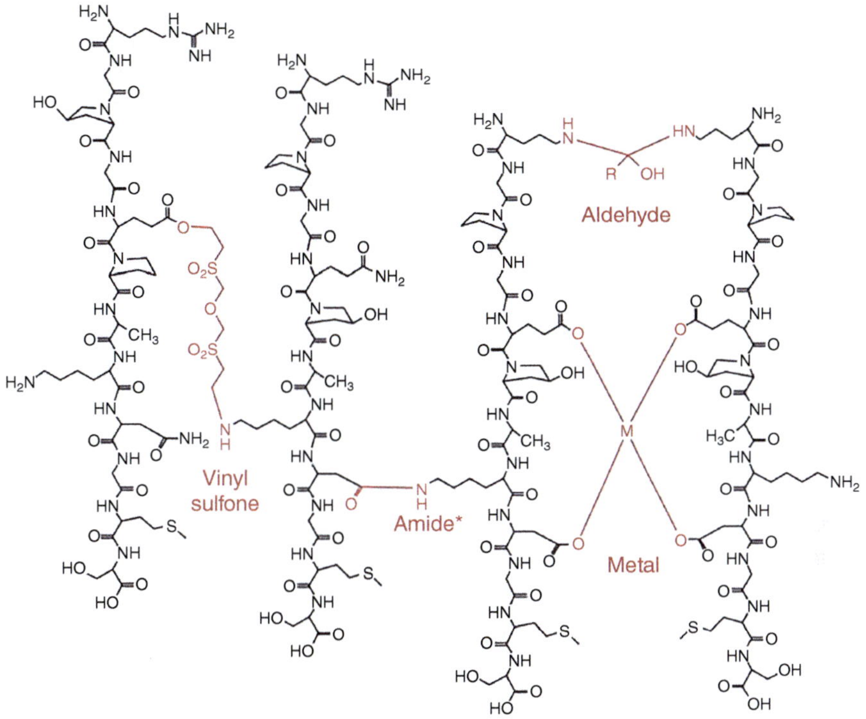

Figure 5.9 Structure of cross-linked gelatins with various reagents. *Can be formed by carbonyl activation, enzyme, or aging. Source: Reproduced with permission from Schrieber and Gareis [1]; © 2007, John Wiley & Sons.

"Macromers" with polymerizable pendant chains can be obtained by reaction with active derivatives of acrylic and methacrylic acids. In a second-reaction stage, irradiation, radicals, or redox reactions can induce polymerization, causing the resulted modified gelatins to be transformed into derivatives withstanding boiling water and organic solvents. Compared with conventional cross-linking of gelatin, this way is capable of accurately controlling the product characteristics due to the separated cross-linking and modification processes.

Another kind of chemical modification is dependent on a biological process. The portions of broken linear collagen chains are reconnected by the addition of the enzyme transglutaminase, leading to the increased viscosity and less melting of the gel that is heated to 100 °C. Unfortunately, the gel firmness is only slightly enhanced. Some countries utilize this effect to manufacture luncheon meat or similar products during the meat processing. Some classical reaction mechanisms of normal chemical modification for gelatin are introduced in the following text (Figure 5.10).

5.3.2.1 Phosphorylation Modification

Phosphorylation is a general way to modify the properties of protein. The common phosphorylation reagents contain trisodium trimetaphosphate, phosphorus

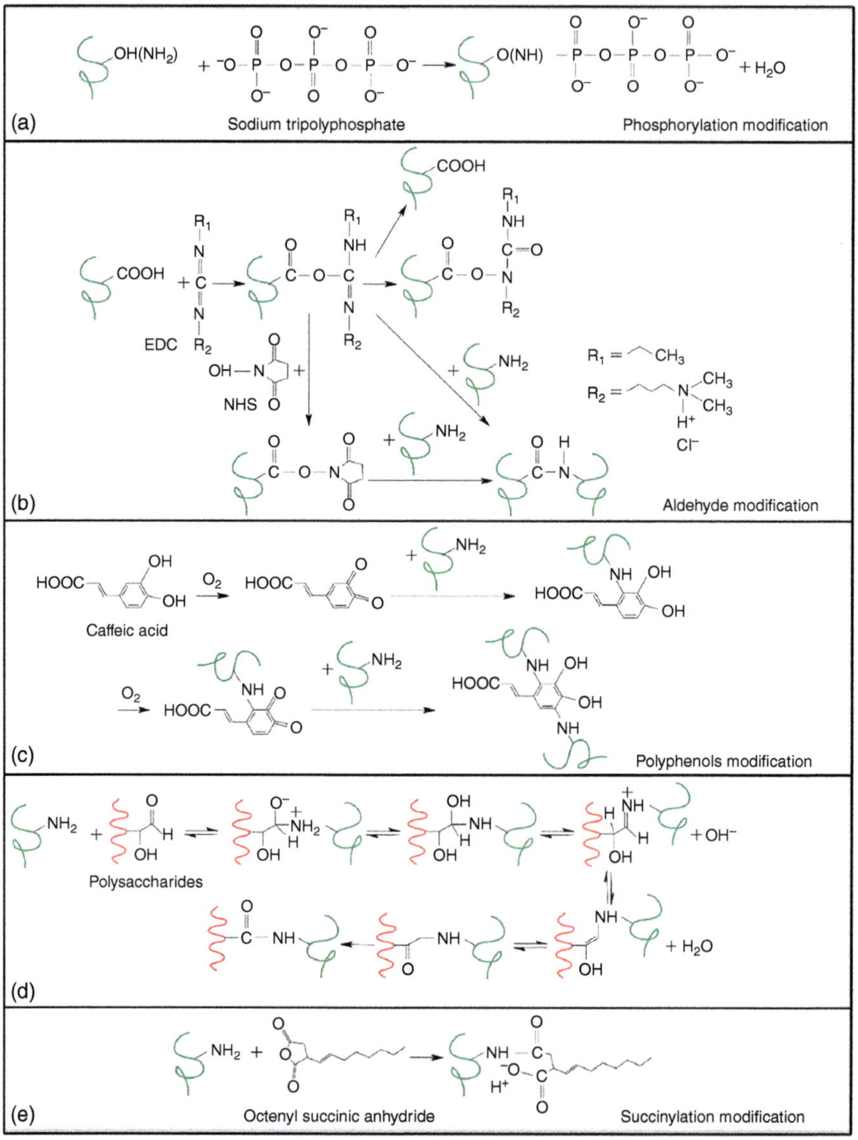

Figure 5.10 Typical reaction mechanisms of common chemical modification of gelatin. (a) Phosphorylation modification. (b) NHS/EDC Aldehyde modification. (c) Polyphenol modification. (d) Glycosylation modification. (e) Succinylation modification. Source: Reproduced with permission from Zhang et al. [20]; © 2020, Elsevier Ltd.

oxychloride, glucose-6-phosphate, sodium tripolyphosphate, and so on [32]. The reaction of the phosphate groups of a phosphorylation reagent with the hydroxyl/amino groups of the protein polypeptide chain enables the phosphate group to introduce to a protein polypeptide chain. Normally, one phosphorylation molecule may chemically cross-link one to three protein molecules, causing the increased surface hydrophobic property of protein, and facilitate hydrophobic interactions of protein with oil droplets and protein molecules at the interface [33].

Phosphorylation modification is able to improve the ionic interaction between the phosphate groups and NH_3^+ of amino acids in the modified gelatin, which enhances the gel strength of gelatin [34]. The linkages between phosphate and gelatin are impacted by a few significant experimental parameters, for example, reaction time and pH, as well as the ratio of the phosphorylation reagent to gelatin. Too longer reaction time is likely to enable gelatin chains to aggregate for the formation of larger bundles, which may have an interfering effect on forming a finer and homogeneous gel network and reducing the gel strength. An alkaline environment may promote the phosphorylation. An excessive phosphorylation reagent may enhance the electrostatic repulsive force between gelatin molecules and thus reduce the gel strength.

It is legitimately forecasted that the microwave-assisted application is likely to improve the phosphorylation degrees of the obtained gelatins and enhance the emulsifying characteristics of gelatins.

5.3.2.2 Aldehyde Modification

Aldehyde modification is one of the general protein modification ways through employing aldehyde-based cross-linking reagents. It plays an important role in enhancing mechanical, thermal, and gel characteristics of gelatins. Depending on whether to turn into a segment of modified gelatin molecule, the aldehyde-based cross-linking agents include two groups. *N*,*N*-(3-dimethylamino propyl)-*N*′-ethyl carbodiimide (EDC) is a typical reagent that cannot turn into a segment of gelatin molecule after modification. EDC is capable of activating the carboxylic acid residues of aspartic and glutamic acids in gelatin chains, resulting in the formation of an active *O*-acylisourea intermediate group. The reaction of activated groups with free amino residues of hydroxylysine and lysine residues in (other) gelatin molecules causes the formation of amide bonds in one gelatin molecule or between gelatin molecules [35]. Due to its instability in water, the intermediate group activated by EDC usually reacts with *N*-hydroxysuccinimide (NHS) to generate the NHS-ester, which increases the cross-linking efficiency. Typical reagents that become a segment of gelatin molecule after modification contain glutaraldehyde, formaldehyde, and terephthalaldehyde. Glutaraldehyde and terephthalaldehyde are easy to react with ε-amino groups to create new covalent bonds in a gelatin molecule or between gelatin molecules. Formaldehyde is easy to react with ε-amino groups to generate new covalent bonds between gelatin molecules.

Due to the fact that they are inexpensive and able to react rapidly with ε-amino groups in proteins, glutaraldehyde and formaldehyde are the most general aldehyde-based cross-linking reagents. Unfortunately, their applications in food industry have been limited by their potential toxicity. As safe or low-toxic reagents, *N*-hydroxysuccinimide adipic-acid derivative and terephthalaldehyde occurred in nature and EDC/NHS shows large application prospects in the field of gelatin modification [36]. A lot of hard work ought to be performed in this direction in order to stimulate the development of food industry [37].

5.3.2.3 Polyphenols Modification

Owing to the hydrogen bonding and hydrophobic interactions, gelatin and polyphenols are able to form nanocomplexes, which can considerably improve the thermal and emulsifying properties as well as the gel strength of gelatin. Unfortunately, these nanocomplexes are just formed and stabilized at specific processing conditions. Thus, in order to enhance the properties of gelatins, it is necessary to explore the chemical cross-linking of gelatin with polyphenols for the formation of covalent bonds between them.

In order to enhance the emulsifying properties of gelatins, oxidized polyphenols have been used to modify gelatin by covalently reacting with amino acids or sulfhydryl side chains of gelatin (such as Lys, Arg, and His) to form C—N and C—S bonds. [38, 39]. Compared with unmodified gelatins, gelatins modified by oxidized polyphenol possessed enhanced antioxidative activity and reduced surface hydrophobicity but similar droplet size distribution of the emulsions. Moreover, gelatins modified by oxidized polyphenol showed a smaller increase in vitro antioxidative activity than non-covalent mixtures of polyphenol/gelatin. However, modification enabled the oil-in-water emulsion to have a stronger stabilization ability, which could restrain lipid oxidation of emulsion within the stored process. The possible reason is that the gelatin after modification may be easy to locate and adsorb at the interface of oil and water. It is necessary to investigate the influences of polyphenols on the molecular structure of gelatin at the interface of oil and water.

5.3.2.4 Succinylation Modification

During the process of typical succinylation modification, the succinylating reagent reacts primarily with the ε-amino group of lysine and minorly with the N-terminal amino acids in proteins or polysaccharides, which convert the macromolecules from residue with positive charge into residue with negative charge by N-acylation. The general succinylating reagents contain succinic anhydride [40] and octenyl succinic anhydride [41]. Nevertheless, succinic anhydride is listed in the 2017 Class 3 carcinogen list by the WHO international cancer research institute, thus causing its application to be limited in food industry. Octenyl succinic anhydride modification has been used to modify starches in food industry over half a century.

Zhang et al. investigated the impact of octenyl succinic anhydride modification on the characteristics of gelatins from bovine bone or fish skin and the fish oil-loaded emulsions [42]. The study indicated that the succinylation degree of gelatin logarithmically increased with the mass ratios of succinylating reagent to gelatin. The succinylation modification had no significant influences on the incipient droplet sizes of emulsion. Increasing the succinylation degree of gelatin from bovine bone could enhance the droplet stabilization ability, creaming index, and the phase transition time of emulsions. However, increasing the succinylation degree of gelatin from fish skin just improved the emulsion droplet stability. This study extended the application of succinylation modification in the field of gelatin science. It is imperative to further analyze the influence of succinylation modification on the molecular structures of gelatins and promote the application of succinylated modified gelatins.

5.3.3 Enzymatic Modification of Gelatin

Enzymatic modification methods are used to improve the functional properties of macromolecules by employing enzymes to modify macromolecules. Enzymatic modification has several important merits, namely larger selectivity and substrate specificity, milder reaction conditions, and sustainable enzymes, than chemical modification. Among them, larger selectivity and substrate specificity endow the resulted products with well-designed and/or stereospecific structures. Thus, the enzymatic modification is a more environmentally friendly method to enhance the properties of macromolecules than chemical modification [43]. Herein, only the enzymatic modification of the resulted pure gelatin after extracting gelatin is discussed.

Some enzymes have been investigated to improve the functional properties of protein by modification, for example, protein kinases [32], tyrosinases [44], laccases [45], a mixture of glucose oxidase and horseradish peroxidase [46], and transglutaminases [47]. Proteins can be phosphorylated by protein kinases by adenosine triphosphate to enhance their emulsion stabilization ability and foam expandability. Unfortunately, this enzymatic phosphorylation may not be suitable for food industry because proteins only introduce few phosphate groups as well as enzyme and adenosine triphosphate are both expensive. Oxidoreductases, such as tyrosinases, laccases, and peroxidases, are capable of cross-linking proteins with the assistance of tyrosine residues or polyphenols. The type of oxidoreductases causes enzymatic modification to have different mechanisms. Conjugates of protein and polyphenol are likely to possess the excellent gel and emulsifying properties [48]. Nevertheless, oxidoreductases are not able to obviously enhance the gel properties of gelatin principally due to the small amount of tyrosine (~0.5%) in gelatin. Tyrosinases result in forming *o*-quinones by the reaction of *o*-diphenols with cysteinyl, tyrosyl, nucleophilic lysyl, and histidinyl residues of proteins. As a result, Tyr-Cys, Tyr-Tyr, or Tyr-Lys are directly cross-linked in proteins. Laccases and peroxidases generate a hydrogen atom from polyphenols and aromatic amines causing the formation of radicals. These radicals are then oxidized to quinones or go through a variety of coupling reactions to catalyze proteins' cross-linking. A mixture of glucose and two oxidases (glucose oxidase and horseradish peroxidase) can be used to cross-link gelatins to polymerized molecules with more well-organized secondary structures owing to the formation of dityrosine. The cross-linked gelatins are able to form gels with higher strength in shorter time. Transglutaminases can particularly cross-link peptide bonds between the γ-amide group of a glutamine residue and the ε-amino groups of a lysine residue to generate ε-(γ-glutamyl)-lysine peptide bonds in proteins via a three-step reaction of acyl transfer, cross-linking, and deamidation. Transglutaminase modification has been applied with success to improve rheological and gel properties as well as melting points of gelatins, owing to the formation of a uniform and dense gel network with quite small voids [30].

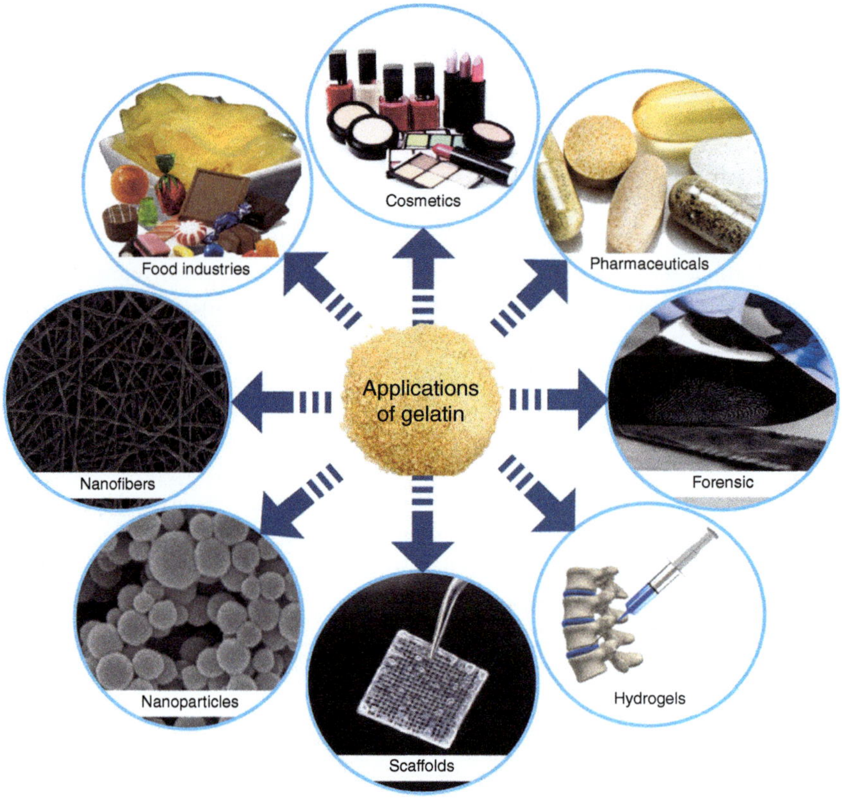

Figure 5.11 Applications of gelatins (pictures taken randomly from google images). Source: Alipal et al. [4]/with permission of © Elsevier Ltd.

5.3.4 Complex Modification of Gelatin

Complex modification refers to the incorporation of chemical, physical, and enzymatic modifications or of different modifications in the same kind of modifications. In theory, the complex modification via reasonable design is more beneficial to improve the functional characteristics of macromolecules than individual modification (Figure 5.11).

In order to modulate lipid digestion of the emulsions stabilized by interface layers based on gelatin, several efforts had been performed to investigate the incorporation of multilayer formation with enzymatic modification. Zeeb et al. explored the influence of laccase (a polysaccharide cross-linking enzyme) on the lipolysis and stability of fish gelatin-sugar beet pectin double-layered oil-in-water emulsions by preparing and comparing the fish gelatin primary, fish gelatin-sugar beet pectin secondary, and cross-linked secondary emulsions [49]. They found that the laccase modification was able to improve the multilayer emulsion stability. In the human gastrointestinal tract models, the interfacial layer composition was a determining factor for the stability of emulsion to coalescence and flocculation. There was no significant impact for cross-linking on the degree and rate of lipid digestion of secondary emulsions.

In fact, the degree and rate of lipid digestion for secondary emulsions, whether or not to cross-link, were quicker than those of primary emulsions. Moreover, the lipid digestion of these secondary emulsions was likely not to be impacted by electrostatic regulation via salts [50]. Thus, complex modification was a promising application for food emulsions.

References

1 Schrieber, R. and Gareis, H. (2007). *Gelatine Handbook*. New York: Wiley.

2 Haug, I.J. and Draget, K.I. (2011). *Handbook of Food Proteins*. Amsterdam: Elsevier Ltd.

3 Abedinia, A., Nafchi, A.M., Sharifi, M. et al. (2020). Poultry gelatin: characteristics, developments, challenges, and future outlooks as a sustainable alternative for mammalian gelatin. *Trends in Food Science & Technology* 104: 14–26.

4 Alipal, J., Pu'ad, N.A.S.M., Lee, T.C. et al. (2021). A review of gelatin: properties, sources, process, applications, and commercialization. *Materials Today: Proceedings* 42: 240–250.

5 Karim, A.A. and Bhat, R. (2009). Fish gelatin: properties, challenges, and prospects as an alternative to mammalian gelatins. *Food Hydrocolloids* 23 (3): 563–576.

6 Lu, L., Garcia, C.A., and Mikos, A.G. (1999). In vitro degradation of thin poly(DL-lactic-*co*-glycolic acid) films. *Journal of Biomedical Materials Research* 46: 236–244.

7 Gómez-Guillén, M.C., Giménez, B., López-Caballero, M.E., and Montero, M.P. (2011). Functional and bioactive properties of collagen and gelatin from alternative sources: a review. *Food Hydrocolloids* 25: 1813–1827.

8 Cao, S., Wang, Y., Xing, L. et al. (2020). Structure and physical properties of gelatin from bovine bone collagen influenced by acid pretreatment and pepsin. *Food and Bioproducts Processing* 121: 213–223.

9 Frushour, B.G. and Koenig, J.L. (1975). Raman scattering of collagen, gelatin, and elastin. *Biopolymers* 14: 379–391.

10 Giménez, B., Gómez-Guillén, M.C., and Montero, P. (2005). The role of salt washing of fish skins in chemical and rheological properties of gelatin extracted. *Food Hydrocolloids* 19: 951–957.

11 Damrongsakkul, S., Ratanathammapan, K., Komolpis, K., and Tanthapanichakoon, W. (2008). Enzymatic hydrolysis of rawhide using papain and neutrase. *Journal of Industrial and Engineering Chemistry* 14: 202–206.

12 Pitpreecha, S. and Damrongsakkul, S. (2006). Hydrolysis of raw hide using proteolytic enzyme extracted from papaya latex. *Korean Journal of Chemical Engineering* 23: 972–976.

13 Ahmad, T., Ismail, A., Ahmad, S.A. et al. (2017). Recent advances on the role of process variables affecting gelatin yield and characteristics with special reference to enzymatic extraction: a review. *Food Hydrocolloids* 63: 85–96.

14 Asomaning, J., Haupt, S., Chae, M., and Bressler, D.C. (2018). Recent developments in microwave-assisted thermal conversion of biomass for fuels and chemicals. *Renewable and Sustainable Energy Reviews* 92: 642–657.

15 Liu, T., Dai, H., Ma, L. et al. (2019). Structure of Hyla rabbit skin gelatin as affected by microwave-assisted extraction. *International Journal of Food Properties* 22: 1594–1607.

16 Kim, T.-K., Ham, Y.-K., Shin, D.-M. et al. (2020). Extraction of crude gelatin from duck skin: effects of heating methods on gelatin yield. *Poultry Science* 99: 590–596.

17 Huang, Y., Li, A., Qiu, C. et al. (2017). Self-assembled colloidal complexes of polyphenol–gelatin and their stabilizing effects on emulsions. *Food & Function* 8: 3145–3154.

18 Karnjanapratum, S. and Benjakul, S. (2020). Asian bullfrog (*Rana tigerina*) skin gelatin extracted by ultrasound-assisted process: characteristics and *in-vitro* cytotoxicity. *International Journal of Biological Macromolecules* 148: 391–400.

19 Mirzapour-Kouhdasht, A., Sabzipour, F., Taghizadeh, M.S., and Moosavi-Nasab, M. (2019). Physicochemical, rheological, and molecular characterization of colloidal gelatin produced from common carp by-products using microwave and ultrasound-assisted extraction. *Journal of Texture Studies* 50: 416–425.

20 Zhang, T., Xu, J., Zhang, Y. et al. (2020). Gelatins as emulsifiers for oil-in-water emulsions: extraction, chemical composition, molecular structure, and molecular modification. *Trends in Food Science & Technology* 106: 113–131.

21 Kozlov, P.V. and Burdygina, G.I. (1983). The structure and properties of solid gelatin and the principles of their modification. *Polymer* 24: 651–666.

22 Galant, S. (1983). Formation of collagen-like structure of gel after radiation modification of gelatin. *Colloid and Polymer Science* 261: 445–451.

23 Benbettaïeb, N., Karbowiak, T., Brachais, C.H., and Debeaufort, F. (2016). Impact of electron beam irradiation on fish gelatin film properties. *Food Chemistry* 195: 11–18.

24 Devi, A.F., Buckow, R., Hemar, Y., and Kasapis, S. (2014). Modification of the structural and rheological properties of whey protein/gelatin mixtures through high pressure processing. *Food Chemistry* 156: 243–249.

25 Kulisiewicz, L. and Delgado, A. (2009). Network structure of gelatin gel at high pressure determined by rheological measurements. *High Pressure Research* 29: 67–71.

26 Walkenström, P. and Hermansson, A.-M. (1997). High-pressure treated mixed gels of gelatin and whey proteins. *Food Hydrocolloids* 11: 195–208.

27 Bhat, R. and Karim, A.A. (2009). Ultraviolet irradiation improves gel strength of fish gelatin. *Food Chemistry* 113: 1160–1164.

28 Otoni, C.G., Avena-Bustillos, R.J., Chiou, B.S. et al. (2012). Ultraviolet-B radiation induced cross-linking improves physical properties of cold- and warm-water fish gelatin gels and films. *Journal of Food Science* 77: E215–E223.

29 Sung, W.-C. and Chen, Z.-Y. (2014). UV treatment and γ irradiation processing on improving porcine and fish gelatin and qualities of their premix mousse. *Radiation Physics and Chemistry* 97: 208–211.

30 Huang, T., Tu, Z., Shangguan, X. et al. (2019). Fish gelatin modifications: a comprehensive review. *Trends in Food Science & Technology* 86: 260–269.

31 Saber, M.M. (2019). Strategies for surface modification of gelatin-based nanoparticles. *Colloids and Surfaces B: Biointerfaces* 183: 110407.

32 Li, C.-P., Enomoto, H., Hayashi, Y. et al. (2010). Recent advances in phosphorylation of food proteins: a review. *LWT- Food Science and Technology* 43: 1295–1300.

33 Xiong, Z., Zhang, M., and Ma, M. (2016). Emulsifying properties of ovalbumin: improvement and mechanism by phosphorylation in the presence of sodium tripolyphosphate. *Food Hydrocolloids* 60: 29–37.

34 Kaewruang, P., Benjakul, S., and Prodpran, T. (2014). Characteristics and gelling property of phosphorylated gelatin from the skin of unicorn leatherjacket. *Food Chemistry* 146: 591–596.

35 Kuijpers, A.J., Engbers, G.H.M., Krijgsveld, J. et al. (2000). Cross-linking and characterisation of gelatin matrices for biomedical applications. *Journal of Biomaterials Science, Polymer Edition* 11: 225–243.

36 Shen, L., Tian, Z., Liu, W., and Li, G. (2015). Influence on the physicochemical properties of fish collagen gels using self-assembly and simultaneous cross-linking with the *N*-hydroxysuccinimide adipic acid derivative. *Connective Tissue Research* 56: 244–252.

37 Mohtar, N.F., Perera, C.O., and Hemar, Y. (2014). Chemical modification of New Zealand hoki (*Macruronus novaezelandiae*) skin gelatin and its properties. *Food Chemistry* 155: 64–73.

38 Aewsiri, T., Benjakul, S., Visessanguan, W. et al. (2009). Antioxidative activity and emulsifying properties of cuttlefish skin gelatin modified by oxidised phenolic compounds. *Food Chemistry* 117: 160–168.

39 Aewsiri, T., Benjakul, S., Visessanguan, W. et al. (2010). Antioxidative activity and emulsifying properties of cuttlefish skin gelatin-tannic acid complex as influenced by types of interaction. *Innovative Food Science & Emerging Technologies* 11: 712–720.

40 Lawal, O.S. (2005). Functionality of native and succinylated Lablab bean (*Lablab purpureus*) protein concentrate. *Food Hydrocolloids* 19: 63–72.

41 Sweedman, M.C., Tizzotti, M.J., Schäfer, C., and Gilbert, R.G. (2013). Structure and physicochemical properties of octenyl succinic anhydride modified starches: a review. *Carbohydrate Polymers* 92: 905–920.

42 Zhang, T., Ding, M., Tao, L. et al. (2020). Octenyl succinic anhydride modification of bovine bone and fish skin gelatins and their application for fish oil-loaded emulsions. *Food Hydrocolloids* 108, 106041.

43 Karaki, N., Aljawish, A., Humeau, C. et al. (2016). Enzymatic modification of polysaccharides: mechanisms, properties, and potential applications: a review. *Enzyme and Microbial Technology* 90: 1–18.

44 Jus, S., Stachel, I., Fairhead, M. et al. (2012). Enzymatic cross-linking of gelatine with laccase and tyrosinase. *Biocatalysis and Biotransformation* 30: 86–95.

45 Azarikia, F., Wu, B., Abbasi, S., and McClements, D.J. (2015). Stabilization of biopolymer microgels formed by electrostatic complexation: influence of enzyme (laccase) cross-linking on pH, thermal, and mechanical stability. *Food Research International* 78: 18–26.

46 Han, Y.-P. and Zhao, X.-H. (2016). Properties of bovine gelatin cross-linked by a mixture of two oxidases (horseradish peroxidase and glucose oxidase) and glucose. *CyTA - Journal of Food* 14: 457–464.

47 Liu, Y., Weng, R., Wang, W. et al. (2020). Tunable physical and mechanical properties of gelatin hydrogel after transglutaminase crosslinking on two gelatin types. *International Journal of Biological Macromolecules* 162: 405–413.

48 Quan, T.H., Benjakul, S., Sae-leaw, T. et al. (2019). Protein–polyphenol conjugates: antioxidant property, functionalities and their applications. *Trends in Food Science & Technology* 91: 507–517.

49 Zeeb, B., Lopez-Pena, C.L., Weiss, J., and McClements, D.J. (2015). Controlling lipid digestion using enzyme-induced crosslinking of biopolymer interfacial layers in multilayer emulsions. *Food Hydrocolloids* 46: 125–133.

50 Zeeb, B., Weiss, J., and McClements, D.J. (2015). Electrostatic modulation and enzymatic cross-linking of interfacial layers impacts gastrointestinal fate of multilayer emulsions. *Food Chemistry* 180: 257–264.

6

Applications of Gelatin in Electrochemical Energy Storage and Conversion

6.1 As Binders for Battery Electrodes

As a significant component in rechargeable batteries, binders endow the electrodes with high mechanical strength and interconnected structures to guarantee the electronic/ionic transfer during the charge–discharge process of batteries [1]. However, the conventional binders are undesired candidates for batteries with high performance owing to their relatively weak adhesiveness, low mechanical strength, and poor functionalities. Therefore, it is necessary to explore advanced binders to improve the performance of batteries. A binder with high performance should not only possess great adhesion property for the electrode materials to the current collector, but also provide multifunctional properties, e.g. conductivity, anode/cathode–electrolyte interphase (solid electrolyte interphase [SEI] and CEI) stabilization, and polysulfide anchoring. Currently, different binders are being investigated for electrode materials of batteries, e.g. lithium–sulfur (Li–S) batteries and lithium-ion batteries (LIBs).

6.1.1 As Binders of Electrodes for Lithium–Sulfur Batteries

For Li–S batteries, the cathode is normally produced by coating the blend of binder, conductive carbon, and elemental sulfur on the Al current collector followed by drying at the increased temperature in vacuum condition [2]. The drying process may result in the loss of active sulfur material because of the weak adhesion ability of binder. Additionally, sulfur in cathode is converted into a long-chain polysulfide in the discharge process. Subsequently, long-chain polysulfide is dissolved into the liquid electrolyte and then reduced, which causes high reduction of soluble polysulfide and solid products, e.g. Li_2S_2 and Li_2S. The reactions are shown as below [3]:

$$2Li + S_8 \leftrightarrow Li_2S_8 \tag{6.1}$$

$$2Li + Li_2S_8 \leftrightarrow 2Li_2S_4 \tag{6.2}$$

$$2Li + Li_2S_4 \leftrightarrow 2Li_2S_2(\text{insoluble}) \tag{6.3}$$

$$2Li + Li_2S_2 \leftrightarrow 2Li_2S(\text{insoluble}) \tag{6.4}$$

Collagen-Derived Materials: Synthesis and Applications in Electrochemical Energy Storage and Conversion, First Edition. Feng Wang, Yaqin Huang, and Jin Niu.
© 2022 WILEY-VCH GmbH. Published 2022 by WILEY-VCH GmbH.

Due to the dissolved lithium polysulfide and insoluble Li_2S in electrolyte, the cathode has a considerably poor sulfur utilization. Furthermore, sulfur and conductive carbon are easy to aggregate, and the electrical contacts between them become worse if the binder is unable to make cathode material form a good electric network [4] (Figure 6.1).

Among polymer binders, gelatin is a main binder used for Li–S batteries due to its following merits for the preparation of the cathode [6]: (i) Gelatin is a kind of ampholytic polymer with ionizable groups (e.g. COOH and NH_2). Thus, gelatin has good hydrophilicity and is basically insoluble in generally organic electrolyte solvents (e.g. carbonic acid esters and ether), leading to the great stability of electrodes. (ii) The large viscosity enables gelatin to be appropriate as an adhesion

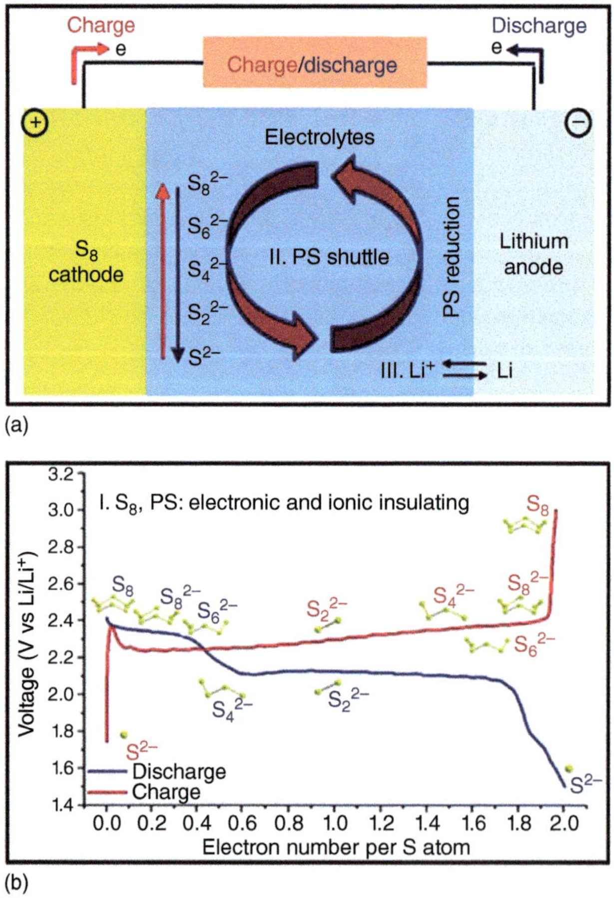

Figure 6.1 (a) Working mechanism of a Li–S battery involving formation of lithium polysulfides, (b) charge/discharge curve indicating contribution of various states of sulfur during lithiation and delithiation. Source: Reproduced with permission from Saroha et al. [5]; © 2021 Springer Nature Switzerland AG.

agent for bonding various kinds of small granules onto substrates. (iii) The excellent dispersibility of gelatin is beneficial to the preparation of electrodes.

Sun J. et al. prepared two kinds of cathodes for Li–S batteries, with gelatin and poly(ethylene oxide) (PEO) as binder, respectively, and compared their structure as well as electrochemical performance in 1 M $LiClO_4$ dimethoxy ethane/1,3-dioxolane ($V/V = 1/1$) electrolyte [6]. The results demonstrated that the gelatin binder not only possessed the effective adhesive and dispersion abilities for the cathode materials, but also could act as an electrochemically stable binder. The cathode using gelatin binder exhibited a large original capacity of 1132 mAh g^{-1}, as well as a reversible capacity of 408 $mAhg^{-1}$ over 50 cycles, indicating higher capacities than the cathode using PEO binder in the same conditions.

The cathode performance can be further improved via using modified gelatin binder. For example, gelatin binder modified with L-cysteine not only was used as an excellent adhesion and dispersion agent for the cathode materials, but also had the ability to form a steady electric network between the sulfur and the conducting carbon [7]. Thus, the cathode showed smaller transfer resistance, less polarization, and improved cycle performance during cycles, leading to higher original capacity of 1135 mAh g^{-1} and reversible capacity of 754 mAh g^{-1} over 50 cycles.

Additionally, apart from the great adhesion and dispersion abilities for the prepared cathode, acylated gelatin binder was also capable of inhibiting effectively the dissolution of polysulfide into the electrolyte due to the large affinity of carbonyl groups for polysulfide [8]. The studies showed that more polysulfide Li_2S_n ($4 \leq n \leq 8$) were restricted in the cathode with the increase of carbonyl groups in the acylated gelatin binder. Therefore, compared with sulfur cathode using gelatin binder, the cathodes show smaller transfer resistance, less polarization, and larger capacities over 100 cycles at 0.1 and 0.5 C, respectively.

Composite binders combining gelatin with other materials, such as gelatin–polyethylenimine composite (GPC), are also able to effectively improve the cathode performance [9]. Polyethylenimine in GPC binder offered amounts of amino groups and branched network architecture, causing GPC binder to possess high adsorption ability for polysulfide intermediates via electrostatic interaction, and hence "shuttling effect" of sulfur cathode is confined during discharge–charge process (Figure 6.2). Meanwhile, GPC binder enables cathode to maintain stable and homogenous structures due to the great adhesion and dispersion capability of gelatin. The electrochemical measurements indicated that GPC binder considerably improved cyclic lifespan of Li–S batteries without obvious capacity decay at rate of 1 C, which is 27–35% larger than batteries using gelatin or polyethylenimine binder (Figure 6.3).

6.1.2 As Binders of Electrodes for Lithium-Ion Batteries

The water-soluble gelatin also shows tremendous potential as binder of the electrodes for LIB. For instance, Zhao et al. reported a green gelatin binder cross-linked by polyethylene glycol diacrylate (PEGDA) for the $LiFePO_4$ cathode [10]. Benefiting from the cross-linked network and unique groups (e.g. carboxyl and amino groups)

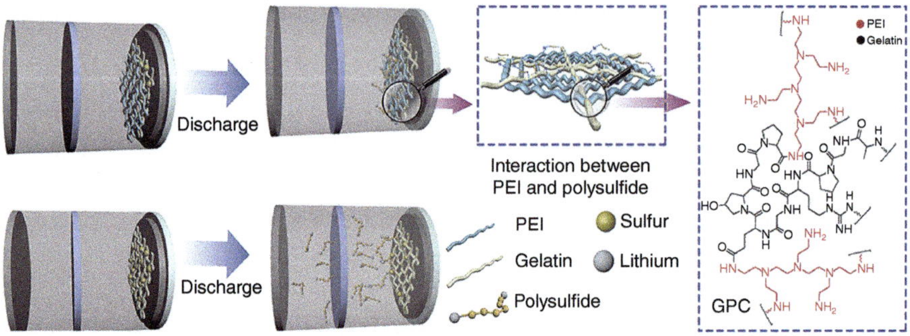

Figure 6.2 The polysulfide-trapping mechanisms of batteries with GPC binder. Source: Reproduced with permission from Akhtar et al. [9]; © 2018 Elsevier Ltd.

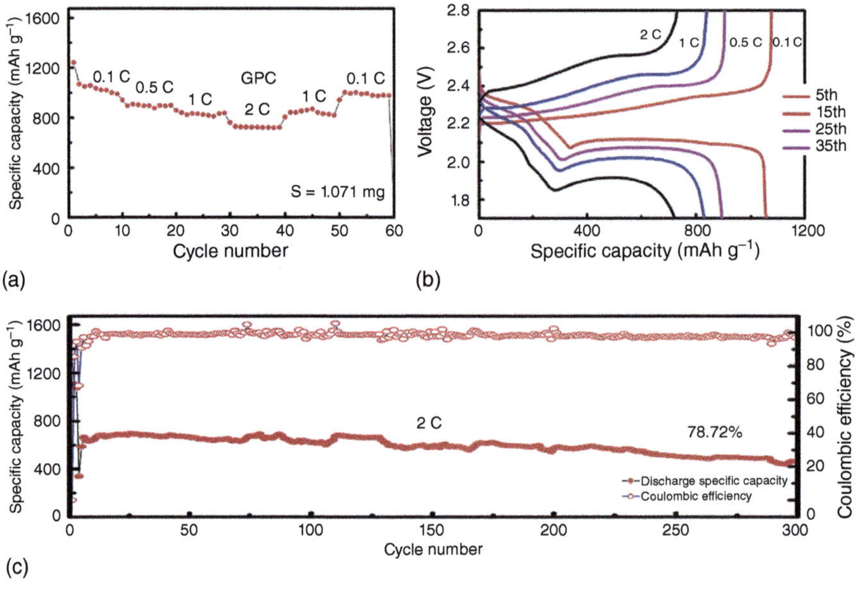

Figure 6.3 (a) The rate capacity of GPC-based battery and (b) corresponding charge–discharge voltage curve, (c) long cycling performance of GPC-based battery at 2 C. Source: Reproduced with permission from Akhtar et al. [9]; © 2018 Elsevier.

in the gelatin backbone, the cross-linked gelatin binder (named as GP binder) was conducive to stabilizing the structure of cathode after long-term cycles via decreasing the formation of the crack and inhibiting the exfoliation of electrode materials from the Al foil current collector. Moreover, the GP binder with 3D network contributed to the formation of interconnected conductive bridges between the binders and $LiFePO_4$ granules to facilitate the Li^+ and electrons transport.

The superior performance of LIBs was largely associated with the multiple chemical interactions among the binder, collector, and active materials (Figure 6.4b). Firstly, the carboxyl and amino groups in the GP binder were capable of forming hydrogen bonding with the Al current collector, making the $LiFePO_4$ granules and

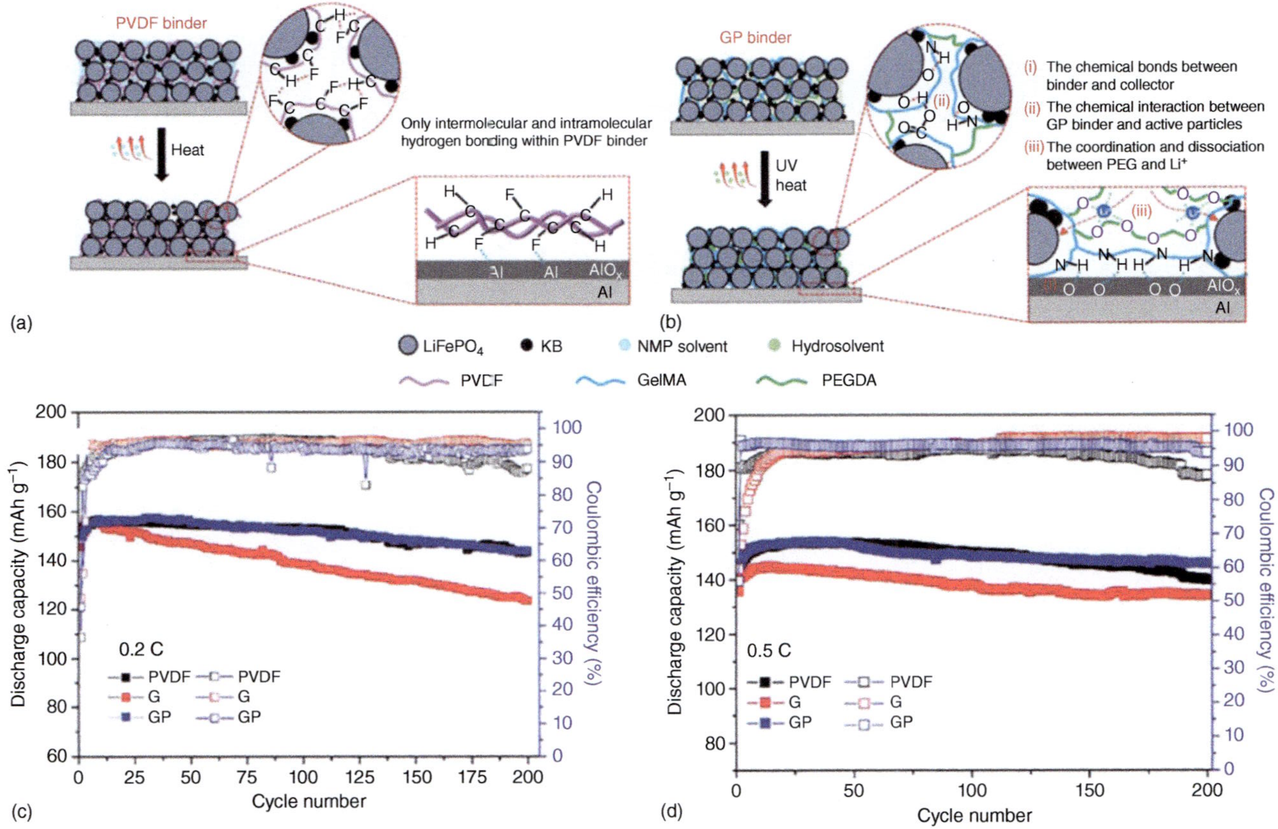

Figure 6.4 Schematic illustrations of the preparation and binding mechanisms of the $LiFePO_4$ electrodes using (a) PVDF and (b) GP binders. Cycling performances of the $LiFePO_4$ cathode with PVDF, G (gelatin), and GP binders at (c) 0.2 C and (d) 0.5 C. Source: Reproduced with permission from Zhao et al. [10]; © The Royal Society of Chemistry 2020.

conducting materials closely bond onto the current collector and then considerably confine the detachment of electrode material. Secondly, the strong interaction of hydrogen bonding between GP binder and $LiFePO_4$ granules enabled the $LiFePO_4$ granules to be wrapped tightly by the GP binder to reduce the electrode polarization, which ensured a highly homogenous distribution of the electrode materials and the structural integrity of electrode. Thirdly, a number of complexation sites were formed via the chemical interaction of coordination and dissociation between the ether linkages in the PEGDA chain and Li^+. The complex Li^+ could move as the segmental motion of the polymer matrix with the assistance of these complexation sites, which facilitates the Li^+ migration in the cathode. By contrast, the commercial polyvinylidene fluoride (PVDF) binder dependent on weak van der Waals interaction usually was inclined to intramolecular and intermolecular agglomeration, causing inferior conduction and detachment of electrode (Figure 6.4a). Therefore, the $LiFePO_4$ electrode using GP binder showed a larger capacity of 140.3 mAh g^{-1} with high retention of 90.1% over 300 cycles at a rate of 0.5 C than that with PVDF binder (114.4 mAh g^{-1} and 74.8%) (Figure 6.4c,d).

6.2 As Modifying Agents for Battery Electrodes

Apart from being a binder for electrodes, gelatin is also used for the modification of electrodes via adding it into the electrode or coating onto the electrode surface to enhance the properties of LIB or other electrochemical energy-storage devices.

6.2.1 As Modifying Agents of Electrodes for Lithium–Sulfur Batteries

In Li–S batteries, SEI might be generated at the interface of Li metal anode and electrolyte due to spontaneous chemical reactions and electrochemical reductions of the electrolyte. SEI is beneficial to inhibit the consecutive parasitic reactions between the Li metal anode and electrolyte and effect the stripping and deposition of Li metal [11, 12]. A desired SEI should have remarkable Li^+ conductivity, great elastic strength, large mechanical modulus, homogenous component, and suitable thickness to restrict the initiation and growth of Li dendrites and keep the significant interface fluctuation in the charge–discharge process [13] (Figure 6.5).

Gelatin-based fibers (GFs) were used as an artificial SEI coated on the surface of Li anode to inhibit uncontrollable growth of Li dendrites and the shuttling of polysulfides in Li–S batteries (Figure 6.6) [14]. The prepared GFs as a bifunctional interlayer exhibited interconnected 3D network structure, great adhesion, and polarity, leading to the tight contact with Li anode. The many homogenously distributed polar moieties (N, O) of GFs layer enabled uniform Li deposition via adjusting the Li-ion flux, inhibiting the growth of Li dendrites. Meanwhile, the GFs layer restrained the undissolved Li_2S_2/Li_2S from depositing the surface of Li anode, enabling the shuttle effect to be inhibited and the interface impedance to be decreased in the charge–discharge process. Moreover, the polysulfide was capable of returning to the cathode and keep on participating in following reactions during cycling, resulting in the decreased superfluous loss of Li and S.

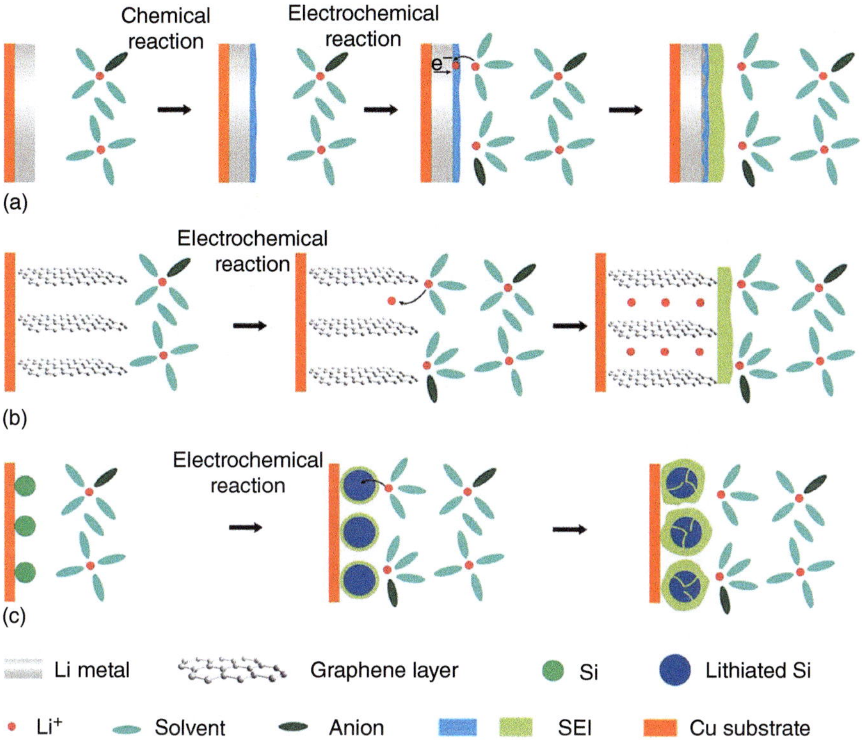

Figure 6.5 Graphical illustration of SEI formation processes on (a) Li, (b) Gr, and (c) Si negative electrodes. Source: Reproduced with permission from Wu et al. [12]; © 2020 Wiley-VCH GmbH.

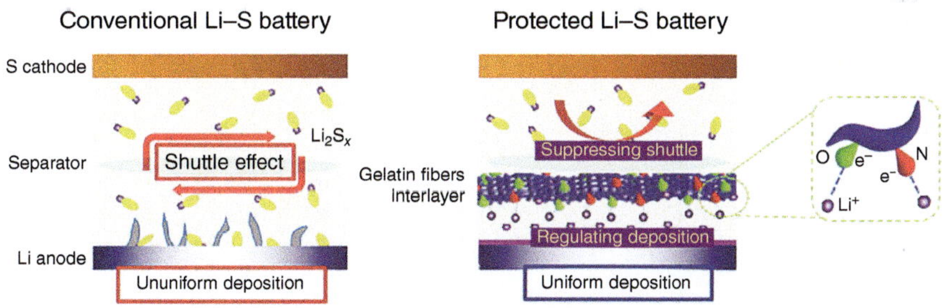

Figure 6.6 Graphical illustration of gelatin-based artificial SEI's effects: simultaneous regulation of lithium deposition and suppression of shuttle effect for Li–S batteries. Source: Reproduced with permission from Akhtar et al. [14]; © 2020 Science Press and Dalian Institute of Chemical Physics, Chinese Academy of Sciences.

Thus, the Li–Cu half cells exhibited average coulombic efficiency of up to 96% over 160 cycles. The Li–Li symmetric cells showed long cycle life of 800 hours with low-voltage hysteresis below 10 mV. Consequently, the assembled Li–S full cells exhibited higher capacity retention of about 14% than the cell without GFs layer over 100 cycles at the rate of 0.5 C (Figure 6.7). This research demonstrated that

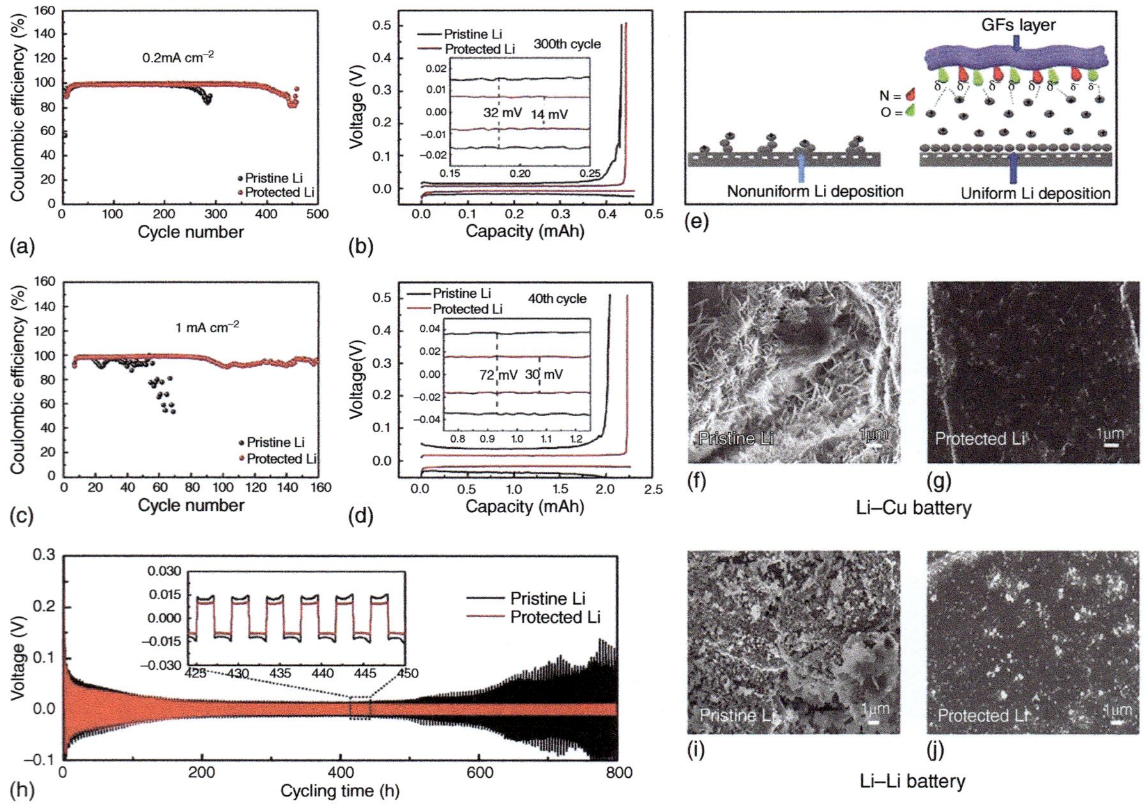

Figure 6.7 Li–Cu cells: (a, c) Coulombic efficiency at various current densities (0.2 and 1 mA cm^2), (b, d) voltage–capacity profiles, (e) schemes of Li deposition, and (f, g) surface. Morphology of cycled pristine and protected Li anodes after removing fibers layer. Symmetric Li–Li batteries: (h) The galvanostatic cycling diagram of the pristine and protected batteries and (i, j) surface morphology of cycled pristine and protected Li anodes after removing fibers layer. Source: Akhtar et al. [14]. © 2020 Science Press and Dalian Institute of Chemical Physics, Chinese Academy of Sciences.

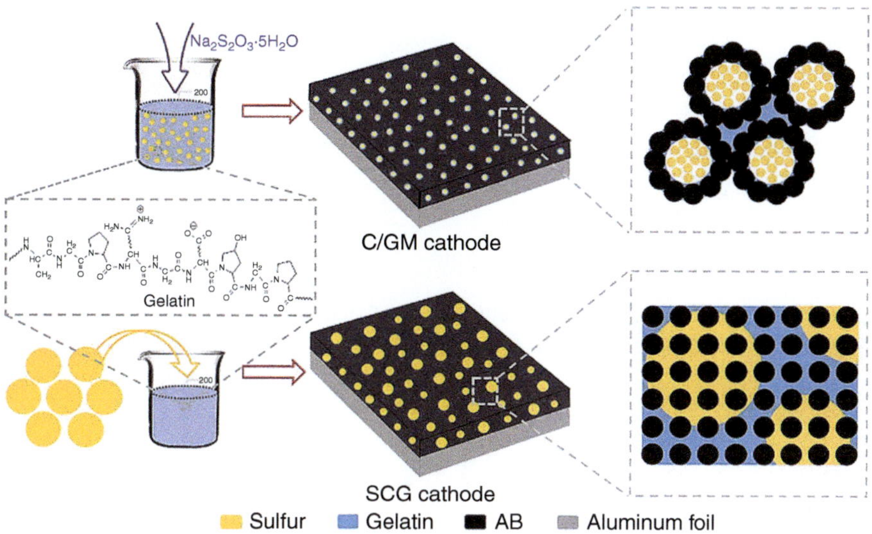

Figure 6.8 Graphical illustration of the experiment process. Source: Reproduced with permission from Sun et al. [15]; © 2019 The Electrochemical Society.

GFs were promising modified agents for electrode to improve the workability and cycle life of Li–S batteries.

Additionally, gelatin can also be used to encapsulate the sulfur to a limited region to suppress the polysulfide diffusion in Li–S batteries. Sun et al. reported a kind of carbon/gelatin microcapsule (GM) reactor to enhance electrochemical performance of Li–S batteries (Figure 6.8) [15]. Gelatin could be used as not only a binder with high adhesion and electrochemical stability for the cathode materials, but also a significant dispersion agent to offer abundant nucleation sites for the sulfur growth and dispersion of the submicron-sulfur granules, and the carbon/gelatin microcapsule kept the shape and size of submicron-sulfur granules and restrained the shuttle effect of polysulfide via preventing the electrochemical reaction on the cathode during cycling.

Beneficial from the carbon/gelatin microcapsule reactor, the Li–S cells achieved large capacities and prolonged lifespan. The original specific capacity exhibited was as high as 1259.0 $mAh\,g^{-1}$, and it was still able to maintain 961.0 $mAh\,g^{-1}$ over 100 cycles (Figure 6.9). Moreover, the specific capacity at 2 C is approximately five times as much as that of cells with SCG (mixtures of sulfur, carbon, and gelatin) cathode (Figure 6.10).

6.2.2 As Modifying Agents of Electrodes for Lithium Batteries

The irreversible capacity for most Li batteries is usually related to irreversible structure changes or undesired reactions between electrolyte and electrode. The undesired reactions are able to be effectively restrained via forming a protective SEI on the surface of the reactive electrode. It was reported that gelatin dissolved in water had a positive effect on the decrease of the irreversible capacity for graphite anode [16].

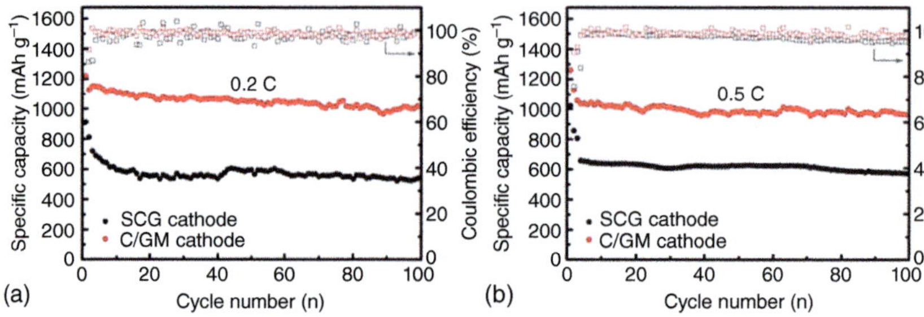

Figure 6.9 Cycling stabilities of two cells at (a) 0.2 C and (b) 0.5 C. Source: Reproduced with permission from Sun et al. [15]; © 2019 The Electrochemical Society.

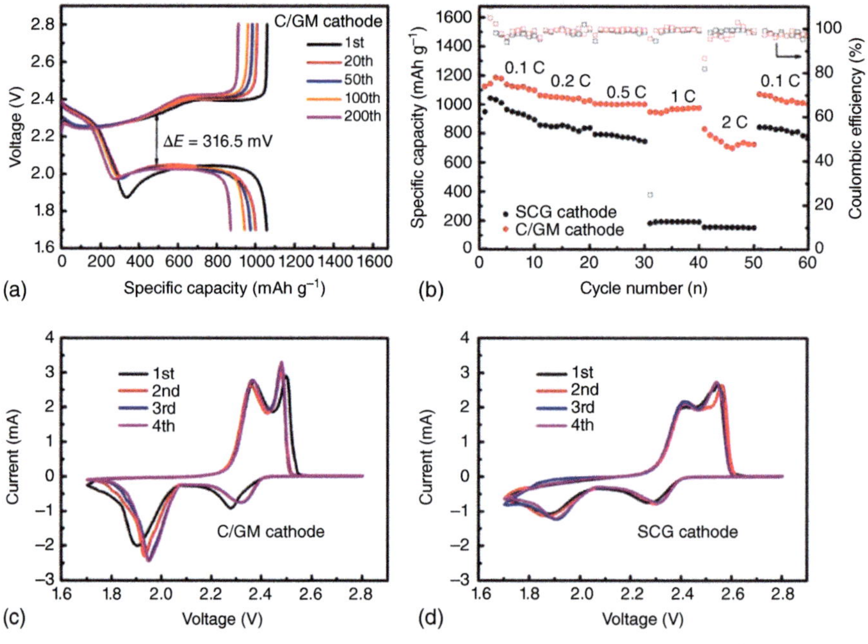

Figure 6.10 (a) Voltage–capacity profiles of cell with C/GM cathode at 0.5 C. (b) Cycling stabilities of two cells at various rates. Cyclic voltammetry (CV) profiles of cells with (c) C/GM cathode and (d) SCG cathode. Source: Reproduced with permission from Sun et al. [15]; © 2019 The Electrochemical Society.

Benefiting from ionizable groups, e.g. NH_2 and COOH, gelatin can react with a few metals, such as Pt, Pd, Au, and Al [17], leading to unique physicochemical characteristics of electrodes in Li-ion batteries.

Montoro prepared a pyrite composite electrode via adding gelatin into natural pyrite ($FeS_{1.9}$, n-type semiconductor) under nonaqueous treatments [18]. The gelatin coated with the pyrite grains was beneficial to prevent the products formed during the Li stripping/deposition process from being degraded by the electrolyte, leading to the increase of the specific capacities of pyrite composite electrodes under room temperature. Compared with the pyrite electrodes without adding gelatin, pyrite composite electrodes showed higher specific capacities and superior capacity retention. They showed a reversible exchange of 1.2 e/FeS_2, causing the higher specific capacity of up to 279 mAh g^{-1} over 15 cycles at 0.4 mA cm^{-2} than that of reported $Li_{2-x}FeS_2$ ($0 < x < 0.8$) electrode.

6.3 As Modifying Agents for Battery Separators

The separator in batteries is an important component that promotes ions migration but blocks electrons diffusion between cathodes and anodes [19]. Furthermore, the separator avoids the direct contact of cathodes and anodes to guarantee the safety of batteries. If separators are unable to avoid the contact between two electrodes, the internal short circuit happens. A few factors result in the separators failure. For example, separators may rupture due to extrusion and puncture under electrical and mechanical misuses conditions. Currently, polyolefin microporous membranes, e.g. polypropylene (PP) and polyethylene (PE), are the most commonly applied separators. Unfortunately, the poor physical interaction between the fibers results in nonwoven membranes with inferior mechanical strength, and hence dendrites can easily penetrate these membranes, bringing about internal short circuits. Thus, it is paramount to explore unique separator materials to satisfy the demands for high safety and superior electrochemical performance of batteries. The separators with multifunctional characteristics, e.g. dendrite suppression, dendrite detection, flammability, and thermal shutdown, are fabricated to intelligently control the safety of batteries via preventing the thermal runaway phenomenon and other hazards (Figure 6.11).

6.3.1 As Modifying Agents of Separators for Lithium–Sulfur Batteries

Plenty of methods to modify separators have been investigated and reported for Li–S batteries. Using a separator modified by conducting carbon interlayer is considered as a useful strategy to enhance the cycle stability of Li–S batteries [20]. Generally, the separator modified by conducting carbon interlayer not only can suppress the

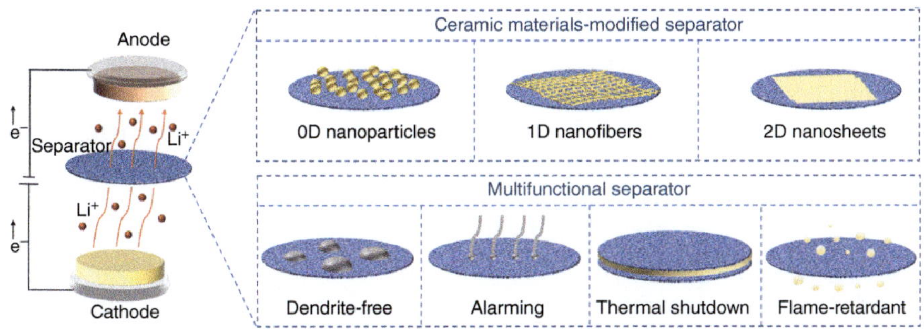

Figure 6.11 Graphic illustration of separator construction for LIBs with high safety. Source: Reproduced with permission from Liu et al. [19]; © 2020 Zhengzhou University.

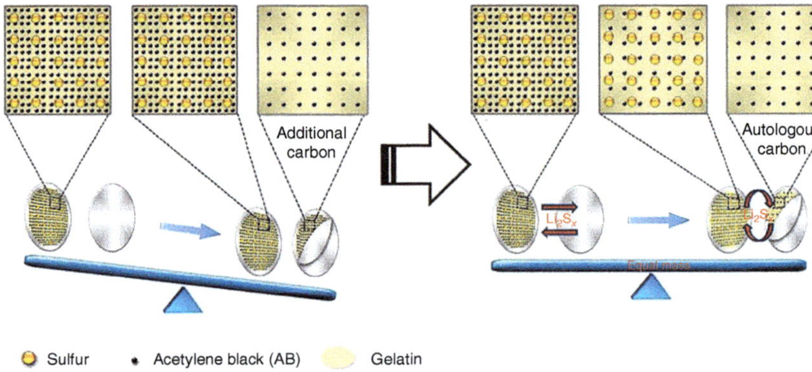

Figure 6.12 Graphic illustration of thin carbon layer-modified separator obtained from its cathode (IM). Source: Reproduced with permission from Liu et al. [20]; © 2016 American Chemical Society.

dissolution and shuttling of polysulfides in the charge–discharge process, but also act as absorption reagent for confining the electrolyte, including the dissolved polysulfide in the cathode space to reutilize.

A novel modified separator, one side of which was coated with conducting carbon and gelatin, was designed and prepared to anchor polysulfides with effect as well as to improve reversible capacity and cycle stability of Li–S batteries (Figure 6.12) [20]. For this modified separator, the gelatin including $-NH_3$ and $-COOH$ could chemically bond with polysulfides, further enhancing the ability of the carbon layer to trap the dissolved polysulfides. Meanwhile, the gelatin could offer strong adhesion to bind the separator and conducting carbon. The gelatin with water solubility was also able to increase the surface hydrophilic property of the modified separator, which was useful for the electrolyte permeation. Moreover, this modified separator was completely isolated from the cathode, avoiding the extra weight involved.

Thus, the modified separator combined the physical barrier of conductive carbon and chemical interaction of gelatin blocked the diffusion and shuttling of the dissolved polysulfides, and hence reduced the loss of sulfur material during cycling. Furthermore, the modified separator was beneficial to increase the energy density without any additional material. Benefiting from the modified separator, the Li–S batteries achieved a higher specific capacity of 660 mAh g^{-1} over 100 cycles than the common Li–S batteries (460 mAh g^{-1}) (Figure 6.13).

Chen et al. developed a protein-based Janus nanofabric as separator, which was prepared by incorporating a gelatin-coated conducting nanofabric (G@CNF) and a gelatin nanofabric (G-nanofabric) into a heterostructure (Figure 6.14) [21]. The G@CNF as a layer of blocking polysulfide considerably increased the ability of trapping polysulfide benefiting from powerful interactions between polysulfide and gelatin. The G-nanofabric as an ion-regulating layer possessed remarkable

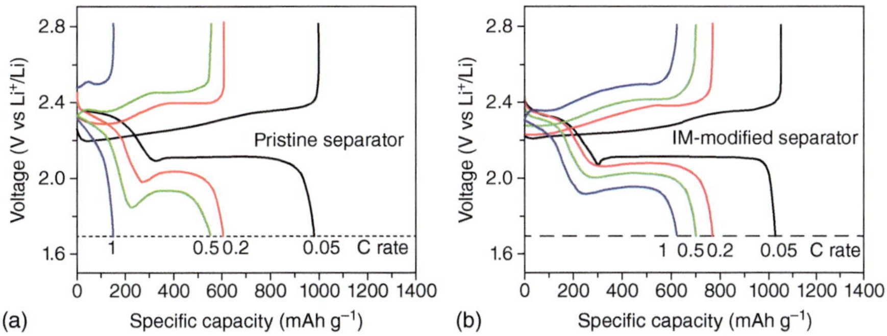

Figure 6.13 Voltage curves of the Li–S cells with (a) pristine separator and (b) IM-modified separator at various discharge/charge rates. Source: Reproduced with permission from Liu et al. [20]; © 2016 American Chemical Society.

wettability, great ionic conductivity (4.9 mS cm^{-1}), and a large lithium-ion transference number (0.73), which were beneficial to homogenize the Li deposition, and hence inhibited the lithium dendrites growth (Figures 6.15 and 6.16). Due to the bifunctionality of the Janus nanofabric, the Li–S cell showed large capacity, superior rate capability, and long cycle lifespan. The first specific capacity of 890 mAh g^{-1} had been achieved with a fading rate of 0.117% after 300 cycles at the current density of 0.5 A g^{-1}.

6.3.2 As Modifying Agents of Separators for Other Batteries

Research showed that collagen hydrolysate (CH) uniformly coated on absorbed glass mat (CH@AGM) between the AGM separator and the anode could significantly stabilize metal anodes and restrain dendrites growth in Li and Zn metal batteries with a large areal-capacity cathode [22]. CH with a surface electronegativity in either organic or aqueous electrolytes endowed the surface of CH@AGM with conductivity and enabled a deionization shock in randomly charged porous media under the case of overlimiting current, which facilitated homogenous deposition of Li^+ and Zn^{2+}. The combination of CH with Li^+ or Zn^{2+} further enabled cations to distribute on the entire surface of the Li or Zn metal anode, which guided metal deposition along the surface or horizontal direction of the anode and hence considerably stabilized Li and Zn formations. This kind of superposed moderating effects in metal electrodeposition could realize metal deposition without dendrite at large current densities and construct Li and Zn metal batteries with excellent cycle stability, which employed a cathode with high loading ($LiMn_2O_4$ [LMO]).

5-Ah LMO|AGM + CH@AGM|Zn full cell showed remarkable stability after 600 cycles under fully deep discharge. The 65-Ah and 200-Ah Zn metal batteries with high capacity were also fabricated and exhibited over 400 stable cycles with <10% decay rate.

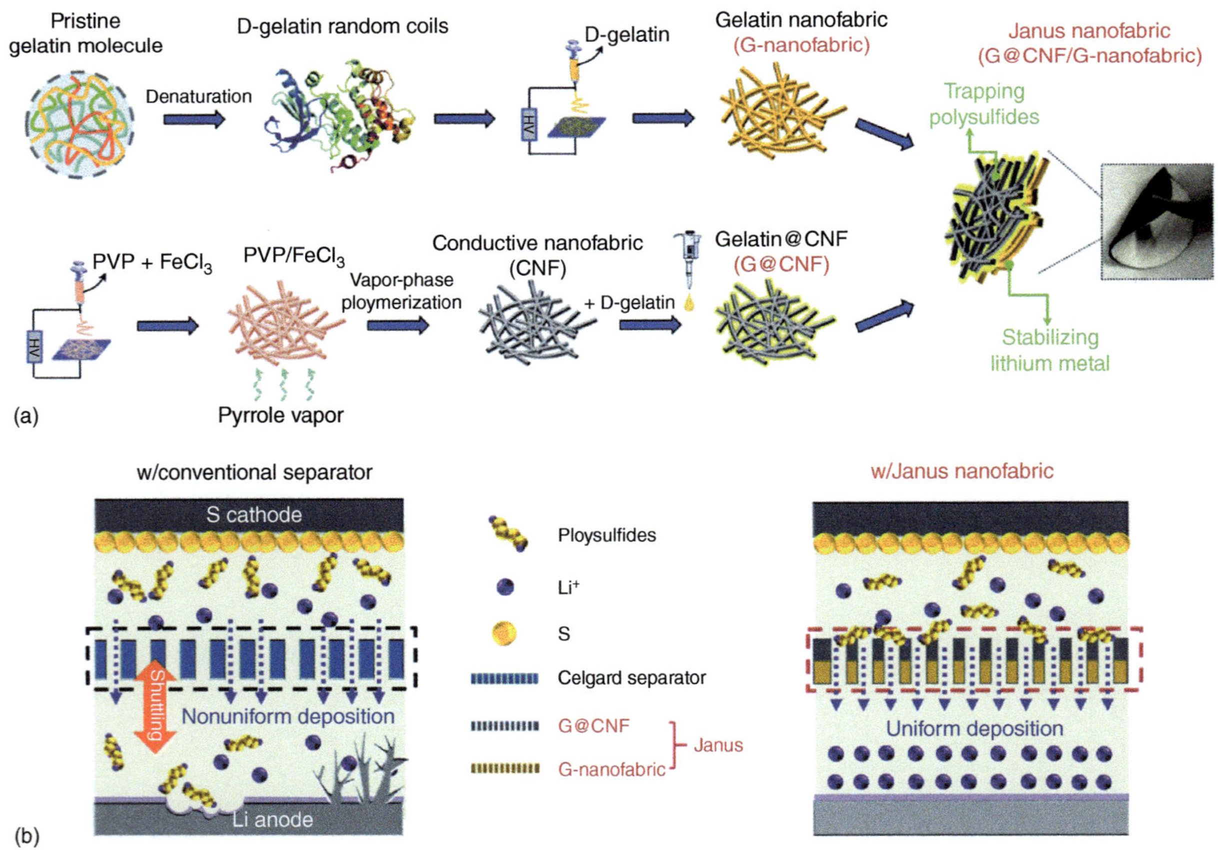

Figure 6.14 Fabrication of the protein-based Janus nanofabric and its effect on trapping polysulfides and stabilizing ion deposition. (a) Schematic illustration of the fabrication process of the Janus nanofabric. The Janus nanofabric consists of an upper layer of gelatin-functionalized conductive nanofabric and a lower layer of gelatin nanofabric. (b) Schematic illustration of the contribution from a conventional separator and Janus nanofabric to polysulfide diffusion and Li-ion deposition in Li–S batteries. Source: Reproduced with permission from Chen et al. [21]; © The Royal Society of Chemistry 2020.

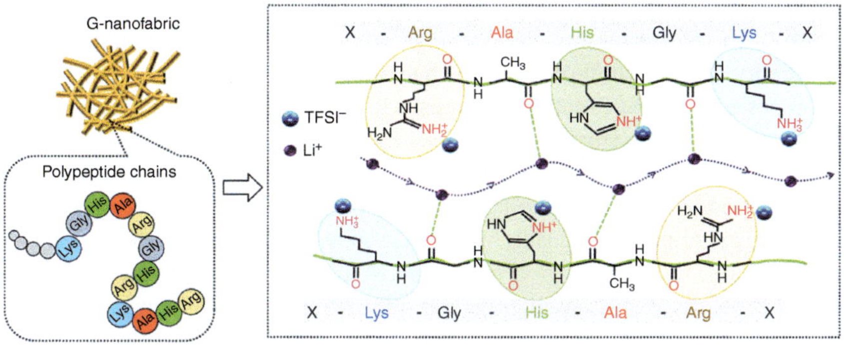

Figure 6.15 Schematic illustration of the possible ion-transport situation for the G-nanofabric. Source: Reproduced with permission from Chen et al. [21]; © The Royal Society of Chemistry 2020.

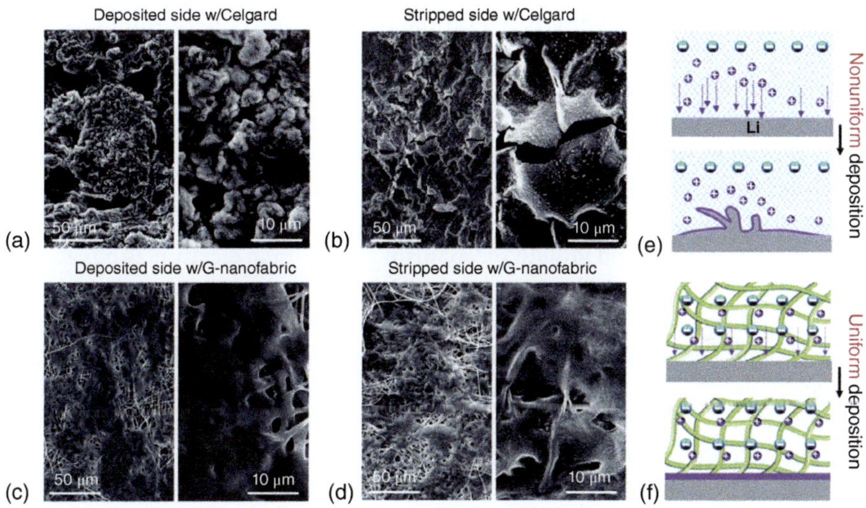

Figure 6.16 Morphological observation of postmortem Li metal. (a–d) Morphology of Li metal over 26 cycles of plating or stripping of Li in Li/Li symmetric cells with (a, b) Celgard® and (c, d) the G-nanofabric. Schematic illustration of the Li deposition processes with (e) Celgard or (f) gelatin separators. Source: Chen et al. [21]/with permission of Royal Society of Chemistry.

6.4 As Electrodeposition Additives for Electrode Preparation

Electroplated metal films are commonly used as electrodes and applied in several energy-storage and conversion devices, such as Li-ion batteries. Adding various additives into electroplating media is an effective method to improve the electrodeposition efficiency and form the metal films with required quality [23]. The additives in the electroplating solution have an extreme impact on the nucleation and crystal growth as well as the mechanical and physicochemical characteristics of the electrodeposit. It is reported that the addition of an organic additive with low concentration in the electroplating electrolyte greatly influences texture, morphology, surface quality, and electrodeposition mechanism of electrodeposit.

Gelatin possesses polar functional groups, e.g. $-NH_2$ and $-COOH$, which enable it to be adsorbed on the surface protrusions in priority and navigate the uniform deposition of metal ions in the following electrodeposition process. Thus, gelatin is capable of acting as an additive to modify the electrochemical reactions and crystal growth of metal electrodeposits. When gelatin as additive is applied in copper electrodeposition, the cathode overvoltage becomes high in a wide concentration range, which results in decreased grain size and increased nucleation rate. During the deposition process of Zn–Ni coating, gelatin additive is capable of modifying the deposit properties as well as enhancing the efficiency of electrodeposition and corrosion resistance. The gelatin added to the metal-coating matrix causes the compositional and morphology variation of the electrodeposited alloy.

Panat's group employed hydroquinone (HQ) and gelatin as organic additives to control the film microstructure based on the bath chemistry [24]. Adding gelatin individually not only inhibited the hydrogen gas evolution and influenced the film nucleation process that controls its grain morphology, but also generated pinholes in the film and sludge in the bath. However, incorporation of HQ and gelatin created a quite homogenous grain architecture with little or no dendrite formation and delayed the evolution of hydrogen gas, which was considered necessary for the electroplating process. The results demonstrated that the addition of two additives changed the nucleation and growth mechanism of tin film in electroplating process. Gelatin inhibited the tin nucleation rate, which controlled the grain and prevented the large dendrites' formation, while HQ acted as an antioxidant to restrain the formation of pinholes and sludge as well as generate a homogenous film (Figure 6.17).

Tabuchi's group synthesized successfully ternary Sn–Sb–Co alloy film via the co-electroplating way utilizing aqueous solution bath including gelatin, which acted as a viscosity control agent to generate films with uniform surfaces [25]. Compared with binary Sn–Co and Sn–Sb film electrodes, the Sn–Sb–Co film electrode possessed higher capacities and more superior cycle stability.

The copper substrates with porous structures for Li-battery electrodes were prepared via the electrodeposition of copper particles using gelatin as additive [26]. Gelatin or hydrazine monohydrate was found to be necessary for the formation of porous structures. The larger gelatin concentration was beneficial to form the copper substrates with smaller pore diameters.

Wang's group reported a direct current electroplating method using composite gelatin and poly-2-sulfur-2-propane sodium sulfonate as additive to prepare Cu with microstructure of fine equiaxed-grained nanotwin for application of ultrathin Cu foil used in LIBs [27]. It was supposed that the dynamic competitive adsorption of gelatin and poly-2-sulfur-2-propane sodium sulfonate avoided the grain growth along with preferential orientation, which made columnar-grained microstructure transform successfully into equiaxed-grained microstructure. As a result, the prepared Cu with equiaxed-grained microstructure showed a superhigh tensile strength of 764 MPa, approximately 40% higher than that of columnar-grained

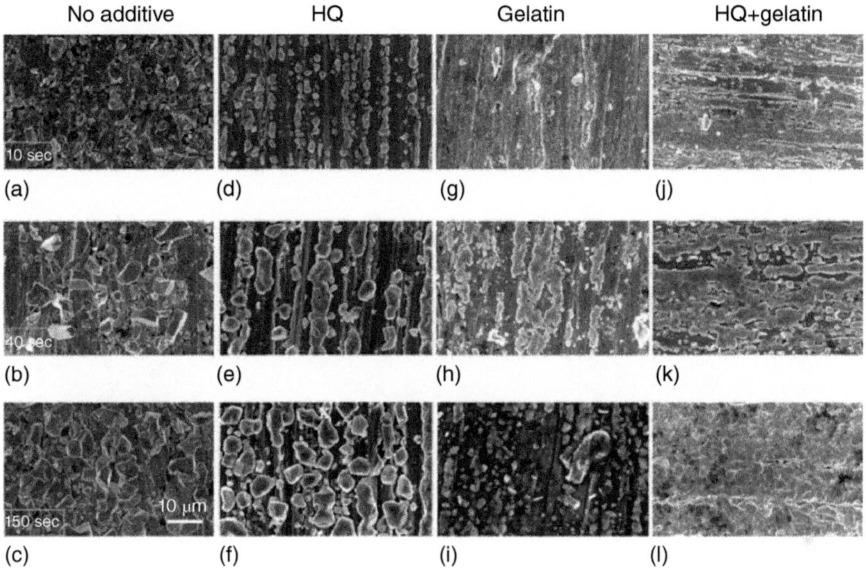

Figure 6.17 Electrodeposition of Sn from different solutions with varying time durations – 10 seconds (first row), 40 seconds (second row), and 150 seconds (third row) with solutions with different chemical compositions. (a–c) No additives present, (d–f) only HQ present as additive, (g–i) only gelatin as additive, and (j–l) both HQ and gelatin present from solutions Ia, IIa, IIIa, and VI′a, respectively. A common scale bar for all the images is shown in (c). Source: Collazo et al. [24]/with permission of © The Electrochemical Society.

nanotwinned Cu. It indicated that a microstructure of fine equiaxed-grained nanotwin enabled Cu to achieve the superhigh tensile strength rather than large columnar-grained nanotwinned structure.

Additionally, electrodeposition technique is also applied to prepare electrocatalytic electrode materials with the assistance of gelatin additive. A novel Sn–Ni alloy coating was prepared on the Cu surface via an efficient and facile electrodeposition technique [28]. Gelatin as additive was added into the bath to modify the morphology, composition, and phase structure of Sn–Ni alloy coatings via changing the electro-crystallization mechanism. The results indicated the electrocatalytic activity of coatings for oxygen evolution reaction (OER) and hydrogen evolution reaction (HER) was dependent on the morphology, composition, and phase structure of coatings deposited on different current densities. The coatings deposited at $4.0\,A\,dm^{-2}$ (with 37.6 wt pct Ni) and $1.0\,A\,dm^{-2}$ (with 19.6 wt pct Ni), respectively, showed the best electrocatalytic activity for OER and HER in 1.0 M KOH solution. The electrochemical corrosion measurements demonstrated that the coatings were stable at working conditions of electrolysis. It is found that the coating with high wt pct Ni favored cathodic reaction (HER) but had a negative impact on anodic reaction (OER).

Apart from being used as an additive to synthesize electrodes via electrodeposition technique, gelatin can also be added into electrolyte to modify the electrodeposition behavior of electrodes during electrochemical reactions. The Zn corrosion and

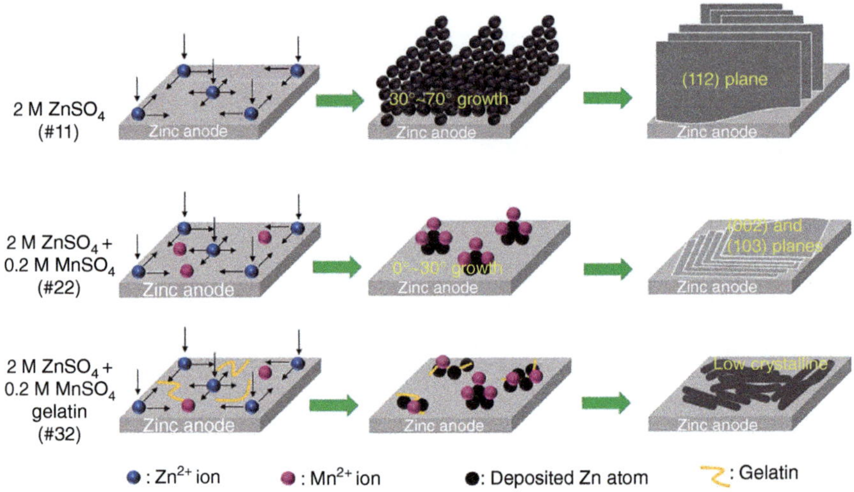

Figure 6.18 The mechanisms of the Zn deposits growth on the Zn surface in the different electrolytes. Source: Reproduced with permission from Liu et al. [29]; © 2020 Elsevier B.V.

dendrite growth are the critical factors of limiting the electrochemical performance of Zn-ion batteries. It is a promising approach of introducing additives into electrodes or electrolytes to address these issues.

As a widely used grain refining and leveling reagent in the electroplating/electrodeposition field, gelatin can be applied to modify the morphology of electrodeposits [29]. Research reported that the preferential adsorption of gelatin on active growth sites of Zn surface caused much more even current distribution around the tip, which increased the polarization of electrode surface and hence restrained significantly the Zn dendrites growth. Namely, the larger electrode polarization resulted in lower current density, and hence reduced the Zn deposition rate and alleviated the concentration polarization. Gelatin could form a prohibitive layer on the Zn surface to inhibit the rapid growth of Zn dendrites, and hence slowed down or prevented the consecutive formation and growth of dendrites (Figure 6.18).

The assembled Zn//Zn symmetric cells suggested that the Zn dendrite formation time (>500 h) in sand was considerably increased due to the coexistence of Mn^{2+} ions and gelatin. The Zn//polyaniline/MnO_2 cell with electrolyte using Mn^{2+}/gelatin additives showed 80% capacity retention over 1000 cycles under the current density of 500 mA g^{-1}. Moreover, the float charge current significantly reduced from about 11.8 to 4.3 ~ 7.6 μA at a constant voltage of 1.9 V.

6.5 As Hydrogel Electrolytes for Batteries

With the rapid development of wearable electronics and medical implants, flexible electrochemical energy-storage devices have attracted wide attention. Compared with the conventional devices, flexible electrochemical energy-storage devices encounter even more challenges because electrochemical performances are not the only concern anymore. A desirable flexible energy-storage device should not only be capable of withstanding different degrees of deformation, such as bending, stretching, or even compressing in practical applications, but also have electrochemical stability and durability. Thus, device components (current collectors, electrodes, separator, and electrolyte) are supposed to be soft and compatible with each other in both mechanical and electrochemical characteristics to avoid the electrochemical efficiency from degradation caused by delamination of components during application.

As one of the most significant components, electrolytes are used to transport ions in the energy-storage device and hence are the determining factor for the device performance, such as ionic conductivity, cycling lifespan, and electrochemical stable potential window (ESPW). Unfortunately, traditional liquid electrolytes are unsuitable for applications of flexible wearables due to their potential hazards, for example, leakage problems and rapid evaporation of the electrolytes. Thus, gel polymer electrolytes (GPEs) have been widely developed. GPEs include immobilized liquid electrolytes in the host made up of one or more polymer matrices. They normally have more superior mechanical integrity compared to liquid electrolytes and higher ionic conductivity than solid polymer electrolytes (SPEs). Moreover, they could also function as separator in addition to acting as electrolyte. Besides, the introduction of polymers is able to widen the voltage window of OER or HER via reducing plenty of free water, which is beneficial to develop batteries with high energy density.

Gelatin can be heated and dissolved in water, and the viscosity enhances gradually in the cooling process. The gelatin solution will convert into a gel when the concentration of the gelatin solution reaches high enough and the temperature becomes low enough. Thus, the GPEs with a certain conductivity and mechanical strength can be attained via adding electrolytes into gelatin solution during the dissolution process.

An integrated cathode/GPE for solid-state Li batteries was synthesized using a cheap and natural gelatin via facile ultraviolet curing [30]. The integrated cathode/GPE possesses improved interfacial contact between cathode and GPE, which is beneficial to promote the rapid ions diffusion at the interface (Figure 6.19). The batteries with integrated cathode/GPE exhibited lower potential polarization, smaller interfacial impedance, and better cycling stability with the 91.4% capacity retention over 200 cycles than the nonintegrated batteries.

Utilizing the feature that can be functionalized of hydrogels based on gelatin, a self-powered strain sensor is designed and prepared for portable and wearable electronic devices based on a galvanic cell and a gelatin-based hydrogel [31]. In the hydrogel matrix, hydrogen bonds and hydrophobic interactions between gelatin and tannic acid provided the hydrogel with tremendous elongation potential of up to 1600%. Moreover, the hydrogel was able to self-heal within 0.65 seconds with high efficiency of 95%. The hydrogel could act as a high-efficiency electrolyte and constructed a hydrogel battery coupled to an air electrode with a thin Zn layer. This self-powered strain sensor exhibited superior resistance to large compressional strain under the circumstance of without sacrificing open-circuit voltage. Beneficial

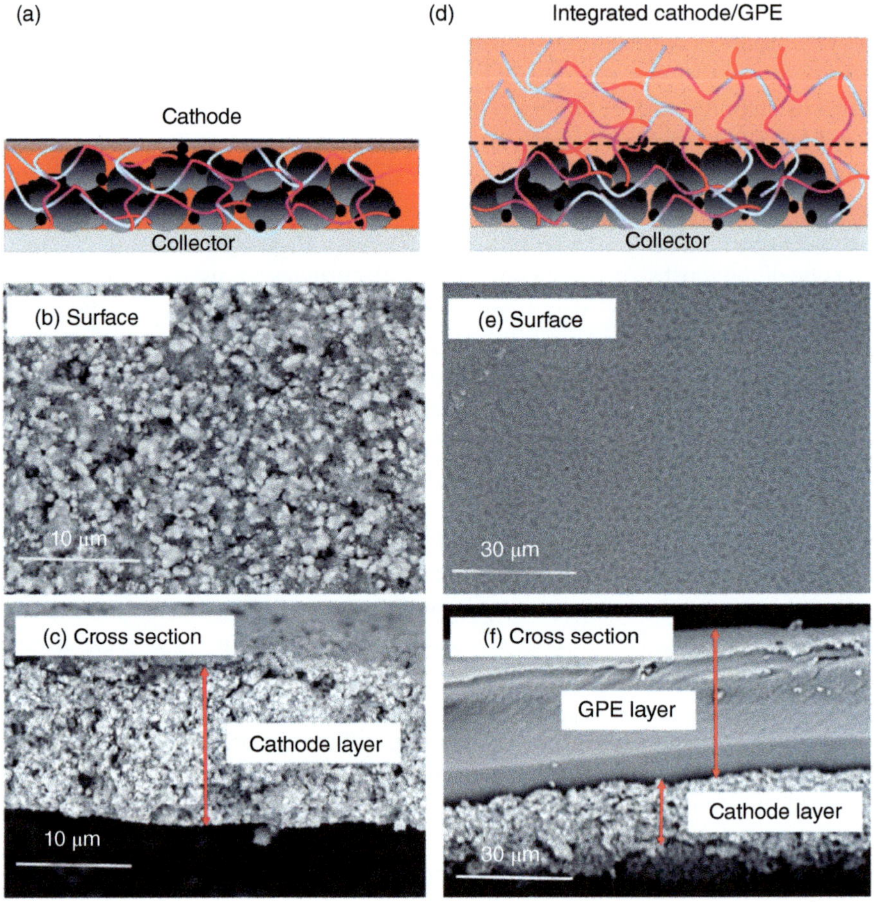

Figure 6.19 The surface and cross section morphology of cathode layer and the integrated cathode-GPE layer by SEM. Scheme of (a) cathode and (d) integrated cathode/GPE (b) and (c) shows the surface and cross section morphology of cathode. (e) and (f) show the surface and cross section of cathode/GPE. Source: Yu et al. [30]/Royal Society of Chemistry.

from the hydrogel stretchability, the devices showed great flexibility and response properties. The ultra-long stability of the device was demonstrated by self-healing and consecutive cycling measurements.

Considering the issue of weak mechanical strength with insufficient cross-linking, Han et al. added appropriate inorganic salt ions into the hydrogel to enhance its strength. A solid-state electrolyte-based gelatin (GSE) was synthesized via a rapid cooling technique [32]. At temperatures over the gelation temperature, gelatin coils were distributed at random in an aqueous solution. After rapid cooling, the gelatin molecules spontaneously formed linkages through the interaction of hydrogen bonds, and the chains were subjected to a conformational transition from disorder to order for producing triple-helix fractions. Furthermore, the helices coiled each other and eventually converted into a GSE (Figure 6.20). The obtained GSE exhibited a high compressive strength of 100 kPa. It was found that

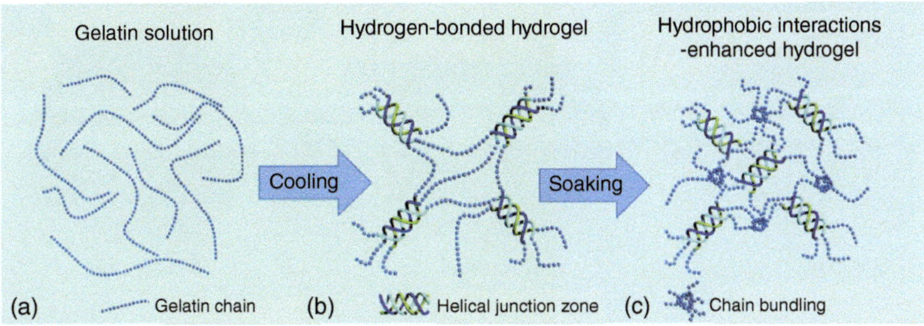

Figure 6.20 The schematic illustration of the molecular conformation of (a) the gelatin solution, (b) the gelatin solid-state electrolyte (GSE-0), and (c) the salt-reinforced GSE-x ($x = 1.0$, 1.5, 2.0, 2.5, 3.0, and 3.5) after soaking the GSE-0 sample in $ZnSO_4/MnSO_4$ solution. Source: Reproduced with permission from Han et al. [32]; © The Royal Society of Chemistry 2019.

the GSE not only possessed good thermal reversibility, but also could tolerate vast inorganic salts. Interestingly, the addition of inorganic salts enabled the GSE to dehydrate. Beneficial from the salt-reinforced effect of the GSE, the obtained GSE underwent further treatment in concentrated inorganic salt solutions, leading to the enhanced hydrophobic interactions between gelatin molecular chains as well as chain bundling in the gel network.

Compared with conventional electrolyte systems, the GSEs synthesized via a simple soaking method contained less free water. Moreover, the GSEs showed the largest fracture tensile strength of 2.78 MPa compared with the reported self-standing single-component solid-state electrolytes for Zn metal batteries. The large fracture tensile strength could successfully suppress the passivation and corrosion reaction, which contributed to the stable cycle of Zn/Zn symmetric cells at the largest current density of $5\,mA\,cm^{-2}$ so far with record large cumulative stripping/plating capacity. Moreover, the solid-state Zn/MnO_2 full cell exhibited a large specific capacity of $285\,mAh\,g^{-1}$, stable cycling performance with the fading of 10% after 500 cycles, and superior safety at extreme working conditions.

Additionally, inorganic salts as additives can also effectively enhance the performance of gelatin. For example, polymer electrolytes based on fish gelatin–added $LiClO_4$ were synthesized by the solution cast way [33]. The ionic conductivity of fish gelatin film would increase from $(9.29 \pm 0.89) \times 10^{-8}\,mS\,cm^{-1}$ to $(3.58 \pm 0.58) \times 10^{-3}\,mS\,cm^{-1}$ after adding 25 wt% $LiClO_4$ into gelatin. Linear sweep voltammetry measurement indicated that the as-prepared electrolyte with $LiClO_4$ had good electrochemical stability even at the high voltage of 1.88 V. The assembled electrical double-layer capacitor using the obtained electrolyte with $LiClO_4$ exhibited an initial specific capacitance of $28.2\,F\,g^{-1}$ and maintained $19.7\,F\,g^{-1}$ over 150 cycles.

Ionogels, a polymer electrolyte introducing ionic liquids, are considered as one of the most excellent electrolyte candidates for next-generation GPE materials due to their high safety and thermal stability, nonflammability, wider electrochemical windows, as well as other merits [34, 35]. Ionogel based on biopolymer (such as

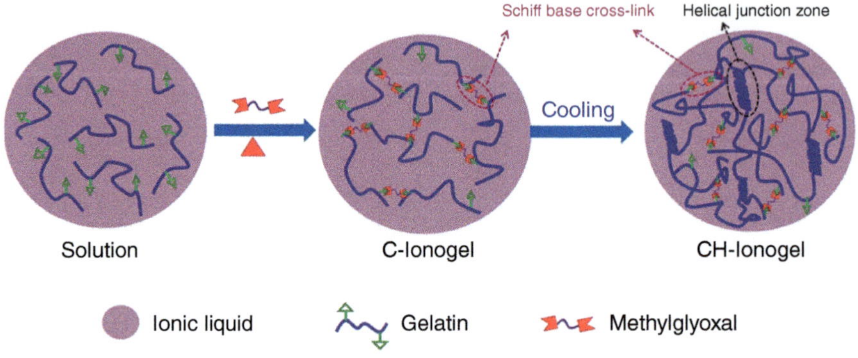

Figure 6.21 Schematic illustration of synthesizing process and structures for self-healing ionogel. Source: Reproduced with permission from Liu et al. [37]; © Springer Science+Business Media, LLC, part of Springer Nature 2019.

gelatin, cellulose, agarose, and guar gum) attracts much more attention than synthetic polymer-based ionogel due to their many advantages, for instance, abundant source, low cost, biodegradability, and environment friendliness [36].

It is noted that hydrogel electrolytes formed through chemical cross-linking have a more stable network structure due to stronger bonds. A unique gelatin-based chemical cross-linked ionogel was prepared through the Schiff-base reaction in ionic liquid of 1-ethyl-3-methylimidazolium acetate (EMIMAc) (Figure 6.21) [37]. Rheological tests indicated that the plateaus modulus in storage modulus increased with the increase of gelatin content. The as-prepared ionogels exhibited great thermal–mechanical stability, and the chemical networks were not broken even at 140 °C. Moreover, the ionogels showed high stretchability and their elongation at break was able to reach up to 368.6%. More importantly, the ionogels not only displayed superior self-healability and reprocessability, but also remarkable ionic conductivity, which was larger than $1.0\,mS\,cm^{-1}$ at room temperature. Consequently, the supercapacitors with this as-prepared ionogel electrolyte exhibited specific capacitance that varied depending on temperature and excellent flexibility (Figure 6.22).

Polyacrylamide (PAM)-based hydrogel prepared with chemical cross-linking is proved to be a promising electrolyte [38]. In the PAM hydrogel matrix, a highly porous framework was formed due to the cross-linking points located among all the chain segments. Additionally, among regions without cross-linking of chain segments, the water molecules were trapped within the PAM frame network by hydrophilic groups, such as amide groups ($-CONH_2$), with the assistance of hydrogen bonding [39]. Thus, Li et al. prepared the hierarchical polymer electrolyte (HPE) via grafting PAM onto gelatin chains filled in the network of a polyacrylonitrile (PAN) electrospun fiber membrane (Figure 6.23) [40]. The grafting of PAM onto a gelatin chain considerably improved the ionic conductivity and mechanical strength of the HPE. Meanwhile, the PAN fiber membrane could further enhance the strength of HPE and significantly decrease the possibility of a battery short circuit. As a result, the HPE film showed an ultrahigh room temperature ionic

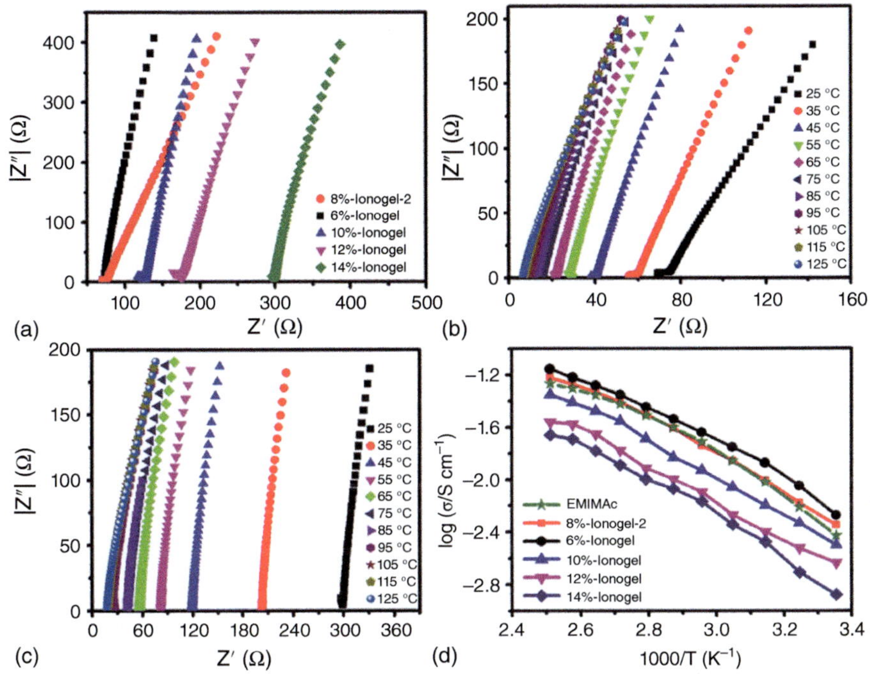

Figure 6.22 (a) The impedance plots of ionogels with different contents of gelatin at 25 °C. (b) The impedance plots of 8%-ionogel-2 at different temperatures. (c) The impedance plots of 14%-ionogel at different temperatures. (d) Temperature dependence of ionic conductivities for EMIMAc and ionogels shown in the Arrhenius convention. Source: Reproduced with permission from Liu et al. [37]; © Springer Science+Business Media, LLC, part of Springer Nature 2019.

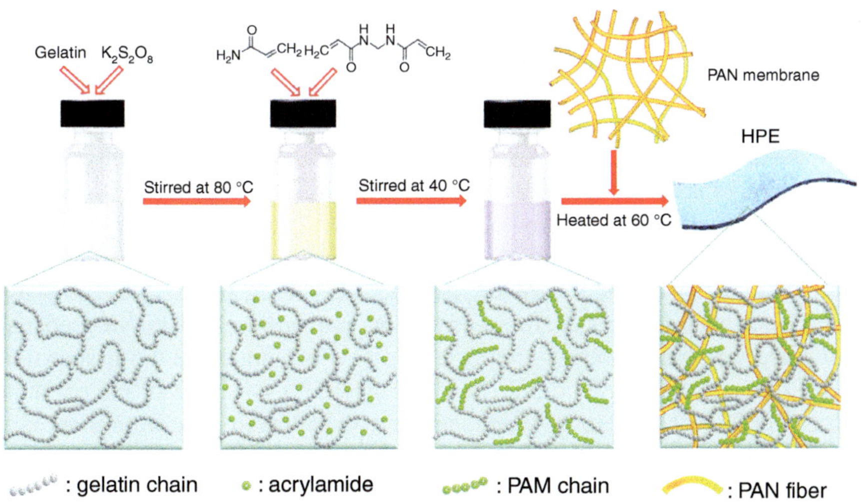

Figure 6.23 Schematic representation of the synthesis route to the HPE. The hierarchical structured HPE was synthesized by grafting PAM on gelatin chains that are filled in the network of PAN electrospun fiber membrane through a facile free radical polymerization approach. Source: Li et al. [40]/with permission of Royal Society of Chemistry.

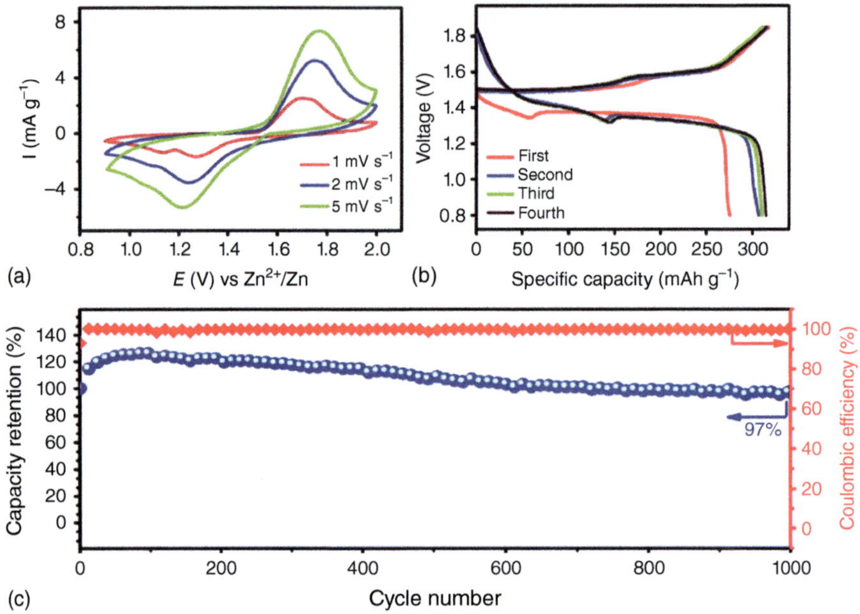

Figure 6.24 Electrochemical performance of the rechargeable solid-state Zn-ion battery. (a) CV curves of the solid-state Zn-ion battery at various scan rates. (b) Galvanostatic charge/discharge profiles at initial four cycles (61.6 mA g^{-1}). (c) Long-term cycling performance and the corresponding coulombic efficiency at 2772 mA g^{-1}. Source: Reproduced with permission from Li et al. [40]; © The Royal Society of Chemistry 2018.

conductivity of 1.76×10^{-2} S cm^{-1} and good mechanical strength as well as excellent flexible performance. The Zn-ion battery with HPE exhibited high reversible capacity of 306 mA h g^{-1} at 61.6 mA g^{-1} and superior cycling stability with capacity fading of 3% over 1000 cycles at 2772 mA g^{-1} (Figure 6.24). Moreover, the areal energy density and large areal power density of the Zn-ion battery reached up to 6.18 mWh cm^{-2} and 148.2 mW cm^{-2} at 2772 mA g^{-1}, respectively.

Sodium alginate (SA), a gel-forming biopolymer, is also regarded as a promising membrane material due to its favorable properties, such as high modulus and concentration of polar groups, as well as easy cross-linking with cations. A gelatin- and SA-based hydrogel membrane electrolyte (GAME) was successfully prepared as the Zn-ion conductor (Figure 6.25) [41]. Beneficial from hydrophilicity of gelatin as well as large dielectric strength and high modulus of SA, an interpenetrated 3D network with abundant polar groups (hydroxyl and carboxyl) of the as-prepared GAME was formed via two cross-linking reactions, which promoted the dissociation of inorganic salts and immobilized the electrolyte. Moreover, the strong covalent bonds as well as reversible ionic bond together with hydrogen bond were coexistent within the dual cross-linked network, which endowed the hydrogel with good toughness, flexibility, thermostability, and self-healing capability. Beneficial from the remarkable properties resulting from gelatin–SA cross-linked reactions, the GAME had the following features: (i) plenty of hydrophilic carboxylic and hydroxyl groups on the polymer chains promoted absorption and dissociation of $ZnSO_4$;

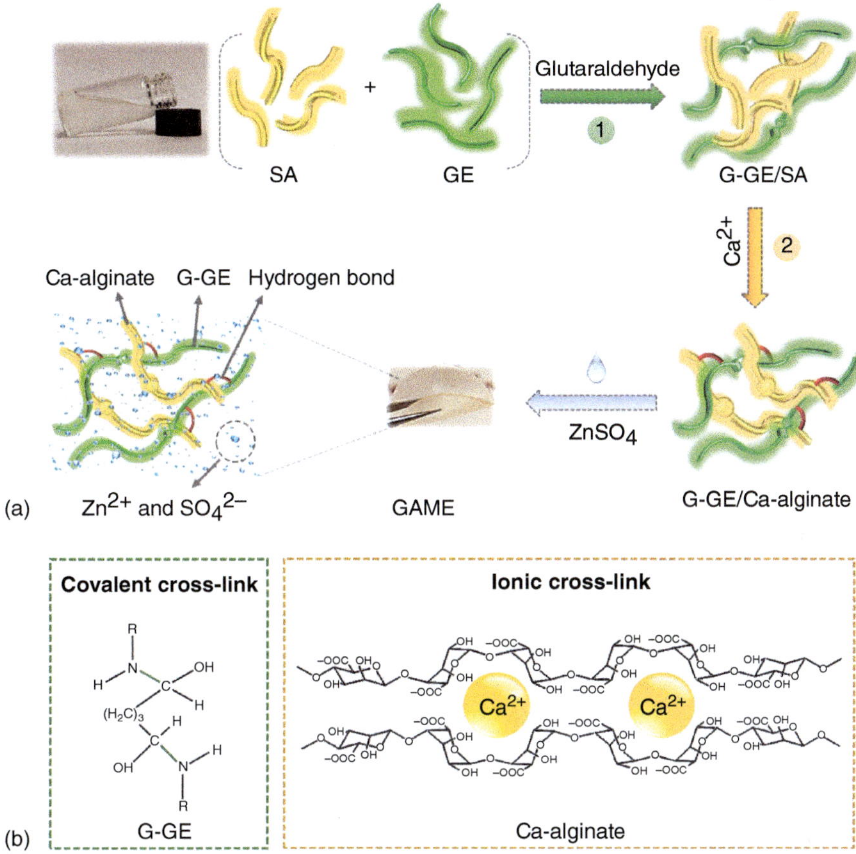

Figure 6.25 (a) Schematic representation of the overall process to make GAME. (b) Cross-linked polymers of G-GE (GE cross-link with glutaraldehyde) and Ca-alginate (SA cross-link with $CaCl_2$). Source: Lu et al. [41]/with permission of American Chemical Society.

(ii) the interpenetrated 3D network offered the GAME with large mechanical strength and enabled more liquid electrolyte to be absorbed; and (iii) the rigid covalent bonds could eliminate mechanical energy caused by deformation, while the reversible hydrogen and ionic bonds as well as chain entanglements between the two polymers permitted fast recovery after deforming. Thus, the novel GAME electrolyte was found to possess both favorable electrochemical and thermal stability as well as mechanical properties.

When absorbed in aqueous solution of $ZnSO_4$, the GAME acted as a superior host for Zn-ion conduction. The electrochemical measurements of symmetrical and full cells demonstrated that the prepared semisolid membrane electrolyte with GAME had a larger reversible capacity, more superior rate performance, and longer cyclic lifespan, indicating better conductivity and resistance to Zn dendrite attack of semisolid membrane electrolyte. The flexible measurements of cell under different bending angles, stress states, and mechanical stimuli also showed that the GAME was excellent electrolyte for Zn-ion batteries-based wearable batteries.

6.6 Other Applications in Electrochemical Energy and Conversion

Additionally, gelatin can be mixed with other chemical substances to prepare gelatin-based particles or membranes for adsorption or reaction [42]. For instance, $Ba(OH)_2$ was encapsulated into microcapsules based on gelatin via a cryogenic-cross-link technique to remove SO_4^{2-} (Figure 6.26). SO_4^{2-} was a main impurity in the manufacture of battery-grade LiOH from acid leaching of ores using H_2SO_4; hence, SO_4^{2-} concentration in aqueous solution must be reduced to around 20 ppm. In the conventional industrial process, SO_4^{2-} was removed via precipitation with Ba^{2+}, which was hard to manipulate stoichiometrically, and excessive Ba^{2+} usually caused secondary contaminations. Furthermore, extra processes were needed to remove fine particles of $BaSO_4$ precipitate, leading to the increased separation cost. However, the gelatin-based microcapsules could effectively remove SO_4^{2-} to below 20 ppm in aqueous solution, and the Ba^{2+} retention rate was about 90% in the entire manufacture process, showing its high efficiency and great mechanical strength for packing.

Gelatin with superior water solubility has numerous advantages in the formation of inorganic–organic templates for controlling the growth of inorganic nanomaterials with various unique architectures. In particular, the gelatin molecule contains amino acids with periodically repeated sequences. For example, the N–H functional groups in the glycine–proline–hydroxyproline segments are inclined to interact strongly with metal ions through multiple nitrogen coordination reactions.

Du et al. reported a simple hydrothermal approach to prepare a mesoporous bowknot-like Co_3O_4 material by utilizing gelatin as the guide agent [43]. During the synthesis process, gelatin as a guiding agent had a significant effect on the formation of mesoporous structure composed of nanoparticles for this bowknot-like Co_3O_4 material (Figure 6.27). Electrochemical measurements indicated the mesoporous bowknot-like Co_3O_4 as anode material of LIB possessed an excellent initial specific capacity of 1427.9 mAh g^{-1} and still maintained 1388.8 mAh g^{-1} after 100 cycles at 0.2 C. Moreover, the Co_3O_4 anode material exhibited a large charge capacity

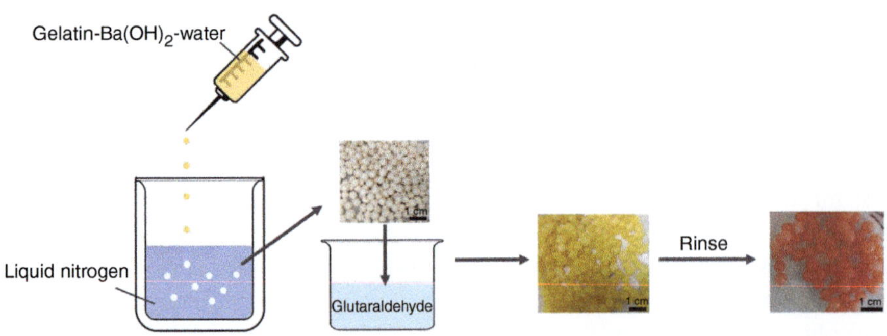

Figure 6.26 Cryogenic-cross-link process for fabricating the $Ba(OH)_2$@gelatin microcapsule. Source: Chen et al. [42]/with permission of © Elsevier Ltd.

Figure 6.27 The morphology of the precursor at different hydrothermal reaction times: (a) 5 hours, (b) 10 hours, (c) 15 hours, and (d) 21 hours. Source: Du et al. [43]/with permission of Royal Society of Chemistry.

of 751.3 mAh g^{-1} over 150 cycles at 2 C. The remarkable performances of the as-prepared Co_3O_4 anode material demonstrated that gelatin was a promising candidate to synthesize the inorganic materials with unique nanostructure.

Electrocatalysts with high activities and low cost for OER have significant practical implications for water splitting. Beneficial from the open structure, widespread source, low cost, and environmental friendliness of iron, iron oxyhydroxide (FeOOH) nanocatalysts are regarded as promising candidates for OER applications. Unfortunately, the low electrical conductivity of the FeOOH (about 10^{-2} mS cm^{-1}) causes slow mass-transfer kinetics and hence reduces the OER efficiency. To address this issue, Zhang et al. prepared high-quality Co-doped FeOOH ($Co_xFe_{1-x}OOH$, $x = 0.23, 0.54, 0.77$) nanomaterials using gelatin as soft template [44]. The obtained $Co_{0.54}Fe_{0.46}OOH$ nanomaterial showed ultrahigh OER activities, fast kinetics, and strong durability. The superior performances of $Co_{0.54}Fe_{0.46}OOH$ nanomaterial could contribute to its branch structure on the surface to provide plenty of active edge sites, which increased charge/mass transport ability, structural stability, and made it easy to release oxygen gas bubbles. Additionally, Co-doping in FeOOH nanostructures provided an ideal four-electron pathway for reversible oxygen evolution and reduction, which was of critical importance for metal–air batteries, fuel cells, and other clean energy devices. Apart from the $Co_xFe_{1-x}OOH$, other electrocatalysts, for example, CoO [45], NiO [46], and Co-doped TiO_2 [47], could also be fabricated via the gelatin-assisted wet chemistry method.

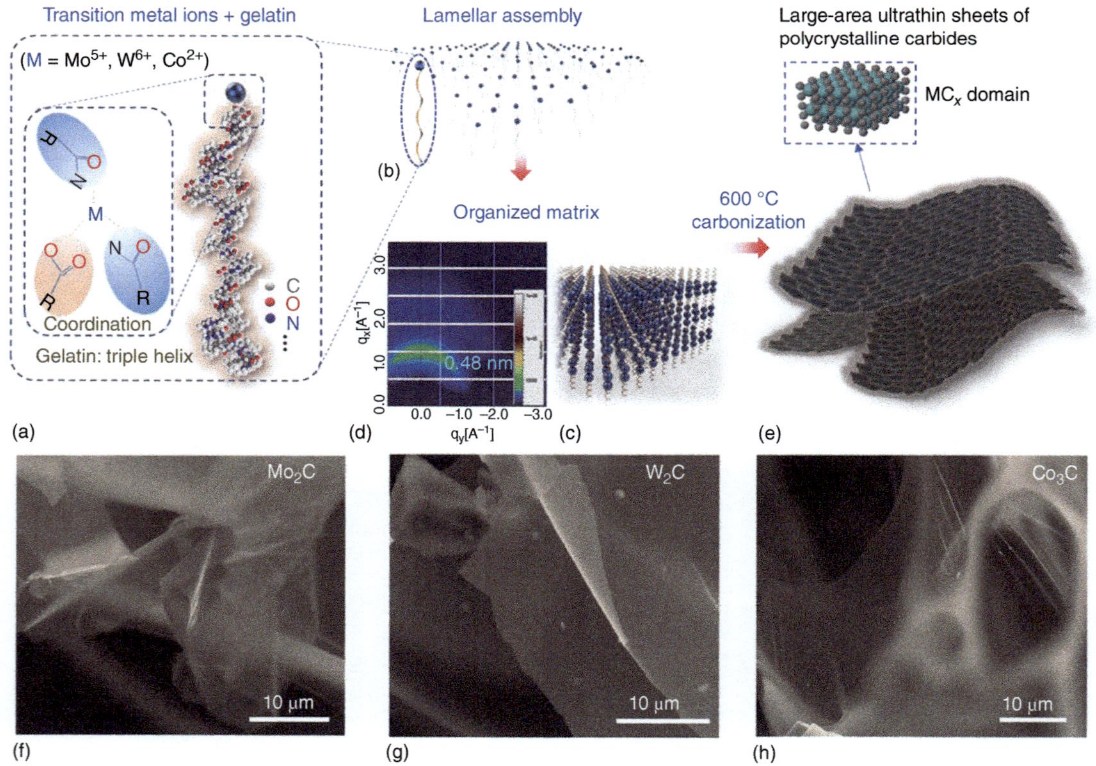

Figure 6.28 Self-assembly process for large-area ultrathin 2D TMCs. (a) Transition-metal ions such as Mo^{5+}, W^{6+}, and Co^{2+} are mixed in the gelatin-mediated hydrogel with rich functional groups as ligands. (b, c) Under an elevated temperature, the ion–gelatin solution takes (b) a lamellar assembly process that results in (c) a layer-by-layer organized matrix. (d) GIWAXS of Mo-gelatin template as an example to show that gelatin constrains the metal ions laterally between the layers. (e) Carbonized in a CH_4/H_2 atmosphere above 600 °C, the gelatin templates constrain the conversion of metal ions laterally between the layers, forming nanocrystalline 2D TMCs. (f–h) SEM images of the resulting ultrathin 2D TMCs of (f) Mo_2C, (g) W_2C, and (h) Co_3C flakes, respectively. Source: Zang et al. [48]/with permission of © John Wiley & Sons.

Due to interesting mechanical, electrical, and electrochemical properties, low-dimensional transition-metal carbides (TMCs) are considered as promising supports for transition-metal catalysts. Zang et al. prepared quite thin TMC sheets with a thickness of about 10 nm and a lateral size of more than 100 μm, decorated by metal nanoparticle ion using a gelatin-assisted bottom-up method [48]. During the preparation process, the self-accumulation of polymeric phases for the metal carbide degradation products occurred at the interface between gelatin hydrogel air and liquid after gelatin hydrogel with embedded transition-metal ions (Mo^{5+}, W^{6+}, Co^{2+}) was heated (Figure 6.28a–e). Therefore, the self-assembly and ordering of gelatin could arrange precursors in macroscopic structures that could be collapsed into dimensionally conserved nanostructures. Mo_2C, W_2C, and Co_3C (Figure 6.28f–h) with lateral size as high as ≈100 μm were synthesized through this method. Among them, ultrathin Mo_2C sheets with Co nanoparticles' decoration (Mo_2C–Co) exhibited superior HER activity with the smallest Tafel slope of 39 mV dec^{-1} and the lowest overpotential of 48 $mV_{j=10\,mA\,cm^{-2}}$ in 0.5 M H_2SO_4. The remarkable performance made Mo_2C–Co a feasible non-noble metal electrocatalyst for the HER and could be competitive with commercial Pt/C. This cheap and scalable preparation method with the assistance of gelatin template provided a novel strategy to obtain low-dimensional TMCs for diverse clean-energy-related applications.

References

1 Zou, F. and Manthiram, A. (2020). A review of the design of advanced binders for high-performance batteries. *Advanced Energy Materials* 10, 2002508.

2 Sun, J., Huang, Y., Wang, W. et al. (2008). Preparation and electrochemical characterization of the porous sulfur cathode using a gelatin binder. *Electrochemistry Communications* 10: 930–933.

3 Wang, Y., Huang, Y., Wang, W. et al. (2009). Structural change of the porous sulfur cathode using gelatin as a binder during discharge and charge. *Electrochimica Acta* 54: 4062–4066.

4 Jung, Y. and Kim, S. (2007). New approaches to improve cycle life characteristics of lithium–sulfur cells. *Electrochemistry Communications* 9: 249–254.

5 Saroha, R., Ahn, J.-H., and Cho, J.S. (2021). A short review on dissolved lithium polysulfide catholytes for advanced lithium–sulfur batteries. *Korean Journal of Chemical Engineering* 38: 461–474.

6 Sun, J., Huang, Y., Wang, W. et al. (2008). Application of gelatin as a binder for the sulfur cathode in lithium-sulfur batteries. *Electrochimica Acta* 53: 7084–7088.

7 Jiang, S., Gao, M., Huang, Y. et al. (2013). Enhanced performance of the sulfur cathode with L-cysteine-modified gelatin binder. *Journal of Adhesion Science and Technology* 27: 1006–1011.

8 Shao, H., Li, C., Liu, N. et al. (2015). Improved electrochemical performance of sulphur cathodes with acylated gelatine as a binder. *RSC Advances* 5: 47757–47761.

9 Akhtar, N., Shao, H., Ai, F. et al. (2018). Gelatin-polyethylenimine composite as a functional binder for highly stable lithium-sulfur batteries. *Electrochimica Acta* 282: 758–766.

10 Zhao, L., Sun, Z., Zhang, H. et al. (2020). An environment-friendly crosslinked binder endowing $LiFePO_4$ electrode with structural integrity and long cycle life performance. *RSC Advances* 10: 29362–29372.

11 Winter, M. (2009). The solid electrolyte interphase – the most important and the least understood solid electrolyte in rechargeable Li batteries. *Zeitschrift für Physikalische Chemie* 223: 1395–1406.

12 Wu, H., Jia, H., Wang, C. et al. (2020). Recent progress in understanding solid electrolyte interphase on lithium metal anodes. *Advanced Energy Materials* 11, 2003092.

13 Hou, Z., Zhang, J., Wang, W. et al. (2020). Towards high-performance lithium metal anodes via the modification of solid electrolyte interphases. *Journal of Energy Chemistry* 45: 7–17.

14 Akhtar, N., Sun, X., Yasir Akram, M. et al. (2021). A gelatin-based artificial SEI for lithium deposition regulation and polysulfide shuttle suppression in lithium-sulfur batteries. *Journal of Energy Chemistry* 52: 310–317.

15 Sun, X., Peng, Q., Ge, Y. et al. (2019). Carbon/gelatin microcapsules for sulfur cathode: a micro-reactor suppressing "Shuttle Effect". *Journal of the Electrochemical Society* 166: A1045–A1050.

16 Drofenik, J., Gaberšček, M., Dominko, R. et al. (2001). Carbon anodes prepared from graphite particles pretreated in a gelatine solution. *Journal of Power Sources* 94: 97–101.

17 Narath, A. and Tiilikka, A. (1961). Investigations of the interaction between gelatin and noble metals and its effect on sensitization by noble metals. *The Journal of Photographic Science* 9: 303–311.

18 Montoro, L. (2003). Gelatin/DMSO: a new approach to enhancing the performance of a pyrite electrode in a lithium battery. *Solid State Ionics* 159: 233–240.

19 Liu, Z., Jiang, Y., Hu, Q. et al. (2020). Safer lithium-ion batteries from the separator aspect: development and future perspectives. *Energy & Environmental Materials* 0: 1–27.

20 Liu, N., Huang, B., Wang, W. et al. (2016). Modified separator using thin carbon layer obtained from its cathode for advanced lithium sulfur batteries. *ACS Applied Materials & Interfaces* 8 (25): 16101–16107.

21 Chen, M., Chen, Z., Fu, X., and Zhong, W.-H. (2020). A Janus protein-based nanofabric for trapping polysulfides and stabilizing lithium metal in lithium–sulfur batteries. *Journal of Materials Chemistry A* 8: 7377–7389.

22 Zhi, J., Li, S., Han, M., and Chen, P. (2020). Biomolecule-guided cation regulation for dendrite-free metal anodes. *Science Advances* 6, eabb1342.

23 Kharmachi, I., Dhouibi, L., Berçot, P. et al. (2018). Electrodeposition behavior, physicochemical properties and corrosion resistance of Ni–Co coating modified by gelatin additive. *Protection of Metals and Physical Chemistry of Surfaces* 53: 1059–1069.

24 Collazo, A., Figueroa, R., Nóvoa, X.R., and Pérez, C. (2015). Electrodeposition of tin from a sulphate bath. *Surface and Coatings Technology* 280: 8–15.

25 Tabuchi, T., Hochgatterer, N., Ogumi, Z., and Winter, M. (2009). Ternary Sn–Sb–Co alloy film as new negative electrode for lithium-ion cells. *Journal of Power Sources* 188: 552–557.

26 Yato, Y., Halid, N.F.B., Chiba, M. et al. (2017). Preparation of Cu foam by electrodeposition of Cu particles. *Proceedings of Chitose International Forum* 18: 19.

27 Li, S., Zhu, Q., Zheng, B. et al. (2019). Nano-scale twinned Cu with ultrahigh strength prepared by direct current electrodeposition. *Materials Science and Engineering A* 758: 1–6.

28 Shetty, S. and Hegde, A.C. (2016). Electrodeposition of Sn-Ni Alloy coatings for water-splitting application from alkaline medium. *Metallurgical and Materials Transactions B* 48: 632–641.

29 Liu, Y., Shi, Q., Wu, Y. et al. (2021). Highly efficient dendrite suppressor and corrosion inhibitor based on gelatin/Mn^{2+} Co-additives for aqueous rechargeable Zn-manganese dioxide battery. *Chemical Engineering Journal* 407: 127189.

30 Yu, F., Zhao, L., Zhang, H. et al. (2021). Cathode/gel polymer electrolyte integration design based on continuous composition and preparation technique for high performance lithium ion batteries. *RSC Advances* 11: 3854–3862.

31 Wang, J., Tang, F., Wang, Y. et al. (2020). Self-healing and highly stretchable gelatin hydrogel for self-powered strain sensor. *ACS Applied Materials & Interfaces* 12: 1558–1566.

32 Han, Q., Chi, X., Liu, Y. et al. (2019). An inorganic salt reinforced Zn^{2+}-conducting solid-state electrolyte for ultra-stable Zn metal batteries. *Journal of Materials Chemistry A* 7: 22287–22295.

33 Kadir, M.F.Z. (2021). Non-faradaic-based supercapacitor fabricated with fish skin gelatin biopolymer electrolyte. *Ionics* 27: 2219–2229.

34 Le Bideau, J., Viau, L., and Vioux, A. (2011). Ionogels, ionic liquid based hybrid materials. *Chemical Society Reviews* 40: 907–925.

35 Tamate, R., Hashimoto, K., Horii, T. et al. (2018). Self-healing micellar ion gels based on multiple hydrogen bonding. *Advanced Materials* 30: 1802792.

36 Guo, S., Zhao, K., Feng, Z. et al. (2018). High performance liquid crystalline bionanocomposite ionogels prepared by in situ crosslinking of cellulose/halloysite nanotubes/ionic liquid dispersions and its application in supercapacitors. *Applied Surface Science* 455: 599–607.

37 Liu, J., Song, H., Wang, Z. et al. (2019). Stretchable, self-healable, and reprocessable chemical cross-linked ionogels electrolytes based on gelatin for flexible supercapacitors. *Journal of Materials Science* 55: 3991–4004.

38 Konarov, A., Voronina, N., Jo, J.H. et al. (2018). Present and future perspective on electrode materials for rechargeable Zn-ion batteries. *ACS Energy Letters* 3: 2620–2640.

39 Huang, S., Zhu, J., Tian, J., and Niu, Z. (2019). Recent progress in the electrolytes of aqueous Zn-ion batteries. *Chemistry* 25: 14480–14494.

40 Li, H., Han, C., Huang, Y. et al. (2018). An extremely safe and wearable solid-state Zn ion battery based on a hierarchical structured polymer electrolyte. *Energy & Environmental Science* 11: 941–951.

41 Lu, Y., Zhu, T., Xu, N., and Huang, K. (2019). A semisolid electrolyte for flexible Zn-ion batteries. *ACS Applied Energy Materials* 2: 6904–6910.

42 Chen, K., Feng, W., Jia, Z. et al. (2020). Cryogenic-crosslink synthesis of $Ba(OH)_2$@gelatin microcapsules for the removal of sulfate in lithium refinery. *Hydrometallurgy* 195: 105342.

43 Du, H., Yuan, C., Huang, K. et al. (2017). A novel gelatin-guided mesoporous bowknot-like Co_3O_4 anode material for high-performance lithium-ion batteries. *Journal of Materials Chemistry A* 5: 5342–5350.

44 Zhang, X., An, L., Yin, J. et al. (2017). Effective construction of high-quality iron oxy-hydroxides and Co-doped iron oxy-hydroxides nanostructures: towards the promising oxygen evolution reaction application. *Scientific Reports* 7: 43590.

45 Chekin, F., Vahdat, S.M., and Asadi, M.J. (2016). Green synthesis and characterization of cobalt oxide nanoparticles and its electrocatalytic behavior. *Russian Journal of Applied Chemistry* 89: 816–822.

46 Chekin, F., Tahermansouri, H., and Besharat, M.R. (2013). Nickel oxide nanoparticles prepared by gelatin and their application toward the oxygen evolution reaction. *Journal of Solid State Electrochemistry* 18: 747–753.

47 Bagheri, S., Chekin, F., and Hamid, S.B.A. (2014). Cobalt doped titanium dioxide nanoparticles: synthesis, characterization and electrocatalytic study. *Journal of the Chinese Chemical Society* 61: 702–706.

48 Zang, X., Chen, W., Zou, X. et al. (2018). Self-assembly of large-area 2D polycrystalline transition metal carbides for hydrogen electrocatalysis. *Advanced Materials* 30: 1805188.

7

Collagen-Derived Carbons

7.1 History of Collagen-Derived Carbons

Collagen is one of the most plentiful biological substances in the environment. Collagen mainly consists of C, N, O, and S. Collagen and its derivatives can be converted into collagen-derived carbons through controlling the pyrolysis conditions. The ordered biological structures and special compositions of precursors endow collagen-derived carbon with unique nanostructures and heteroatom doping, which has a broad application prospect in energy storage and conversion. Collagens-derived carbon is made from the natural products of animal wastes such as bone, fish, and skin by artificial methods like carbonization and activation at high temperatures. Due to the accessibility and abundance of collagen, the collagens-derived carbon is suitable for large-scale production [1]. In 2008, Deng and coworkers prepared ordered hierarchical porous carbon fibers using collagen fiber as a precursor for the first time [2]. The carbon-fibers' pore size can be tailored by changing metal ions, the ratio of metal to collagen, and organic reagent. In the following years, collagen-derived carbons with different structures and compositions have been prepared and widely used in alkali metal-ion batteries [3, 4], supercapacitors [5, 6], electrocatalysis [7–12], and so on.

7.2 Carbonization Mechanism of Collagen-Derived Carbons

Gelatin is a product of the hydrolysis of collagen and is easy to be extracted from animal skin, bones, and fish scales [13]. It contains polypeptides and proteins. By direct carbonization, its composition of amino acids (such as hydroxyproline, proline, alanine, and glycine) can be converted into N,O-doped carbon [14]. Understanding the carbonization process is helpful to control the preparation of gelatin- and collagen-derived carbons [15]. Yang et al. detailedly revealed the carbonization mechanism of gelatin utilizing in situ and ex situ characterizations [16].

Collagen-Derived Materials: Synthesis and Applications in Electrochemical Energy Storage and Conversion,
First Edition. Feng Wang, Yaqin Huang, and Jin Niu.
© 2022 WILEY-VCH GmbH. Published 2022 by WILEY-VCH GmbH.

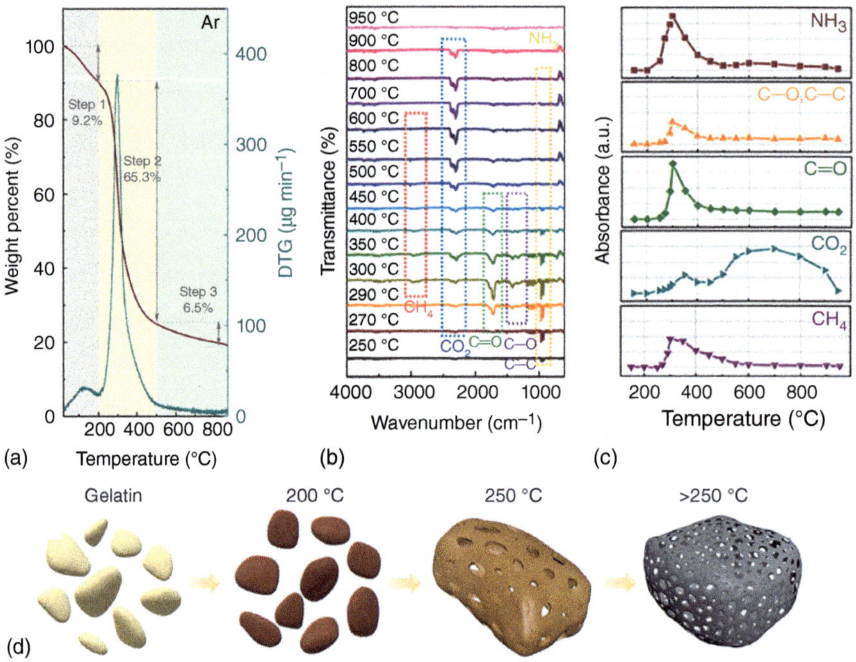

Figure 7.1 (a) TG curves of gelatin. (b)TG–FTIR patterns of gelatin heated in an Ar atmosphere. (c) Gas emission curves recorded by TG–FTIR. (d) Schematic illustration of the morphological evolution during the gelatin carbonization process. Source: Reproduced with permission from Yang et al. [3]; © 2020, Elsevier.

Thermogravimetric (TG) analysis was conducted to explore the formation of gelatin-derived carbons. Three distinct stages of weightlessness were observed in 30–200, 200–500, and 500–900 °C (Figure 7.1a). The slight weight loss in the initial stage is due to the evaporation of water. In the second stage, the weight decreases sharply, and there is a differential TG (DTG) peak at 296 °C, indicating the thermolysis and initial carbonization of gelatin. Further gradual weight loss occurred in the third stage. According to the results of TG analysis, gelatin was pyrolyzed at different temperatures.

Thermogravimetric–Fourier transform infrared spectroscopy (TG–FTIR) was employed to detect the gases produced during pyrolysis. As shown in Figure 7.1b,c, the absorption peaks at 991–766, 2820–2931, 2250–2400, 1475–1000, and 1650–1900 cm^{-1} could be observed, which were the characteristic peaks of NH_3, CH_4, CO_2, C—O/C—C, and C=O, respectively [16, 17]. When the pyrolysis temperature was higher than 250 °C, the absorption intensity of these bands increased considerably, suggesting that numerous NH_3, CH_4, CO_2, and O-containing organic species were produced at 250–500 °C. At about 300 °C, the intensity of NH_3, CH_4, O-containing species achieved the highest, demonstrating that these gases are templates for forming loose and porous structures. The evolution of CO_2 could be observed in two stages. Due to the decarbonylation of alkyl side chains, CO_2 was generated from 250 to 450 °C [18]. When the temperature was above 500 °C, because of the elimination of O-containing functional groups, CO_2 is the main ingredient of the evolving gas. This phenomenon indicated that the carbon structure was first formed at 200–500 °C and developed at 500–900 °C (Figure 7.1d) [19–21].

The graphitization degree and defect density of the samples were demonstrated by X-ray diffraction (XRD; Figure 7.2a) and Raman spectroscopy (Figure 7.2b). The wide peak at 20° was the diffraction of the right-hand triple-helix structure of gelatin. When the temperature was raised to 450 °C, the peak moved to 25°, which was assigned to the characteristic peak of C (002) [22]. This indicated that

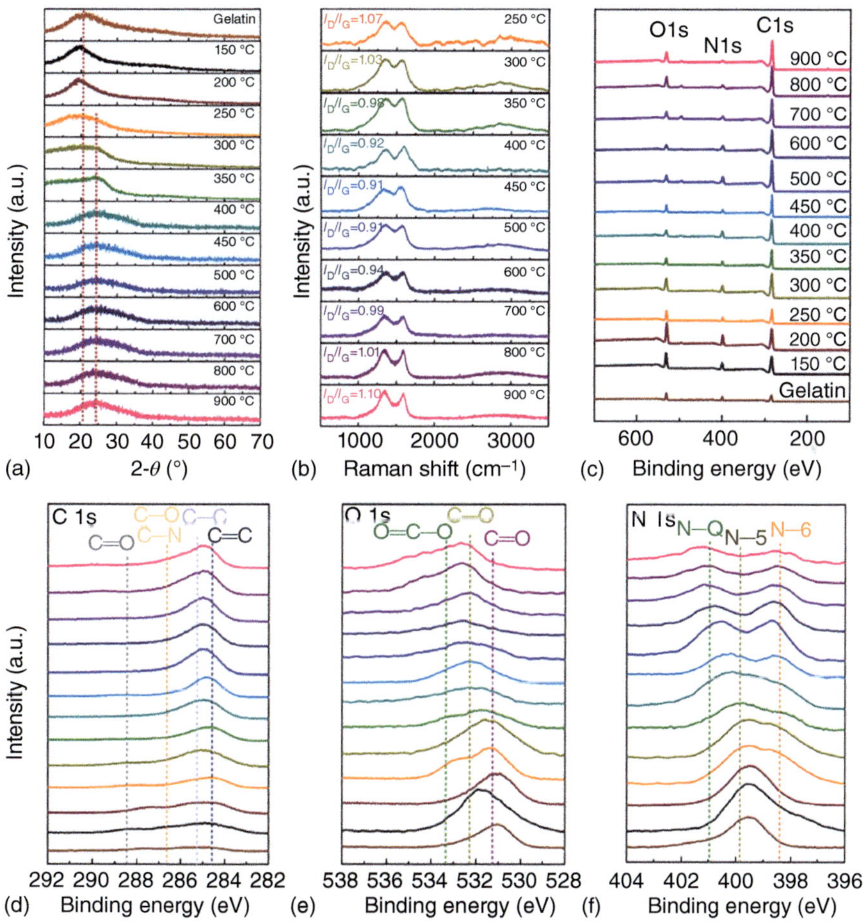

Figure 7.2 (a) XRD patterns and (b) Raman spectra of samples pyrolyzed at different temperatures. (c) XPS spectra, high-resolution (d) C 1s, (e) O 1s, (f) N 1s spectra of samples pyrolyzed at different temperatures. Source: Reproduced with permission from Yang et al. [3]; © 2020, Elsevier.

the gelatin suffered preliminary carbonization. The peak of (002) becomes wider with the increase of temperature, indicating that numerous defects were gradually generated in the carbon structure. Raman spectroscopy further confirmed the existence of these defects (Figure 7.2b). From 250 to 450 °C, the ratio of I_D/I_G decreased, suggesting that the graphitization degree increased. However, when the heating temperatures were above 500 °C, the I_D/I_G intensity ratio increased, indicating that at higher temperatures, due to the removal of heteroatoms and the rearrangement of carbon atoms, defects were produced in the carbon layer.

The degree of heteroatomic self-doping in carbon formation was investigated via elemental analysis and X-ray photoelectron spectroscopy (XPS). Table 7.1 shows the elemental analysis results, indicating that the contents of N and O began to reduce from 200 °C. It was observed by TG–FTIR that due to the breaking of peptide and aliphatic bonds, CH_4, NH_3, CO_2, and O-containing organic gases were generated. Compared with the internal heteroatom, the surface heteroatom can improve the electrochemical performance more effectively, and the surface components were detected by XPS (Figure 7.2c–f). With the increase of pyrolysis temperature, the surface carbon atoms are dominated by C—C, and C=C, while the surface O atoms transformed from C=O to C—O, and O=C—O [23]. In addition, the high-resolution N 1s XPS spectra showed that N atoms altered from pyrrolic N (N-5) to pyridinic N (N-6) and graphitic N (N-Q). The transition from polypeptide chain (five-membered ring) to heteroatom doping and defect-enriched carbon structure (six-membered ring) is consistent [24].

According to the above ex situ and in situ analyses, a graphic illustration of the carbonation mechanism of the gelatin-derived carbons is shown in Figure 7.3: (i) Polypeptide bonds in gelatin were decomposed from room temperature to 200 °C, which were accompanied by water loss, (ii) the carbon started to generate at 200–500 °C. The linear polypeptide chain was cross-linked by cyclization decomposition to form aromatic ring structure, and in the meantime, plentiful CH_4, NH_3, CO_2, and O-containing organics were generated as by-products; (iii) the carbon structure is becoming well developed at 500–900 °C. The oxygen-containing functional groups continue to decompose, generating large amounts of CO_2 and producing abundant defects. The O atoms on the surface transformed from C=O group to C—O, and O=C—O, and the N atoms on the surface changed from N-5 to N-6 and N-Q.

Table 7.1 Elemental analysis [3].

	Gel	150	200	250	300	350	400	450	500	600	700	800	900
C (at.%)	44	46	48	53	61	67	68	69	71	74	76	78	83
O (at.%)	30	29	27	23	16	10	9	9	11	7	6	4	6
N (at.%)	15	16	17	17	16	16	15	15	15	14	13	13	10
H (at.%)	6	6	6	6	5	4	4	3	3	1	1	1	1

Source: Reproduced with permission from Yang et al. [3]; © 2020, Elsevier.

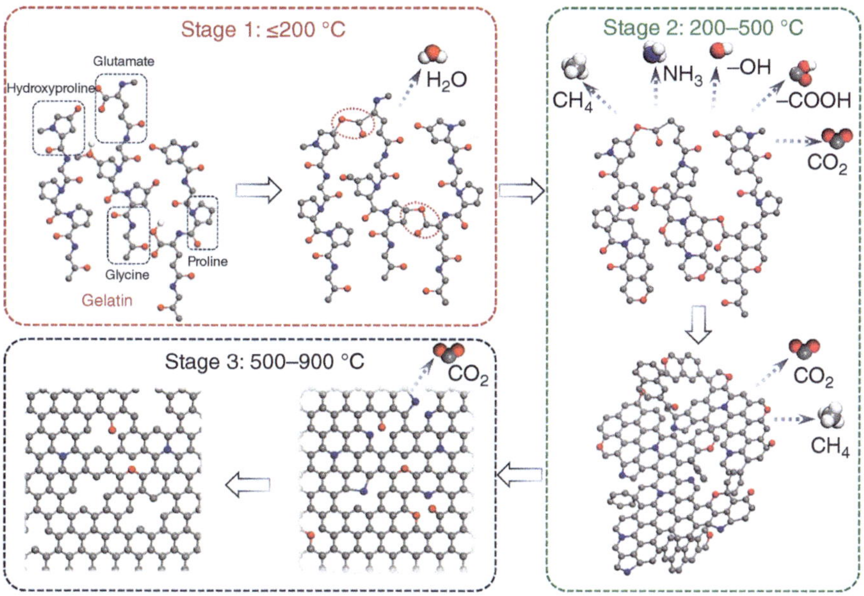

Figure 7.3 Schematic illustration of the carbonization mechanism for gelatin-derived carbons. Source: Reproduced with permission from Yang et al. [3]; © 2020, Elsevier.

7.3 Structures and Compositions of Collagen-Derived Carbons

7.3.1 Morphologies

The diversity of structure is an inherent property of carbon materials (Figure 7.4). The same is true for collagen-based carbon materials as it is known that electrochemical performance is to a great extent morphologically dependent. The collagen-derived carbon materials have various nanostructures with different dimensions [26, 27].

7.3.1.1 Zero-Dimensional Carbon Materials

Zero-dimensional materials are materials within the three-dimensional nanometer scale of 1–10 nm or with it as the basic unit. As a new kind of zero-dimensional carbon materials, carbon dots exhibit ultrafine size and unique physical and chemical properties [28–31]. Since the discovery of carbon dots, because of their quantum size, numerous functional groups, great biocompatibility, environmental friendliness, and photoluminescence properties, they have been largely investigated and gradually used in different fields [32–35]. In recent years, carbon dots also show great potential for energy storage and conversion [36–38].

Through a one-pot hydrothermal reaction, Dehghani et al. prepared graphene quantum dots (GQDs) using collagen as raw materials with 15% quantum yield. Figure 7.5a shows the synthesis step of GQDs [39]. According to the transmission electron microscopy (TEM) images and particle size distribution (Figure 7.5b,c),

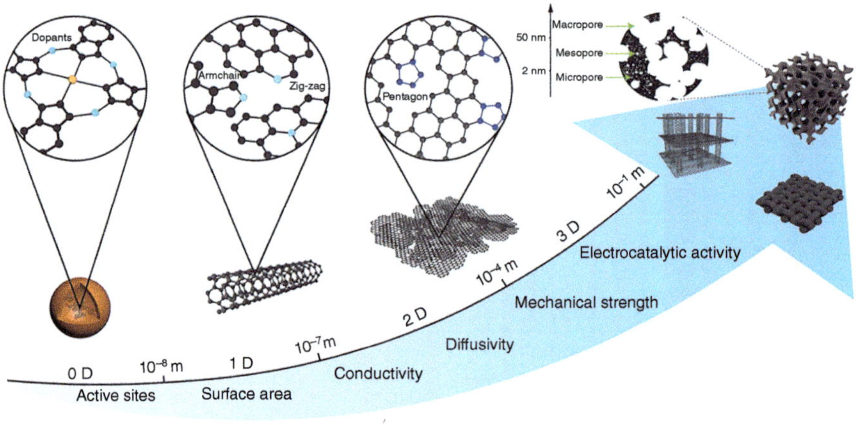

Figure 7.4 Comparison of the structure–property relationship of carbon structures. Source: Reproduced with permission from Jorge et al. [25]; © 2020, John Wiley & Sons.

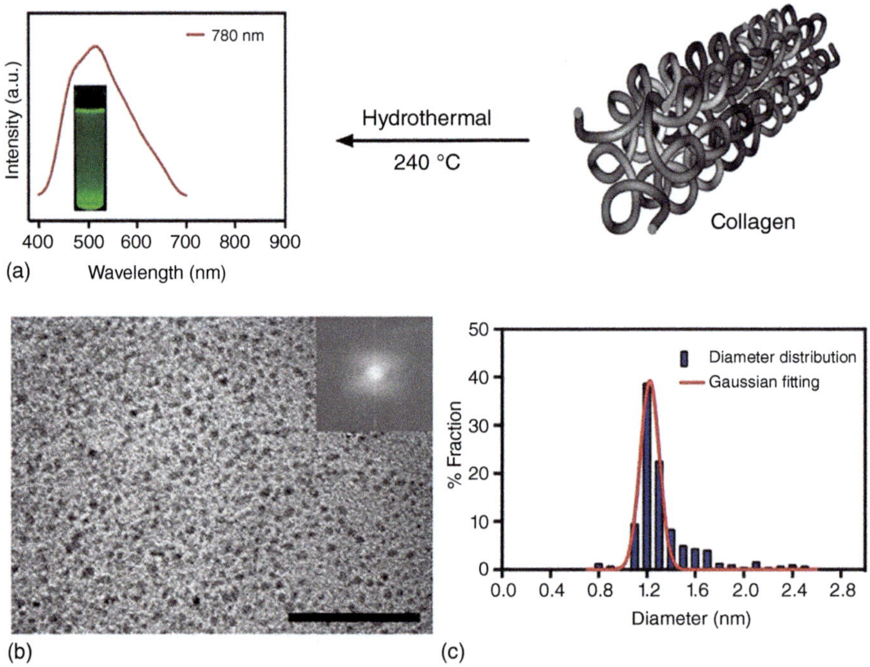

Figure 7.5 (a) Schematic diagram of CQD synthesis, (b) TEM images of CQDs. Inset shows fast Fourier transform (FFT) of nanoparticles with no visible diffraction (scale: 20 nm), (c) particle size distribution of CQDs. Source: Dehghani et al. [39]/with permission of © Elsevier Ltd.

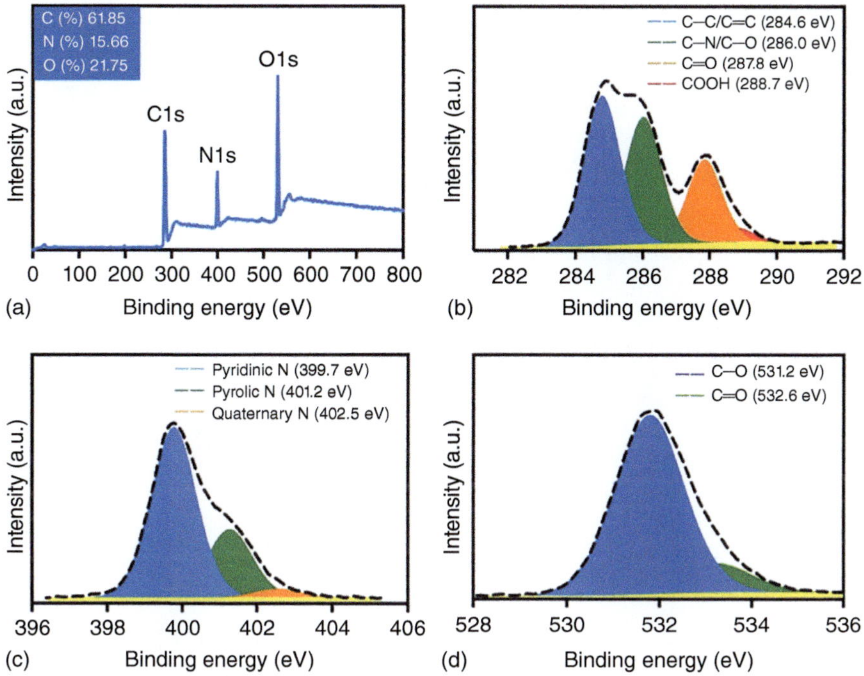

Figure 7.6 XPS spectra of CQDs: (a) Full survey spectrum, (b) C 1s spectrum, (c) N 1s spectrum, and (d) O 1s spectrum (A color version of this figure can be viewed online). Source: Reproduced with permission from Dehghani et al. [39]; © 2018, Elsevier.

the carbon quantum dots (CQDs) can disperse in water homogeneously and show a spherical shape with a particle size of 1.25 ± 0.21 nm.

XPS measurements are implemented to detect the surface chemical composition of CQDs. Figure 7.6a–d shows three peaks can be observed at about 285.1, 400.1, 531.2 eV, indicating the existence of C, N, and O elements. The high-resolution C1s spectrum can be deconvoluted into four peaks (Figure 7.6b), which centered at 284.6 eV (C—C/C=C), 286.0 eV (C—O/C—N), 287.8 eV (C=O), and 288.7 eV (—COOH). Figure 7.6c shows a high content of pyrolysis and pyridine-N. The high proportion of pyridine-N would give an electron, called a delocalized electron, which enhances the possibility of electron excitation at the same excitation energy.

Q.H. Liang and coworkers proposed a facile method to prepare fluorescent carbon dots (CQDs) via the hydrothermal treatment of gelatin solution (Figure 7.7) [40]. In the hydrothermal process, gelatin was first carbonized to small proteins and then hydrolyzed into amino acids rich in amino and carboxyl groups. By further hydrothermal treatment, the amino acids could be partially decomposed and polymerized, and then pyrolyzed into defective CQDs rich in O, S, and -NH_2. At the same time, the hydrophilic group enables CQDs to disperse in water and other polar solvents evenly. The good hydrophilic property of the CQDs is conducive to its application in diagnosis and detection. The prepared CQDs emit strong blue emission at 350 nm excitation wavelength, and the maximum mass yield of CQDs is

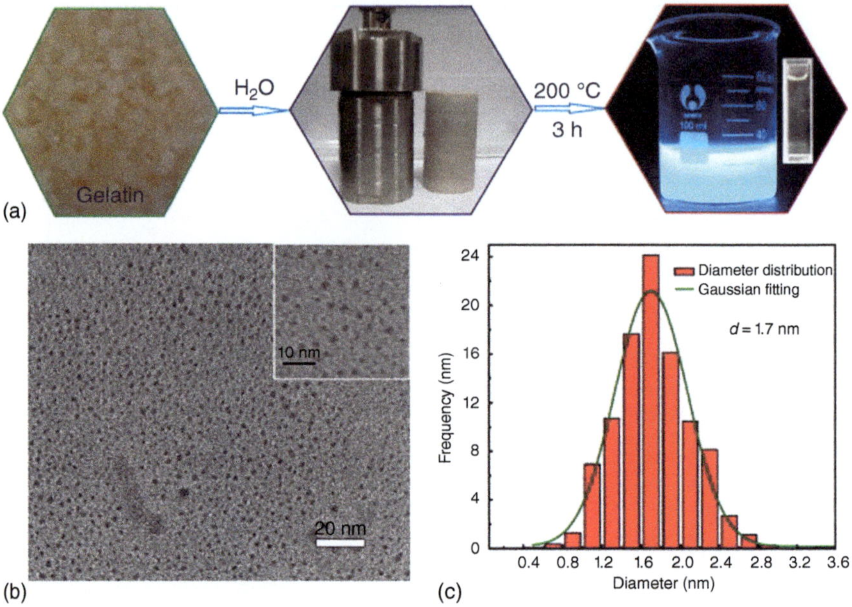

Figure 7.7 (a) Schematic of the synthesis process of photoluminescent CQDs by hydrothermal treatment of commercial gelatin. (b) Low- and high-magnified (inset) TEM images and (c) the diameter distribution of the CQDs. Source: Liang et al. [40]/with permission of © Elsevier Ltd.

31.6%. In addition, these CQDs show pH-sensitive, typical excitation-dependent, and up-converted photoluminescence (PL) characteristics. TEM image and the particle size distribution of CQDs in Figure 7.7b,c show that they have homogeneous size (1.7 nm) as well as an almost spherical shape.

The chemical composition of these CQDs is determined by XPS measurements. In the XPS survey spectra shown in Figure 7.8a, there are four major peaks at 168.2, 284.8, 399.1, and 531.5 eV belonging to S 2p, C 1s, N 1s, and O 1s, indicating the presence of sulfur, carbon, nitrogen, and oxygen elements. Figure 7.8b was the high resolution of the C 1s spectrum. The spectrum can be divided into four peaks centered at 284.1, 285.7, 287.7, and 289.6 eV, which correspond to C—C, C—N, C=O, and O—C bonds. FTIR spectroscopy was performed to obtain further structural insights of CQDs (Figure 7.8c). The characteristic absorption bands of O–H (3500 cm^{-1}), the stretching vibration band of C=O (1700 cm^{-1}), and the stretching vibration bands of C–O (1100 and 870 cm^{-1}) could be clearly seen, which demonstrated the existence of carboxylic acid and other oxygen-containing functional groups. In addition, the vibration and deformation bands of N–H could be observed at 3050–3250 and 1510 cm^{-1}, indicating the existence of amino-containing functional groups. Moreover, there also existed two peaks at 2620 and 680 cm^{-1}, which attributed to stretching vibration bands of S–H and S–C, respectively, suggesting the presence of sulfur-containing functional groups [16]. Furthermore, three distinct peaks at 2950, 1420, and 1330 cm^{-1} are assigned to the stretching vibration C—H, C=C, and C—C, indicating the existence of alkyl and aryl groups.

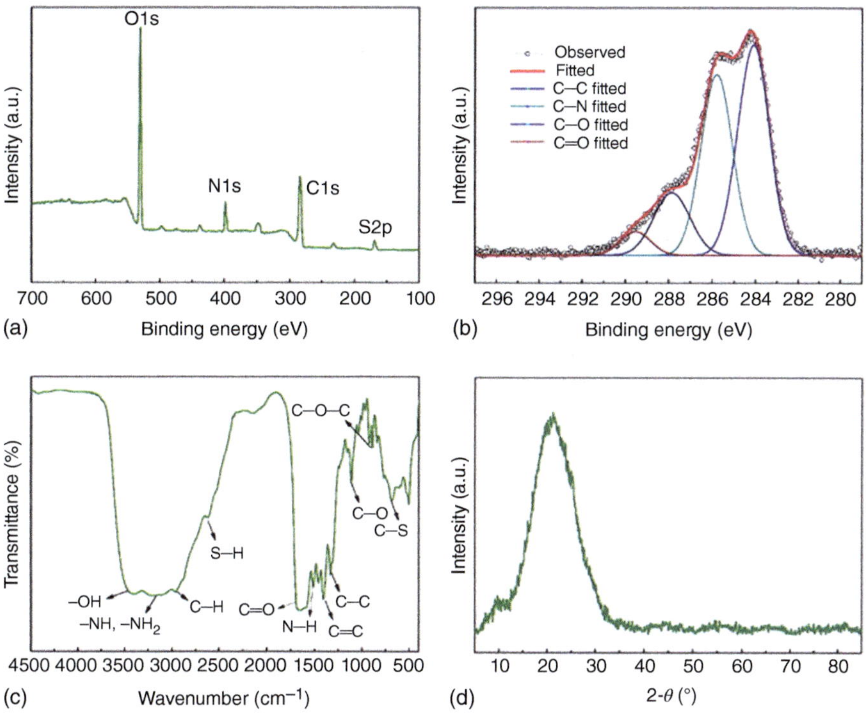

Figure 7.8 (a) XPS survey spectrum, (b) the high-resolution C 1s peaks and the fitting curves, (c) the FTIR spectrum, and (d) the XRD patterns of the as-synthesized CQDs. Source: Reproduced with permission from Liang et al. [40]; © 2013, Elsevier.

The XRD pattern shown in Figure 7.8d exhibits a weakly wide reflection peak with a center of about 22°, indicating that the interlayer spacing of the (002) diffraction peak is 0.41 nm. The layer spacing is larger than that of graphite (0.34 nm) due to the presence of plentiful functional groups. According to the above discussions, the possible CQDs formation mechanism is proposed.

7.3.1.2 One-Dimensional Carbon Materials

One-dimensional nanomaterials are materials that have one of the three dimensions not between 0.1 and 100 nm, such as carbon nanofibers and nanowires. The development of carbon fiber dates back to the eighteenth century, but its use did not materialize until 1960. Natural fibers from plants and animals have been converted to carbon for a long time [41].

As mentioned before, collagen fiber is one of the most abundant natural biomasses, primarily derived from animal skin and traditionally applied as raw material for leather production [42]. Collagen fibers are made up of highly ordered collagen molecules. Collagen is a rod-shaped molecule (length: 300 nm, diameter: 1.5 nm). It consists of three polypeptide chains with helix structures. The three helices are staggered along the long axis of their helix (on the fourth order) and have a corresponding "gap" region with 67 nm in length. Collagen molecules are formed by the fibers and aggregated into microfibers, which are made up of four to eight collagen molecules and further formed into fibrils. These collagen fibrils are organized into fibers that can form larger bundles [43]. Thus, it can be speculated that if collagen fibers are used as carbon precursors, porous carbon materials with definite morphology can be achieved. However, raw collagen fibers have weak mechanical strength and low thermal stability. Therefore, when the collagen fiber is utilized as a carbon source, it needs to be chemically modified to enhance its thermal and mechanical characteristics and "fix" its novel structure before heat treatment.

To this end, Deng et al. reported a new method to prepare controllable porous carbon fiber from collagen fiber [2]. Based on the leather processing principle, collagen fiber is rich in OH, COOH, NH_2, and other functional groups, which can react with metal ions, e.g. Al^{III}, Fe^{III}, Zr^{IV}, and Cr^{III} [44]. As a result, the collagen fiber structure can be fixed via reacting with metal ions.

Porous carbon fiber was prepared from collagen fiber in three stages. Firstly, the structure of collagen fibers was fixed via reacting with metal ions (Al^{III}, Fe^{III}, and Zr^{IV} salts) and organic reagents. Secondly, carbon–metal oxide composite fibers were prepared via vacuum carbonization. Finally, dilute nitric acid solution or H_2SO_4-$(NH_4)_2SO_4$ buffer solution was used to remove the metal oxides formed in the carbonization stage to obtain ordered porous carbon fibers. It should be noted that collagen fibers must be pre-modified with organic agents before reacting with metal ions to improve the structural parameters of collagen fibers. This is because subsequent experiments have shown that the collagen fibers structure cannot be entirely fixed by the metal ions reaction alone. As a result, collagen fibers must be subjected to the treatment with other chemical substances to stabilize their fiber structure and improve their ability to fix metal ions. To synthesize the collagen fiber-Al^{III} precursor, the collagen fibers, after treated with plant tannin (black wattle tannin), were used to react with Al^{III}. Vegetable tannins are rich in *o*-hydroxyphenol rings and therefore have a high affinity for both metal ions and collagen fibers [45]. Actually, the synergistic reaction of plant tannins with collagen fibers and Al^{III} has become an effective method for tanning [46]. For other metal ions, collagen fibers are treated with glutaraldehyde before reacting with Zr^{IV} or Fe^{III}.

The obtained carbon fibers are layered and have multiple layers of structure. SEM image in Figure 7.9 shows that the outer diameter of the carbon fibers is about 1–4 mm, and the length is about 0.5–1 mm. Carbon fibers are composed of smaller ordered fibers with 60 nm average diameter. As a result, the morphology of natural collagen fibers is largely reserved. TEM images show the porous structure of carbon fiber. The pores are 1–10 nm in size. Moreover, the pores have a certain arrangement, corresponding to the smaller carbon-fiber structure (Figure 7.9). These results demonstrate that the elimination of metal oxides from carbon fibers not only preserves the fiber structure but also generates the porous structure of carbon fibers. The specific surface area (SSA) of carbon fiber increases obviously after removing metal oxides. The Brunner–Emmet–Teller (BET) specific surface area of the ZrO_2-treated collagen fibers increased from 177 to 1212 $m^2\ g^{-1}$. These results demonstrate that metal oxides are crucial to increase the carbon-fibers' diameter.

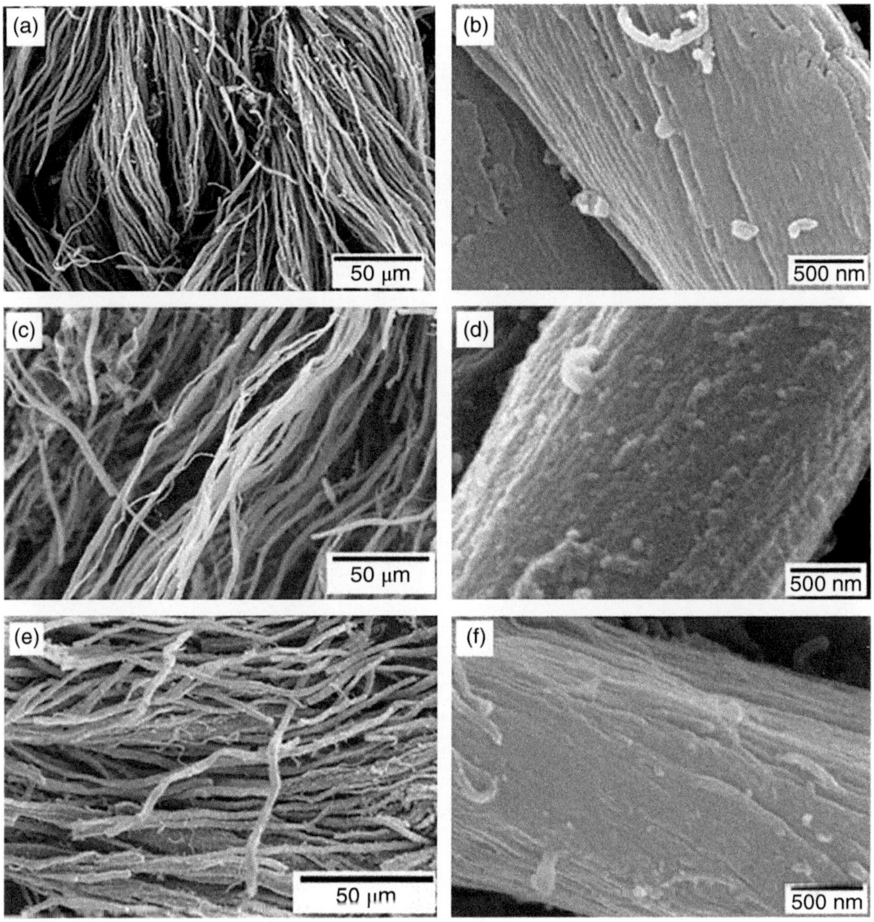

Figure 7.9 SEM images of the carbon fibers obtained from different precursors: (a, b) carbon fiber obtained from the precursor collagen fiber treated with glutaraldehyde and Zr^{IV}; (c, d) carbon fiber obtained from the precursor of collagen fiber treated with black wattle tannin and Al^{III}; (e, f) carbon fiber obtained from the precursor of collagen fiber treated with glutaraldehyde and Fe^{III}. Source: Deng et al. [2]/with permission of © John Wiley & Sons.

7.3.1.3 Two-Dimensional Carbon Material

In the past few years, two-dimensional carbon nanosheet materials were also prepared and widely used in energy storage and conversion due to their large lateral sizes and specific layered structures.

Qiu and coworkers proposed a simple and efficient method to produce flexible B/N-co-doped carbon nanosheets (B/N-CSs) [47]. Through assembling gelatin in a long-range sequence on two-dimensional hard template, the final products could be obtained after carbonization. A schematic of the integrated B/N-CSs strategy is shown in Figure 7.10a. After pyrolysis at 900 °C in an inert atmosphere, the gelatin layer is transformed to carbon nanosheets by thermal condensation and pyrolyzation (Figure 7.10b–d). At the same time, during the carbonization process, the B and N

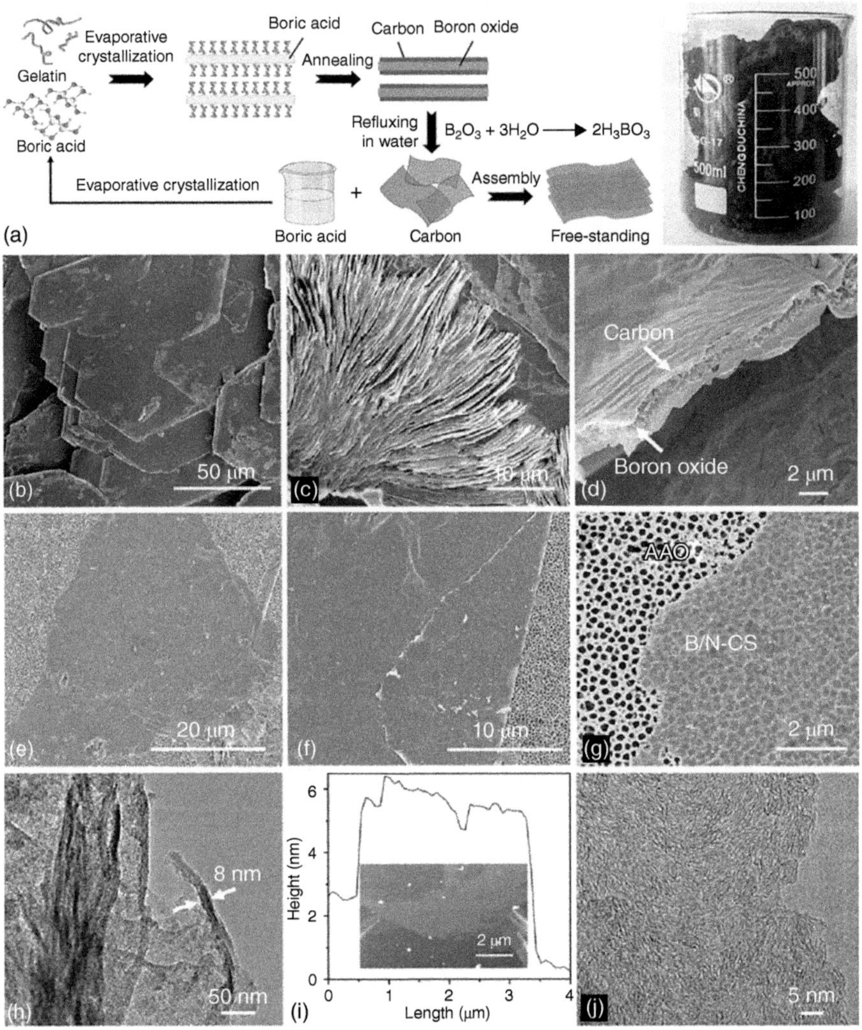

Figure 7.10 (a) Schematic illustration of sustainable synthesis and assembly of B/N-CSs; (b) optical image of doped carbon nanosheets. (b) Top and (c) cross-section SEM view of gelatin-coated boric acid nanoplates; (d) SEM image revealing the formation of carbon nanosheets on boron oxide nanoplate; (e–g) SEM images of B/N-CSs supported on AAO membrane; (h) cross-section TEM image of B/N-CSs; (i) AFM image of a B/N-CS, indicating the uniform thickness of 5.2–6.2 nm across the sheet; and (j) HRTEM image of B/N-CSs. Source: Ling et al. [47]/with permission of © John Wiley & Sons.

heteroatoms are also doped into carbon nanosheets. Boiling water reflux can wash the boron oxide in the sample to form boric acid and finally obtain pure B/N-CSs with gram-grade yield.

SEM images indicate that the B/N-CS has sheet-like morphology with a transverse size of more than 50 μm (Figure 7.10e). Figure 7.10f shows the sheet offers good structural flexibility, which can be folded face to face without cracking. When B/N-CSs are placed on porous anodized alumina (AAO) film, due to the thin thickness of the film, B/N-CSs have greater transparency under the action of electron beam (Figure 7.10g). To calculate the exact thickness of the B/N-CSs, they are embedded in an epoxy resin and then sliced with a slicer. TEM cross-section images of B/N-CSs showed that the thickness was about 5–8 nm (Figure 7.10h). Although the B/N-CSs thickness is thin, they have a flat surface with no noticeable wrinkles. Based on the atomic force microscopy (AFM) measurement, the thickness of the carbon nanosheets can be estimated at 5.2–6.2 nm (Figure 7.10i), consistent with the observation under TEM. Therefore, the B/N-CSs aspect ratio is calculated to be 6200–10 000. As shown in Figure 7.10j, high-resolution transmission electron microscopy (HRTEM) analysis demonstrates that the wafer was composed of many short-range ordered graphite regions at 10–20 nm, indicating its polycrystalline structure.

The combination of the high aspect ratio and B/N-CSs flexibility enables them to be directly assembled through vacuum filtration into individual films that are easy to handle and without the need for any auxiliary additives (Figure 7.11a). With the exception of graphene, this is difficult to achieve for traditional carbon nanosheets with small transverse dimensions and large thickness. The large overlapping area between individual nanosheets helps to keep the internal electrical contact and mechanical properties of the films unaffected during deformation. Thus, the B/N-CS film can be easily bent to 180° or rolled up without detectable damage (Figure 7.11b). The thickness of the flexible B/N-CS membrane can be adjusted via adjusting the filtration volume of the B/N-CS suspension at a certain concentration. As shown in Figure 7.11c, the thickness of the film is about 29 μm. The cross-section SEM images of the film show the neatly arranged lamellar structure of the nanosheets throughout the film (Figure 7.11d), which offers a less tortuous diffusion path for ion transport compared with the particle electrodes. The surface area of the assembled film was slightly lower (393 $m^2 g^{-1}$) than that of B/N-CS powder. In addition, B/N-CSs are able to be assembled into a low-density 3D aerogel with a porosity of over 99% and a density of 4.81 $mg cm^{-3}$ due to the strong interaction between

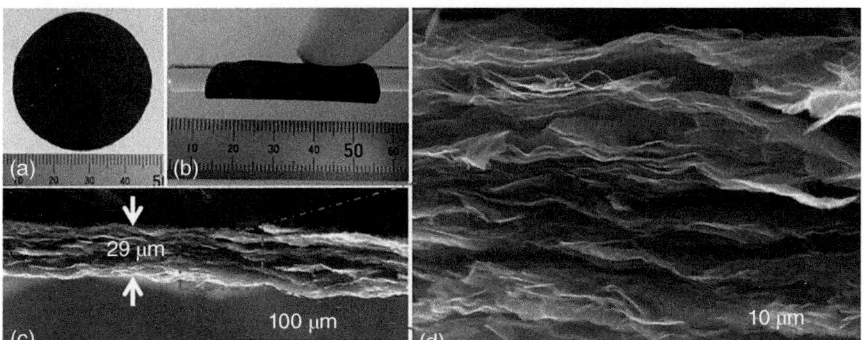

Figure 7.11 (a) Optical image of a freestanding B/N-CS film; (b) a flexible film wrapping around a glass rod; (c, d) cross-section SEM images of a B/N-CS film with a thickness of 29 μm. Source: Reprinted with permission from Ref. [47]; © 2015, WILEY-VCH Verlag GmbH & Co. KGaA.

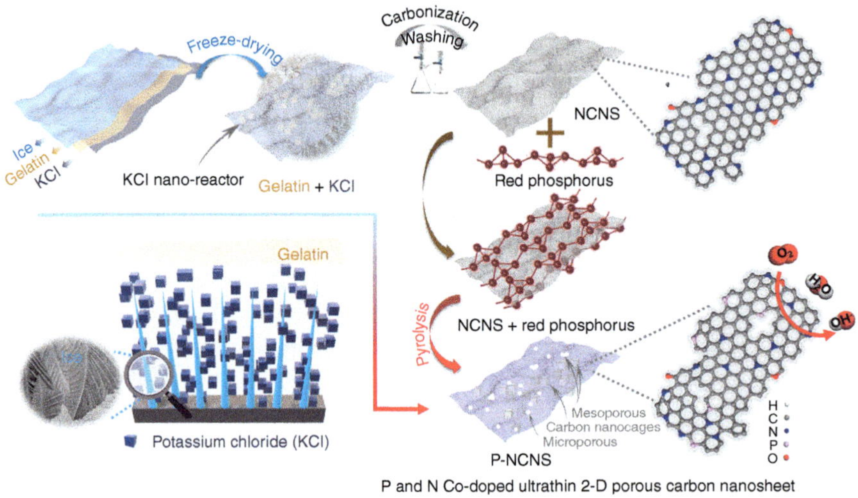

Figure 7.12 Illustration of the synthetic process of P-NCNS electrocatalyst. Source: Ling et al. [48]/© 2020, Elsevier.

them. B/N-CS aerogel has a density comparable to superlight-graphene aerogel (typically 3–5 $mg\,cm^{-3}$).

Zan et al. proposed a facile ice/salt dual-template method to fabricate P- and N-co-doped ultrathin two-dimensional micro/mesoporous carbon nanosheets (P-NCNS) using red phosphorus as P source and gelatin as C and N precursors [48]. Figure 7.12 illustrates the synthesis of ultrathin two-dimensional P-NCNS by dual-template method. The method consists of two steps using red phosphorus (red P) as P source and gelatin as C and N precursors. The first step involves freeze-drying a mixture of gelatin and KCl, where ice/KCl is used as a dual template, and then KCl is used as an independently sealed nanoreactor for restricted pyrolysis to promote high N content (10.38 at%) of the formation of N-doped two-dimensional NCNS. Finally, the obtained NCNS is subjected to a second pyrolysis treatment with red P to convert into P-NCNS. The comparison sample was also fabricated by heat treatment of NCNS at 950 °C without adding red phosphorus, denoted as H-NCNS.

SEM and TEM images show that the prepared P-NCNS and H-NCNS retain the two-dimensional nanosheet structure of ultrathin NCNS, and the carbon nanocages are distributed on the nanosheet (Figure 7.13a,b,d,e,g, h). During the synthesis of two-dimensional carbon nanosheets, KCl crystals precipitate, and carbon nanocages are formed, which is in accord with the results reported in the literature. AFM images indicate the thickness of the P-NCNS nanosheet is about 1.2 nm, similar to that of NCNS (about 1.1 nm) and H-NCNS (about 1.3 nm) (Figure 7.13c,f,i), showing that the nanosheet consists of approximately four graphite layers. To study the influence of KCl and ice on the ultrathin two-dimensional structure of the target during the carbonization process, comparison samples without KCl or ice are prepared by authors. A bulk stacking carbon morphology is observed for the comparison samples, showing that both of them are necessary for forming a unique

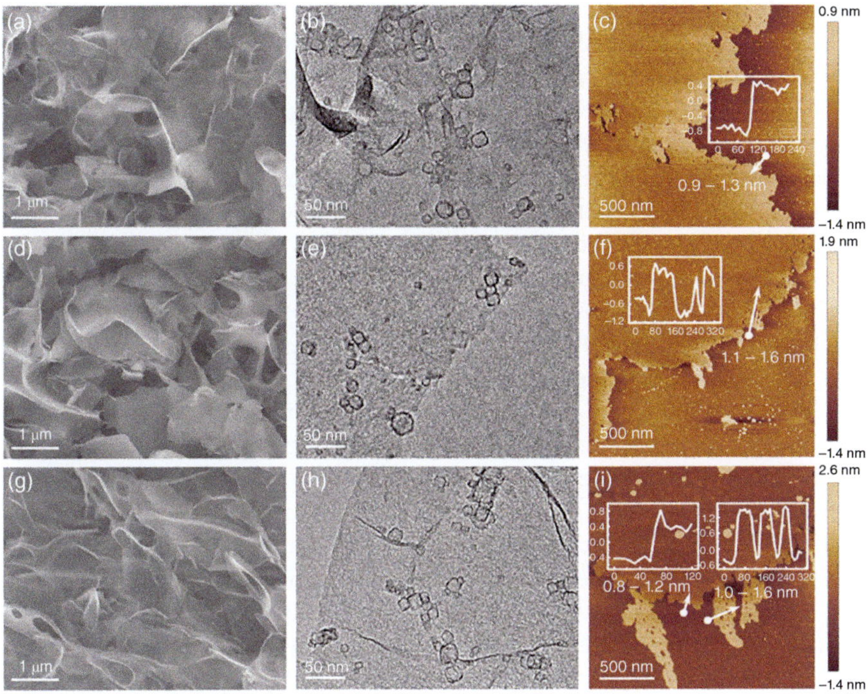

Figure 7.13 Representative SEM and TEM images of (a, b) NCNS, (d, e) H-NCNS, and (g, h) P-NCNS. AFM images of (c) NCNS, (f) H-NCNS, and (i) P-NCNS. Source: Zan et al. [48]/with permission of © Elsevier Ltd.

two-dimensional structure. Further studies have shown that the ice template mainly guides the gelatin precursor to form an ultrathin two-dimensional structure, while KCl template is not only beneficial to keep the two-dimensional structures but also forms a large number of carbon nanocages because of the template of KCl crystals (Figure 7.13b).

7.3.1.4 Three-Dimensional Carbon Materials

Compared to zero-dimensional carbon dots, one-dimensional nanofibers, and two-dimensional nanosheets, there are many obvious advantages for 3D porous carbon materials. The advantages include many exposed catalytically active sites on their large surface, more active sites located on edge compared to the ones on the surface or bulk, excellent electrical conductivity (both in plane and through plane) with a multidimensional conductive network, facilitated electrolyte diffusivity and electrolyte/reactant diffusion process with large volume, and good mechanical strength. From solid-oxide fuel cells with ceramic electrolytes to lithium-ion batteries combining solid-state and liquid diffusion processes, a general goal for electrode design of electrocatalytic energy storage and conversion devices is to maximize the so-called triple-phase boundary size where electrons, reactant, and electrolyte meet, and reaction can occur. In addition, to operate at large current densities and with favorable efficiency in low-temperature fuel cells, electrolyzers, or metal–air batteries, carbon-based electrocatalysts also need a large triple-phase boundary. In conclusion, 3D hierarchical porous network with interconnected porosity and conductive pathways spawned would significantly improve the electrical, structural, and textural characteristics, and promote electron/gas/mass transport in the electrocatalytic process [25].

Typical carbon aerogels are highly porous monoliths and possess large surface area (400–1000 $m^2\ g^{-1}$), controllable pore-size distribution ($\leq$50 nm), and low electrical resistivity ($\leq$40 mΩ cm), which promotes carbon aerogel to become excellent energy storage and conversion materials.

Mao and coworkers proposed a facile method to prepare N- and P-doped carbon aerogels (NPCs) [49]. The NPCs pyrolyzed at different temperatures were named as NPC-T (T was the value of the pyrolysis temperature). The synthesis process, shown in Figure 7.14, involves freeze-drying and direct carbonization of the gelatin and phytic acid gel mixture without an extra activation process. A 3D porous carbon aerogel was obtained after heat treatment.

Figure 7.15a displays the XRD pattern of the NPCs. Two obvious diffraction peaks located at 24° and 44° are ascribed to (002) and (100) planes of carbon, respectively, which indicates the existence of amorphous carbon and graphitic carbon in the NPCs. With the increase of pyrolysis temperature, a rise in the intensities of the (002) and (100) peaks is observed, indicating that the pyrolysis temperature affects the phase structure composition of the catalyst NPCs. Raman spectra of NPCs are displayed in Figure 7.15b. D- and G-bands are observed at approximately 1350 and 1590 cm^{-1}, respectively, suggesting that disordered and graphitic carbon presents in the catalyst materials [35]. As the pyrolysis temperature increases, more defects are generated since the I_D/I_G value, which delivers the graphitization degree and

Figure 7.14 Schematic of the synthesis process of porous carbon frameworks from gelatin and phytic acid. Source: Mao et al. [49]/with Permission of © Elsevier.

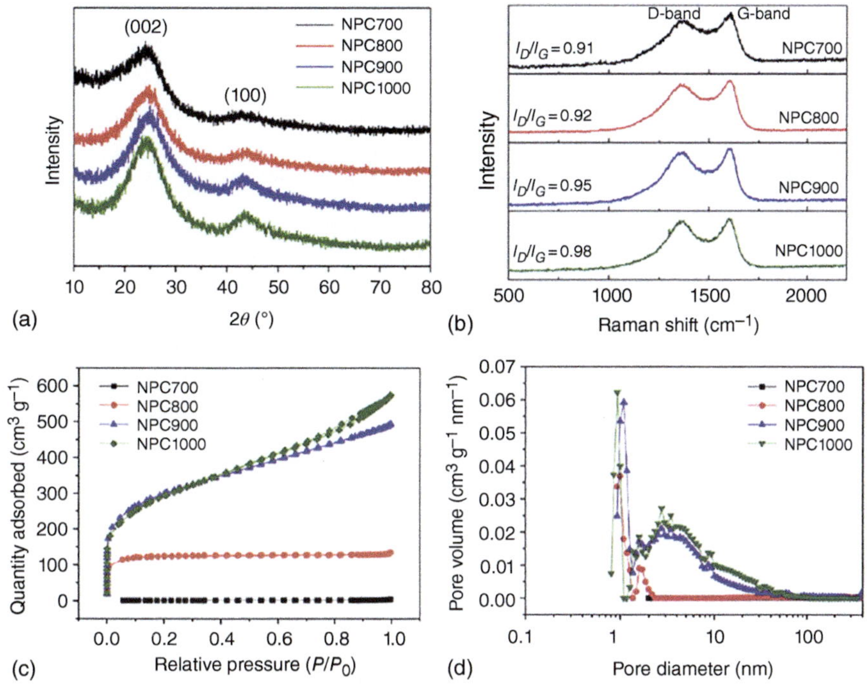

Figure 7.15 XRD patterns (a) and Raman spectra (b) of the NPCs. (c) Nitrogen adsorption/desorption isotherms of the NPCs. (d) Pore-size distributions calculated by the DFT method for the NPCs. Source: Reproduced with permission from Mao et al. [49]; © 2019, Elsevier.

defects information of carbon materials, increases from 0.91 (NPC700) to 0.99 (NPC1000). What's more, the graphitization degree of NPCs is also improved, evidenced by the narrowing half-peak width of D- and G-bands. It is noted that after the P doping, G band shift slightly to a higher wavenumber originated from the increase in the charge density.

Figure 7.15c shows the N_2 adsorption/desorption curves of the NPCs. Both micropores and mesopores can be observed in NPC800, NPC900, and NPC1000, whose adsorption/desorption curves can be attributed as type I. NPC1000 possesses the highest specific surface area of 1056.0 m^2 g^{-1} with an average pore size of 3.37 nm (Figure 7.15d). In contrast, NPC700 possesses the lowest specific surface area and the fewest pores. The pore volume and the specific surface area share the same trend and increase as the pyrolysis temperature rise. Micropores and mesopores coexisting in the hierarchical porous structure offer much more active sites and facilitate mass transfer/diffusion.

Leather wastes (LWs) is a natural polymer material containing >80% collagen. LWs seize unique chemical composition and specific intrinsic structure, e.g. high heteroatom proportion (lots of amino acids) and low ash content, making them good energy storage and conversion precursor material. Liu et al. constructed a carbon aerogel doped with N, O, and S via a one-step activated pyrolysis process (Figure 7.16) [50].

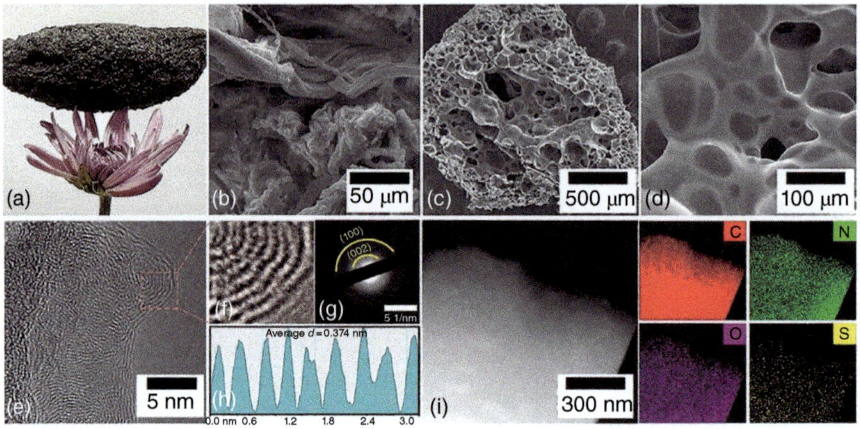

Figure 7.16 (a) Photograph of LWCAs on daisy; (b) SEM image of LWs; (c, d) SEM images of LWCAs; (e) HRTEM image of LWCAs; (f) the magnification of DCNs; (g) SAED pattern for LWCAs; (h) measurement of interlayers distance; (i) the corresponding elemental mappings of LWCAs. Source: Liu et al. [50]/with permission of © Elsevier Ltd.

Figure 7.16a shows that the obtained leather waste carbon aerogels (LWCAs) have a low density. LWs as the precursor is mainly composed of tightly woven natural collagen fibers (Figure 7.16b). Upon carbonization, the unstable organic matter in leather collagen evaporates, leaving voids in the carbon matrix. Thus, hollow vesicle-like cross-linking network structure containing abundant pores presents in LWCA (Figure. 7.16c,d). Layers of distortional carbon nanocrystallites (DCNs) are observed in HRTEM images (Figure 7.16e,f). The selected area electron diffraction (SAED) shows that LWCA is amorphous (Figure 7.16g). The twisted interlayers (Figure. 7.16f–h) indicate these distortions, generating defects and facilitating the capacitive adsorption kinetics. The elemental mapping of the LWCAs suggests heteroatoms N, O, and S are homogeneously distributed, demonstrating that this one-step pyrolysis is efficient in converting LWs into heteroatom-enriched carbon monolith (Figure. 7.16i). Moreover, LWCA displays a large specific surface area (2523 $m^2\ g^{-1}$) with a large number of micropores, which is conducive to store ions.

7.3.2 Porosities

Porous carbons (PCs) are widely used materials with important applications, including energy storage and conversion [10]. The main characteristics of PCs are high surface area and rich porosity [51]. The pores can be divided into micropores, mesopores, and macropores. In accordance with International union of pure and applied chemistry (IUPAC), the diameter of micropores is lower than 2 nm, the diameter of macropores is larger than 50 nm, and mesopores have a pore diameter between that of micropores and macropores [52]. Each type of porous carbon has morphological advantages and unique properties, but their applications are confined by the uniform pore size [53–55]. Hierarchical porous carbon (HPC) breaks through the limitations of the single pore structure of traditional single-stage porous carbon and has drawn widely attention in the last few years. For instance, many studies have demonstrated that HPC is a desirable electrode for supercapacitors because the micropores are storage sites for electrolyte adsorption, while mesopores offer channels for fast diffusion of ions, and macropores reduce the transmission path of ions [56].

Recent researches on collagen-based materials have unlocked their potential applications for the development of high technology. Collagen mainly comprises C, N, O, and S elements. Collagen-derived materials can be converted into collagen-derived PCs by controlled pyrolysis. The natural precursors with specific elemental compositions and ordered biotic structures endow collagen-derived PCs with rich heteroatom doping and unique 3D nanostructures, which have broad application prospects in electrochemical energy storage and conversion. Collagens, e.g. gelatin, bones, and animal skins, have been used as precursors to synthesize PCs [9, 57–59]. Precursors and synthesis methods, especially pyrolysis temperature, have an important influence on the pore structure. To clarify the relationship between precursors and porosity, collagen-rich materials used as PC precursors with porosity of the corresponding PCs are summarized in Table 7.2.

As mentioned above, pore structures differ in PCs produced by different activation processes. Animal derivative gelatin contains a variety of proteins, which possess an average nitrogen content of 16%. Xu et al. have proposed N-doped PC with large surface area using gelatin as carbon precursor [66]. Main steps in this work involve carbonizing gelatin under an inert atmosphere and chemically activating the product using NaOH. The porosity and electrochemistry properties of the product are strongly associated with the weight ratio of NaOH/gelatin-derived carbon.

A series of samples were fabricated with different NaOH/gelatin-derived carbon ratios of 0.5, 1, 1.5, and 2, named as GAC1, GAC2, GAC3, and GAC4, respectively.

Table 7.2 Pore structure of collagen-derived PC.

Sample	Precursor	S_{BET} a) ($m^2\ g^{-1}$)	$S_{micropore}$ b) ($m^2\ g^{-1}$)	V_t c) ($m^3\ g^{-1}$)	$V_{micropore}$ d) ($m^3\ g^{-1}$)	Refs.
PC-1100	Cattle bone	2096	—	1.91	0.084	[22]
FBPC	Fish bone	1517	452	2.53	—	[60]
NHPC–HA/KOH	Cattle bone	2476	273	3.14	0.14	[23]
HPC-550	Fish scale	594	—	0.38	0.21	[61]
HPC-800	Fish scale	3285	—	2.82	0.80	[61]
HPCNS	Cattle bone	890.4	—	0.481	0.289	[62]
PCNS-600	Fish scale	962	880	0.41	0.34	[63]
MPCNS	Fish skin	1017	939	0.422	—	[64]
CCB2	Chicken bone	769	0.35	0.36	0.13	[59]
$CS_{2\text{-}Fe}$	Pig bone	562	0.25	0.2	0.06	[65]

a) Total specific surface area, calculated by the BET method.
b) The specific surface area of the micropore, calculated by the *t*-plot method.
c) Total pore volume, determined at a relative pressure of 0.98.
d) The pore volume of the micropore, calculated by the *t*-plot method.

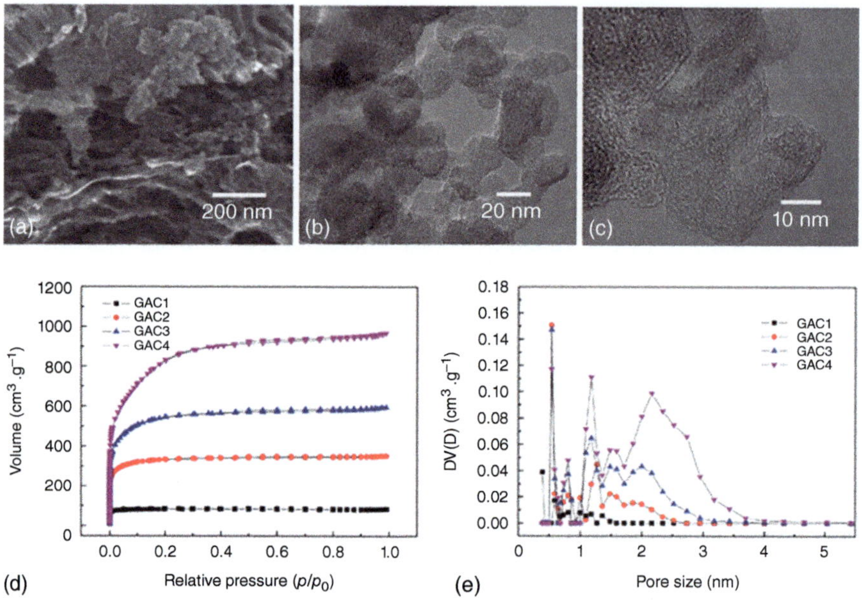

Figure 7.17 FESEM (a), TEM (b) and HR-TEM (c) images of GAC4; nitrogen (77 K) adsorption–desorption isotherms (d) and DFT pore-size distribution (e) of the gelatin-based carbons. Source: Xu et al. [66]/with permission of © The Royal Society of Chemistry.

The morphology and microstructure of NaOH-activated GAC4 are characterized by field emission scanning electron microscopy (FESEM), TEM, and HR-TEM in Figure 7.17. GAC4 possesses a network consisting of compacted carbon particles with interstitial microporosity (Figure 7.17a). Partially aligned graphitic layers are accompanied with amorphous carbon skeleton evidenced by TEM and HRTEM images (Figure 7.17b,c).

The GACs activated at diverse NaOH/char values all show typical type I curves, demonstrating microporous PCs. With the increases of NaOH/char value, the adsorption volume increases dramatically, indicating that the porous structures are more developed; and the wider knee at lower relative pressure suggests wider pores, which demonstrates the key effect of NaOH/char value in chemical activation. High NaOH/char values mean deepened activation and lead to a dramatic increase of BET surface area and pore volume, accompanied by an increase in pore size. Table 7.3 lists the detailed BET surface area and pore volume. The samples prepared at different weight ratios presented surface areas with the range of 323–3012 $m^2\ g^{-1}$, while the nitrogen content varied between 0.88 and 9.26 at%. Figure 7.17d,e presents the N_2 sorption isotherms and pore-size distributions, which indicate that micropores dominate in the prepared samples and the pore size increases as the amounts of activation agent increase.

Zhou and coworkers systematically studied the influence of carbonization temperature on porosity of PCs. They chose carp skin as a carbon precursor to prepare carbon foams (CFs). The preparation process of CFs is illustrated in Figure 7.18.

Table 7.3 Porosity parameters, nitrogen, and content of the gelatin-based carbons in 6 mol l^{-1} KOH aqueous electrolytes [66].

Samples	Elemental analysis N wt%	C wt%	H wt%	S wt%	O wt%	XPS (nitrogen) N at%	N-6 at%	N-5 at%	N-Q at%	N-X at%
CPC-600	10.4	64.7	1.5	0.13	23.25	7.06	1.68	4.32	0	1.06
CPC-700	6.2	84.8	0.5	0.27	8.21	1.05	0.47	0.17	0.23	0.18
CPC-800	2.6	86.8	1.8	0.94	7.90	1.84	0.38	1.29	0.14	0.03
CPC-900	1.6	91.9	1.2	0.87	4.45	0.93	0.32	0.37	0.15	0.09

Source: Reproduced with permission from Xu et al. [66]; © 2012, Royal Society of Chemistry.

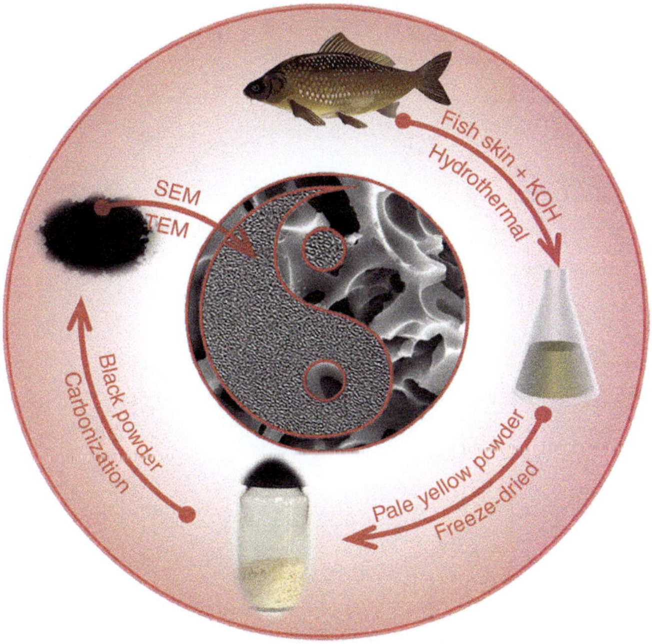

Figure 7.18 Schematic illustration of the preparation process of carbon foams decorated by N and O atoms and the concise characterization of SEM and TEM. Source: Zhou et al. [67]/with permission of © Elsevier Ltd.

Carbon precursor is from the back of grass carp, which is carbonized at different temperatures. The as-prepared products were named as CFs-T, where T represents the carbonization temperature.

The interconnected channels can be observed in Figure 7.19, which are generated by the diffusion of gases (such as H_2O, CO, and CO_2) during the decomposition process of gelatin. The gas production would accelerate as pyrolysis temperature increases, thereby widening the size of the channel. The macropore size of CFs is about 1 μm from SEM image.

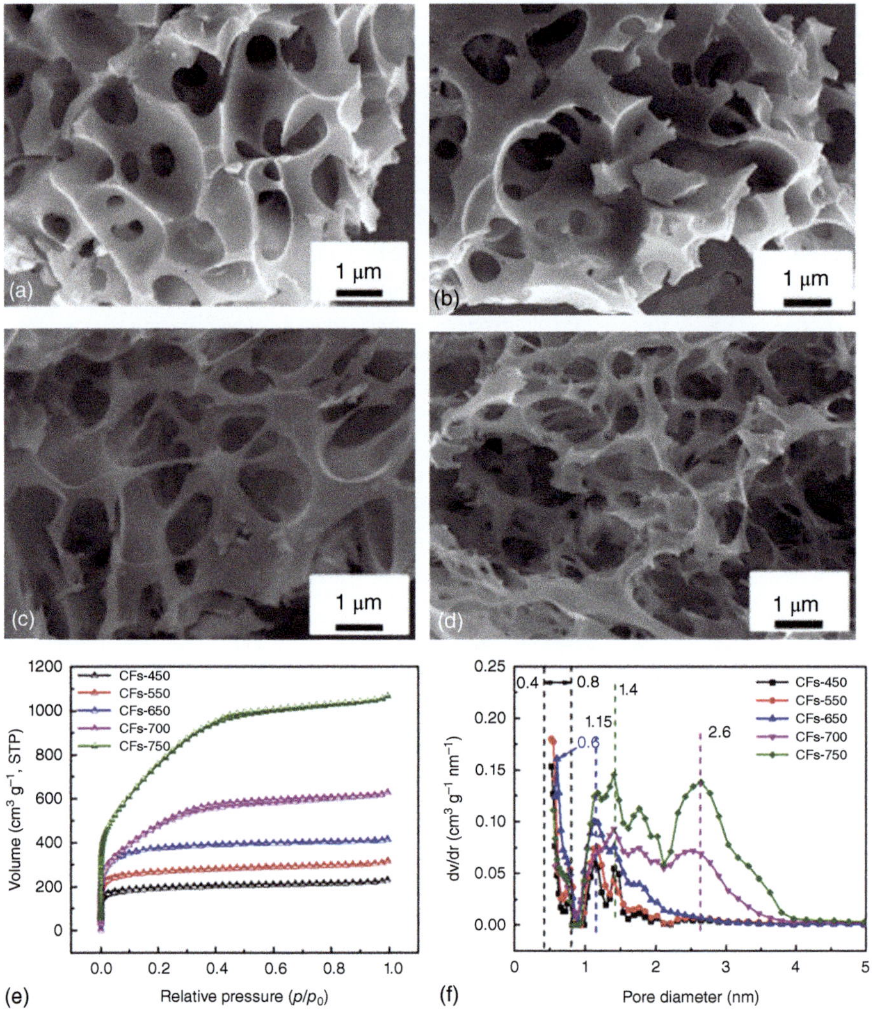

Figure 7.19 SEM images of (a) CFs-450, (b) CF-550, (c) CFs-700 and (d) CFs-750, (e) N_2 adsorption–desorption isotherms, and (f) plots of pore-size distribution of CFs samples. Source: Zhou et al. [67]/with permission of © Elsevier Ltd.

As shown in Figure 7.19, CFs exhibit a non-closed structure. The carbon skeleton has abundant porous channels. These irregularly shaped channels are interconnected to each other to form a high-porosity 3D network. Particularly, with the increase of carbonization temperature, CFs samples show a trend of increasing porosity (Figure 7.19a–d). The BET method was applied to confirm the pore-size distribution of CFs samples. As shown in Figure 7.19e, CFs-450, CFs-550, and CFs-650 present characteristics similar to type I at wide relative pressures (P/P_0) range, which indicates that all three samples have a microporous structure, while CFs-700 and CFs-750 exhibit a typical type IV curve with an obvious hysteresis loop at the P/P_0 range of 0.0–0.4, suggesting the presence of mesoporous structure.

Table 7.4 Specific surface area (SSA) and total pore volume (V_{pore}) of samples calcined at different temperatures [67].

Characteristic	CFs-450	CFs-550	CFs-650	CFs-700	CFs-750
SSA ($m^2\ g^{-1}$)	724.1	1014.2	1369.3	1416.4	2222.7
V_{pore} ($cm^3\ g^{-1}$)	0.3576	0.4927	0.644	0.9715	1.651

Source: Reproduced with permission from Zhou et al. [67]; © 2019, Elsevier.

The pore-size distribution in Figure 7.19f further demonstrates the coexistence of microporous and mesoporous structures in CFs-700 and CFs-750. Table 7.4 lists the detailed values of SSA and pore volume (V_{pore}) of the samples. It is found that the SSA and V_{pore} of the samples increased as the calcination temperature increases.

He et al. employed bovine bones to synthesize a mesoporous N,O-co-doped carbon network via optimizing the pyrolysis method [68]. The authors used SEM and TEM to observe the morphology and structure of the prepared N,O-HPC-x. All samples have a typical 3D porous network with rich mesopores and macropores as shown in Figure 7.20a–d. TEM images show a highly interconnected porous structure, forming ultra-thin sheet-like pore walls (Figure 7.20e–h). The unique porous network facilitates the electrochemical contact between the electrolyte and the electrode, provides fast ion and electron transmission channels, and can enhance the rate capability of the supercapacitor electrode. The nitrogen adsorption/desorption measurement was employed to determine the pore structures of the N,O-HPC-x samples. All the N,O-HPC-x samples display a combination of type I and IV isotherm characteristics (Figure 7.20i). The pore-size distribution plots for all the N,O-HPC-x samples are shown in Figure 7.20j. N,O-HPC-850 has a high mesopore volume of 2.682 $cm^3\ g^{-1}$ involved much more mesopores with the pore size centered at 8 nm than others, thereby enabling a desirable electrode material for supercapacitor with high performance.

Niu et al. used hydroxyapatite/KOH to synergize pyrolysis of cheap bovine bones at 550 °C and fabricated tremella-like hierarchically porous carbon nanosheets (HPCNSs) with a carbon interlayer distance of 0.41 nm [62]. Using bovine bone as a precursor, HPCNS was fabricated by a two-step process of pre-carbonization and carbonization combined with KOH activation. The heat treatment temperature in this work is only 550 °C, which is much lower than previously reported KOH activation method. SEM images show that HPCNS presents a 3D tremella-like structure consisting of two-dimensional nanosheets (Figure 7.21a,b). HPCNS differs from other activated carbons in morphology, primarily retaining the irregular shape of the precursor. According to AFM and TEM images (Figure 7.21c,d), the thickness of HPCNS can be estimated in the range of 10–30 nm. Energy dispersive X-ray spectroscopy (EDX) mapping images (Figure 7.21e–g) show that many N and O atoms are evenly distributed on the carbon nanosheets of HPCNS. HRTEM image (Figure 7.21h) demonstrates that HPCNS is nearly in an amorphous state, but there are few laminated carbon layers. According to the intensity distribution

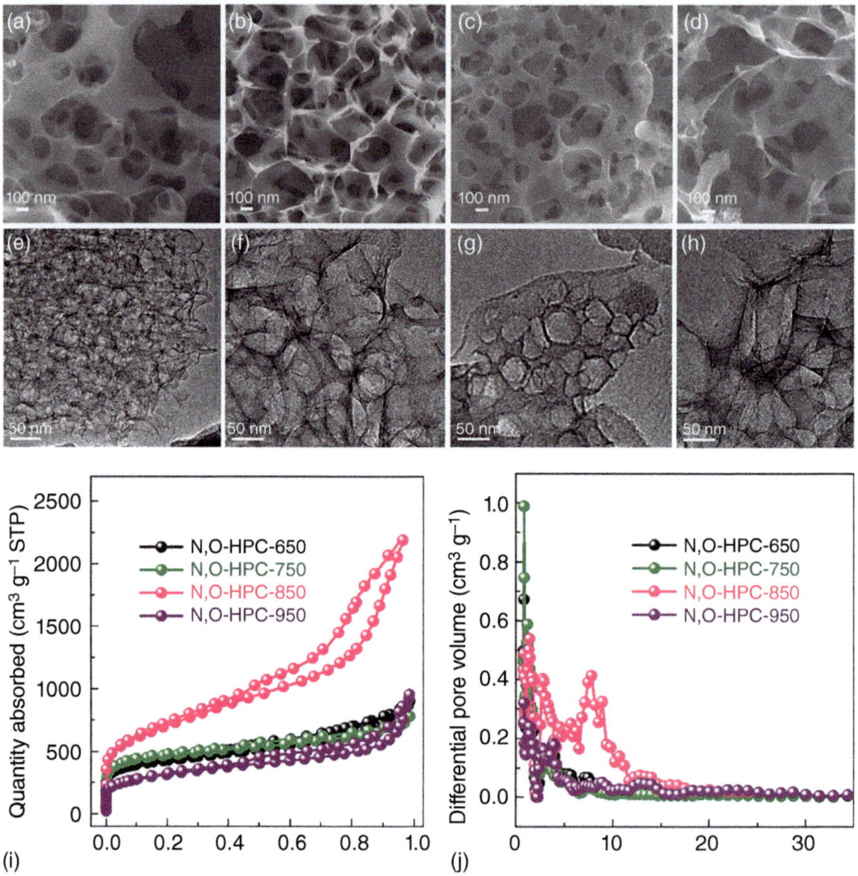

Figure 7.20 SEM images of (a) N,O-HPC-650, (b) N,O-HPC-750, (c) N,O-HPC-850, (d) N,O-HPC-950. TEM images of (e) N,O-HPC-650, (f) N,O-HPC-750, (g) N,O-HPC-850, (h) N,O-HPC-950. (i) N_2 absorption–desorption isotherms, (j) pore-size distributions N,O-HPC-x. Source: He et al. [68]/with permission of © Elsevier Ltd.

of the line scan through the lattice fringes, the carbon interlayer distance of HPCNS is calculated to be ~0.41 nm (Figure 7.21h, inset).

HPCNS exhibits an incorporation of type I and type IV isotherm with a small hysteresis loop at a relative pressure of more than 0.5, showing the co-existence of micropores and mesopores (Figure 7.22a, upper) [69]. The hierarchical porosity of HPCNS is confirmed by pore-size distribution (Figure 7.22b). The SSA of HPCNS was 890.4 $m^2\ g^{-1}$ with 40% of pore volume contributed from mesopore (Table 7.5).

Apart from hierarchically porous carbon nanosheets, hierarchically porous carbon fibers can also be prepared by collagen-rich materials. Liu and coworkers utilized fish bone as a precursor to fabricate fibrous-rich porous carbons [70]. Unlike carbon foams reported in the previous literature, a large number of unwoven and crosslinked carbon nanofibers constitute the entire foam with hierarchical porosity. Figure 7.23a shows the typical manufacturing process. The cleaned tuna bones were

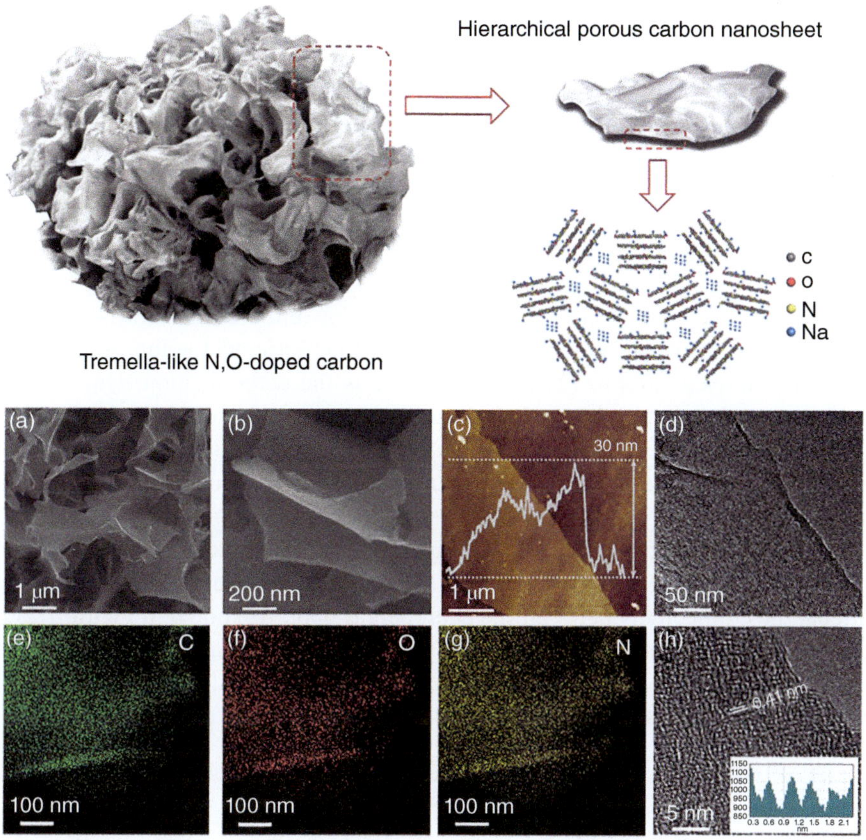

Figure 7.21 (a, b) SEM images, (c) AFM image, (d) TEM image of HPCNS. (e–g) EDX elemental mapping images of HPCNS. (h) HRTEM image of HPCNS (inset: corresponding intensity profile for the line scan across the lattice fringes). Source: Niu et al. [62]/with permission of © Elsevier Ltd.

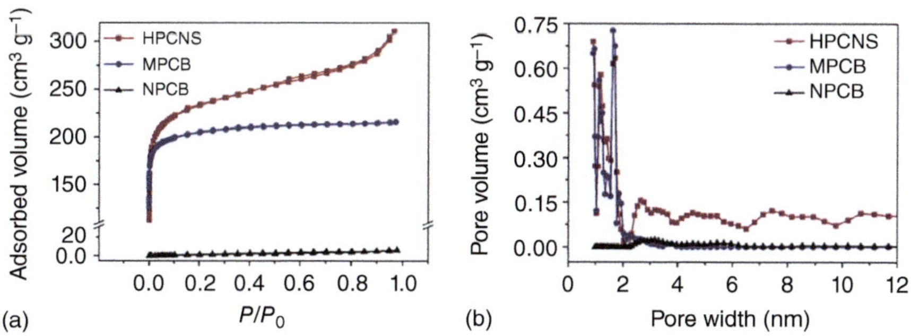

Figure 7.22 (a) Nitrogen adsorption–desorption isotherms and (b) plots of the pore-size distribution for HPCNS, microporous carbon bulk (MPCB), and non-porous carbon bulks (NPCB). Source: Reproduced with permission from Niu et al. [62]; © 2017 Elsevier.

Table 7.5 Pore structural parameters.

Samples	S_{BET} (m^2 g^{-1})	V_t (cm^3 g^{-1})	Pore volume (cm^3 g^{-1})	
			Micropore	Mesopore
NPCB	6.523	0.008	—	0.008
MPCB	810.8	0.304	0.294	0.010
HPCNS	890.4	0.481	0.289	0.192

first heated to 500 °C in N_2 flow for pre-carbonization and heated for carbonization. The inorganic residue was removed with a hydrochloric acid solution to obtain tuna bone-derived carbons (TBCs-T), where T is the carbonization temperature. An electron microscope was used to observe the structural evolution of fish bones during carbonization. As shown in Figure 7.23b, the products have a layered, chamber-like quasi-periodic microstructure separated by lamellae. The resulting chamber is 20–40 μm in size. The enlarged SEM image (Figure 7.23c) reveals the internal structure of the lamella, showing a relatively loose composite of collagen and hydroxyapatite (HAP) nanocrystals. Nevertheless, due to the presence of the lamella, the connectivity between the chambers is poor. As shown in Figure 7.23d, after pre-carbonization, owing to the strong shrinkage of organic mass during the heating process, a dense organic/inorganic hybrid is formed, thereby forming a more ordered porous structure.

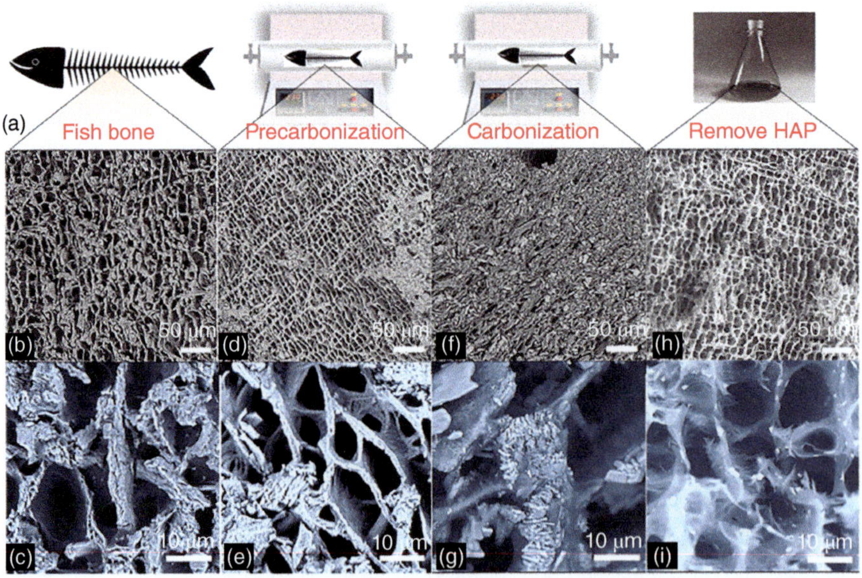

Figure 7.23 (a) Schematic of the fabricating process for the TBC samples. The structural and morphological evolution from (b, c) tuna bones, (d, e) precarbonized sample, (f, g) carbonized samples, and (h, i) the sample after removing HAP. Source: Shan et al. [70]/with permission of © American Chemical Society.

It should be noted that the lamellas between the chambers completely disappear after pre-carbonization, forming an intersecting channel system (Figure 7.23e). Figure 7.23f shows the rapid growth of HAP nanocrystals is induced by the subsequent carbonization treatment, which fills the space of the chamber and decreases the porosity of the sample. Magnified SEM image (Figure 7.23g) shows numbers of large Ca_3PO_4 (CP) crystals surging out of the walls. Fortunately, the carbon scaffold was not destroyed by the growth and accumulation of these crystals. After removing inorganic crystals, foam-like carbon network remains, as shown in Figure 7.23h, indicating that the obtained biocarbons have favorable flexibility and stability. The magnified SEM image (Figure 7.23i) indicates the inner structure of the carbon foam, which has an obvious intersecting porous system.

The microstructure of TBCs was further studied using SEM and TEM. Unlike carbon foam reported in previous works, carbon-fiber network was discovered for the first time to construct the entire foam (Figure 7.24a). As shown in Figure 7.24b, most carbon fibers are isolated and have a homogeneous diameter (~80 nm). In addition, several neighboring carbon fibers adhere to each other to form carbon belts. Both fibers and belts intersect and cross-link each other, generating the walls of the carbon foam. Owing to this novel structural configuration, the prepared TBCs have an interpenetrating channel system like the blood vessels of our body. If the microscale channels in the foamed carbon are regarded as arteries, then the channels in the walls with a diameter of 50–100 nm are regarded as the capillary-like subchannel system. Further investigation by TEM (Figure 7.24c) suggests that the fiber is not solid, which consists of a loose structure with many wrinkles and pores in the size of 10–20 nm.

HRTEM (Figure 7.24d) reveals that carbon fibers contain partially ordered graphene domains, forming pseudographitic arrays. As shown in Figure 7.24e, the curled graphene-like flakes can be found at the end of fibers, which suggests that their general thickness is 2–5 atomic layers. The intensity curve for the line scan across the lattice fringes is revealed by Figure 7.24f, and then the dilated intergraphene spacing is calculated to be about 0.41 nm. Owing to the abundant porosity and graphene-like sheets, the carbon fiber has a novel 1D foam-like porous structure, and the hierarchical structure of porous carbon foam is further constructed by it. As shown in Figure 7.24g, the fish has a porous and hard skeleton. The skeleton of fish bone is composed of mineralized fibrous organic-based tissue with a hierarchical structure, which is similar to many marine biomasses. As basic units, collagen molecules are wrapped by protein in the form of long and narrow crystalline units to form nanofibers. Clusters of these nanofibrils are arrayed in parallel to form the long collagen-based fibers, while the intervals among the fiber clusters are filled up with biominerals, which mainly exist in the form of HAP nanocrystals. After calcination, the organic-based molecules are pyrolyzed and adhere to each other to form carbon fibers with a large amount of HAP nanocrystals. Due to the favorable thermal stability of HAP, these inorganic nanocrystals give the fiber more superior rigidity and support the entire porous skeleton during heating process. When removing them, the mesopores are retained in carbon fibers. Consequently, these porous carbon fibers are entangled and connected to each other to generate the foam-like structure mentioned above.

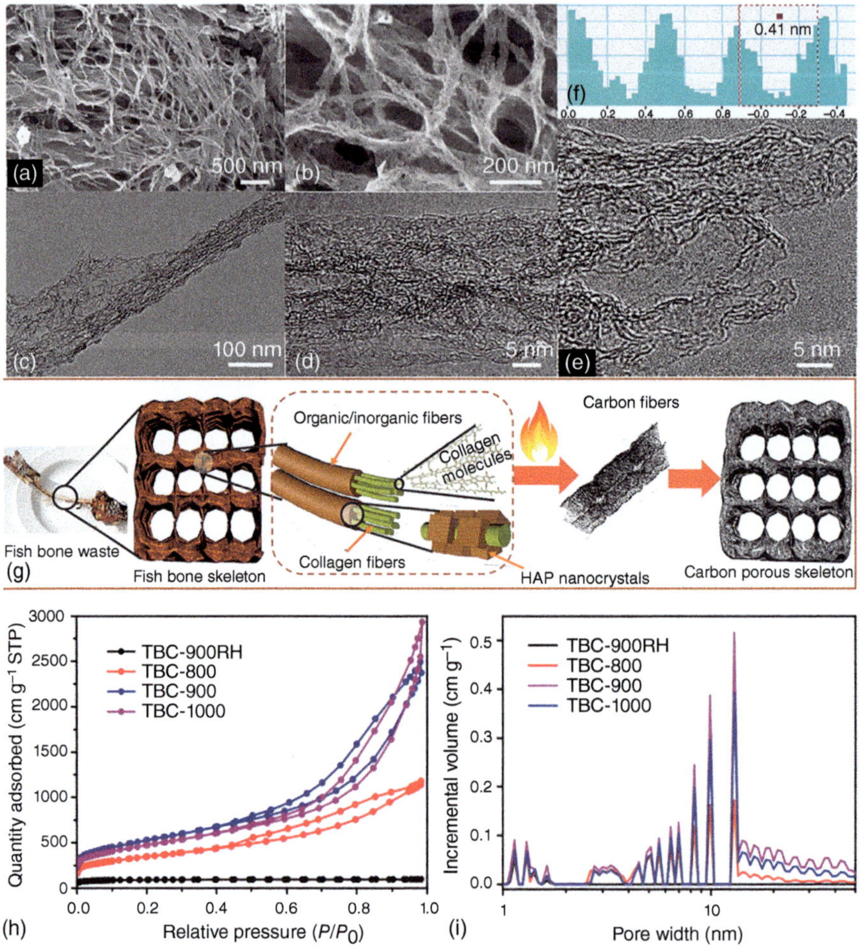

Figure 7.24 (a, b) SEM images of the TBC samples. (c) TEM image of one single carbon fiber of TBCs. (d, e) HRTEM images of the TBC samples. (f) The intensity profile for the line scan across the lattice fringes. (g) Schematic of the microstructural change from fish bones to TBCs. (h) Nitrogen adsorption–desorption isotherms; (i) the pore-size distribution. Source: Shan et al. [70]/with permission of © American Chemical Society.

Figure 7.24h shows the isotherms curves of samples carbonized at different temperatures. All samples present a distinct rise of the N_2 isotherm at the low-pressure range ($P/P_0 < 0.01$), indicating the characteristic of micropore filling. A distinct hysteresis loop is found at 0.5–0.8 P/P_0, confirming the presence of abundant mesopores. The isotherms of TBC-900 and TBC-1000 samples show an obvious increase of the adsorption at high relative pressures ($P/P_0 > 0.9$), indicating the formation of macropores in the carbon. For comparison, the sample named as TBC-900RH is fabricated at the same circumstances except for removing HAP at the beginning. The TBC-900RH sample shows a quite low porous structure, which indicates that HAP nanocrystals have a significant effect on the formation of the porous architecture of TBCs. The measurement results show TBC-900 and TBC-1000 have high specific surface area (1468 and 1670 $cm^2\ g^{-1}$) and pore volume (2.98 and 3.86 $cm^3\ g^{-1}$), respectively, much larger than those of TBC-800 (1209 $cm^2\ g^{-1}$ and 1.69 $cm^3\ g^{-1}$) and TBC-900RH (285 $cm^2\ g^{-1}$ and 0.13 $cm^3\ g^{-1}$). Figure 7.24i shows the pore volume is mainly attributed to the mesopores, which is beneficial from the existence of HAP nanocrystals. In addition, for TBC-900 and TBC-1000, the ratio of pore volume (more than 20 nm) increased, which is perhaps ascribed to the fast growth of these nanocrystals to macroscale crystals as the heating temperature rises, hence generating more macropores in carbon matrix.

In addition, N-doped highly porous carbon material with hierarchical porous structure was synthesized via carbonizing pig bones [65]. Apart from animal bones, other collagen-rich materials are widely used as precursors of porous carbon as well. N.V. Salim et al. reported that collagen could be used as a precursor for renewable fiber formation to produce porous carbon fibers [71].

Obviously, carbon materials with customized pore structures have shown excellent performance in specific energy storage/conversion applications [72, 73]. However, the fundamental relationship between pore structure and performance remains undefined and needs further study. Given the diversity of applications, a common perspective is necessary to depict the role of pore structure. Specifically, in the pore structure that provides channels, spaces, and interfaces, the spaces and channels are crucial for the storage and transportation of electroactive species, as well as the occurrence of electrochemical reactions and electrostatic adsorption on the interface. As for electrochemical energy storage and conversion, the role of the pore structure varies with specific application requirements. A well-made pore structure with rich micropores and mesopores can assist in constructing a larger accessible interface, thereby obtaining more active sites and higher capacitance/capacity, while the hierarchical pore structure with consecutive channels will facilitate the mass transfer of electroactive species. Thus, a comprehensive understanding of the modulation of the pore structures and the functioning mechanism of performance will contribute to realizing more excellent electrochemical performance and thorough comprehension of the specific structure–property relationship [74].

7.3.3 Heteroatoms and Defects

Carbon materials with collagen as the precursor have unique electronic and chemical structures and rich surface feature. The large amounts of N, S, and P in collagen molecules make the collagen-derived carbon materials rich in heteroatoms and defects. This kind of carbon material has great advantages in energy storage and conversion [75, 76].

7.3.3.1 Types and Functions of Defects

Introducing defects is one of the important means of material modification. Defect modification can directly change the chemical and physical structure, thereby improving the intercalation ability of materials and greatly improving the properties related to energy storage and conversion. Defects directly affect the energy conversion process of materials by changing thermodynamics and improving kinetics. The existence of defects increases the energy and entropy of the system, and they can be used as active sites for electrochemical phase changes. It can be known from the surface reaction limit kinetic calculation that the phase boundary extends from surface to surface along planes of rapid ionic diffusion in which facet edges or defects can serve as nucleation sites. The same result was obtained in the experiment. Defects accelerate the phase change process by lowering the energy barrier and changing the material transport path [77]. Defects may also decrease the stress and electrostatic repulsion between neighboring oxygen layers, directly altering the migration energy and diffusion barriers that alkali ions need to overcome during intercalation.

The essential defect is the defect introduced at the same time in the actual preparation process of the material, including point defects, line defects, and Stone–Wales defect, and so on. In addition, to produce intrinsic defects, the properties of two-dimensional materials are often adjusted by artificially introducing some external defects during the experiment, mainly including displacement doping, adsorbing atoms, adsorbing molecules, etc. [78]. Sometimes, introducing defects does not directly affect the performance of the material. This chapter mainly introduces the defects induced by heteroatom doping.

7.3.3.2 Types and Functions of Heteroatoms

Figure 7.25a illustrates different types of heteroatom-doped carbon [75]. Nitrogen (N) is the most commonly used and most effective heteroatom in the doping strategy of carbon materials (Figure 7.25b). The enhancement of the performance for carbon materials by N atoms is based on the following reasons: (i) N and C are neighboring elements in the same period, which can adjust the number of electrons in carbon materials; (ii) N as electron donor provides more electrons to the delocalized carbon network, resulting in the increased electronic conductivity; (iii) the atomic radius of N is close to that of C, which decreases lattice mismatching; (iv) the configuration of C—N bond works similar to the n-type semiconductor materials that can be applied for different electronic applications [75]. N atoms in carbon materials prepared from nitrogen-containing biomass precursors are uniformly distributed. Generally, N atoms can form four different chemical bonds in the sp^2 grid of C atoms, namely pyridine-N, pyrrolic/pyridone-N, graphitic/quaternary-N, and oxidized pyridine-N. Pyridine-N is bonded to two sp^2 carbon atoms in the six-membered ring and offers electron lone pairs to the carbon framework. Pyrrolic-N is similar to pyridine, while it bonds to the sp^2 carbon in five-membered carbon ring with hydroxyl group. Graphitic/quaternary-N replaces sp^2 carbon atom with all three valence electrons bonded to neighboring carbon atoms with three σ-bonds by which, one electron of N fills the π-state of carbon, and the other electron enters the conduction band (π^*-state), leading to a strong N-doping effect.

The introduction of nitrogen into the carbon material helps to enhance the performance during the electrochemical reaction. Cheng et al. researched the involvement of N doping in redox reactions during the electrochemical processes both in experiment and theory. They found through calculations that oxidized pyridine-N with

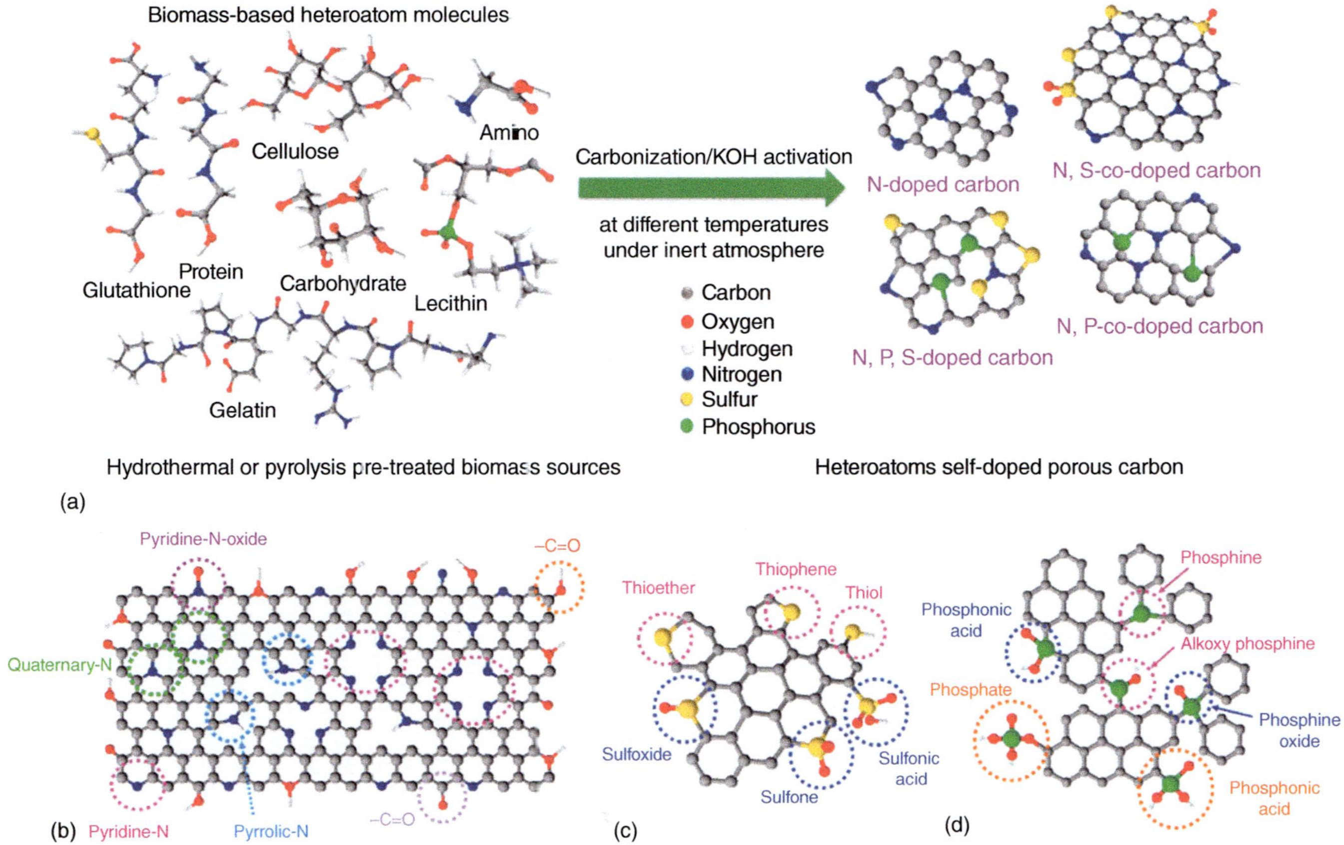

Figure 7.25 (a) Schematic illustration of heteroatoms doped carbon; types of (b) nitrogen, (c) sulfur, and (d) phosphorus functional groups in the carbon matrix. Source: Reproduced with permission from Gopalakrishnan and Badhulika [75]; © 2020, Elsevier.

strong electron affinity and positively charged pyridine-N produced a redox pair. This clear mechanism is found between pyridine-N and pyrrole-N as well, which enhances the storage of H atoms and promotes the attachment of protons.

The introduction of sulfur (S) atoms with a radius much larger than that of carbon will cause the destruction of the sp^2 planar structure and produce defects, which serve as the active point of the redox reaction [79]. S-doping into carbon architecture leads to the opening of bandgap and inclusion of localized states leading to electronegative diversities via inducing polarization effect. In addition, S-doping is able to significantly change the spin/charge density distribution and distort the carbon lattice via generating larger electroactive sites. As shown in Figure 7.25c, the introduction of sulfur atoms into the carbon sp^2 plane can form a variety of sulfur-containing chemical bonds, e.g. thiol, sulfonic acid, thiophene (aromatic sulfur), sulfone, and sulfoxide.

Zhao et al. used mesoporous carbon element to study the effect of doping S atoms in the carbon frame on the capacitor performance [80]. They attributed the performance improvement to (i) sulfur atoms are rich in valence electrons, which can be activated synergistically with carbon atoms to produce some positions with higher electron cloud density on the carbon sp^2-conjugatied surface; (ii) sulfone and sulfoxide substances can undergo a variety of redox Faraday reactions. The Faraday reaction includes the following steps: the sulfone group is reduced to a sulfoxide group, the sulfoxide group is reversibly reduced to sulfinic acid, and the sulfinic acid ionizes and absorbs protons.

Phosphorus is an n-type doping element of the same main group as nitrogen, and it has the same doping characteristics. However, the atomic radius of P is larger than that of N (0.070 nm) and S (0.104 nm), which can increase the interlayer spacing of carbon materials. Most of the doped P atoms in the carbon material form P–O groups and can have a higher doping amount, forming a thin layer on the surface of the carbon grid, hence enhancing the electrochemical performance. P atoms are sp^3 hybridized to form bonds, leading to distortion and open edges of the sp^2-conjugated sheet of C, offering many active sites for energy storage and transformation. It is found that doping heteroatom with larger radius into a carbon framework leads to a larger carbon–heteroatom bond distance protruding out of the sheet with ripple effect and lifting of neighboring carbon atoms [81]. This modification/deviation of the sp^2 carbon means the reactivity of these atoms with larger bandgaps.

The process of doping P into the carbon material first forms an unstable reduced state. Then P is gradually oxidized by oxygen groups and humidity to form a variety of P-containing oxidized functional groups, as shown in Figure 7.25d. These functional groups include phosphine, alkoxyphosphine, phosphonic acid, phosphine oxide, and phosphate. Therefore, charge delocalization/asymmetric spin density is enhanced due to the presence of P, thereby improving electrochemical properties.

7.3.3.3 Collagen-Derived Carbons with Heteroatoms and Defects

Currently, there are two major methods for doping heteroatoms in carbon materials. One is "in situ doping," that is, selecting precursors containing heteroatoms, e.g.

polymers, biomass, and N-containing organics. The other is "after treatment doping," that is, reagents containing heteroatoms such as urea, ammonia, and amines are added during the preparation process. "In situ doping" can provide a higher doping amount and introduce heteroatoms into the sp^2-conjugated ring of carbon. "After treatment doping" can only introduce surface functional groups, so the amount of doping is small. The content of heteroatoms is impacted by factors such as preparation temperature and active agent. Therefore, choosing an appropriate activation temperature and appropriate raw materials to maintain an appropriate amount of doping is the most critical issue that needs to be resolved. In the process of in situ doping, biomass is a good precursor. These resources are abundant and environment friendly, and are ideal raw materials for preparing doped carbon. Biomass can be divided into animal, plant, microbial, and other types. This chapter focuses on the application of N-doped carbon materials from collagen in energy storage and conversion.

Lee et al. synthesized N-doped porous carbons from collagen. Through vacuum heating at different temperatures (400–800 °C), highly porous N-doped carbon was prepared from paraformaldehyde cross-linked collagen [82]. It is shown that the carbonization temperature is an important factor affecting in situ doping. Based on Figure 7.26, Table 7.6 summarizes the total N content and the amounts of their individual N-doped structures. From these data, three main characteristics of N doping can be summarized. Firstly, the total N content reduces with the increase of carbonization temperature. At the same time, different types of nitrogen-doped structures are gradually lost as the temperature increases. Secondly, the N atoms of pyridine-N and pyrrole-N at the edge of the sp^2-conjugated sheet are relatively unstable. At higher carbonization temperatures, these N atoms are removed or

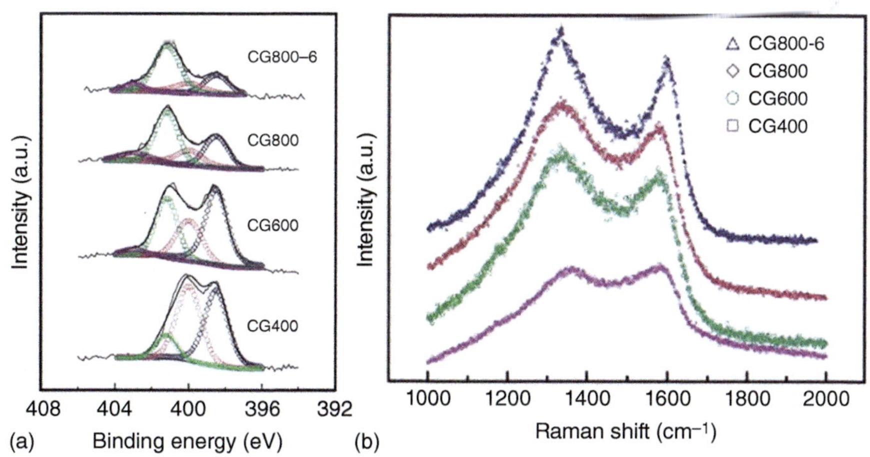

Figure 7.26 (a) XPS core-level spectra of N 1s for CG400, CG600, CG800, and CG800-6. The black line is the raw spectrum; the gray, pink, green, and purple line correspond to pyridine-N, pyrrolic-N, quaternary-N, and pyridine-N-oxide, respectively. (b) Raman spectra of CG400, CG600, CG800, and CG800-6 obtained at 633 nm excitation. Source: Reproduced with permission from Lee et al. [82]; © 2012, Elsevier.

Table 7.6 Deconvolution results (at%) for the XPS core-level spectra of N 1s and O 1s for four collagen-derived carbons [82].

	S_{BET} ($m^2\ g^{-1}$)	Intercept at Y axis ($cc\ g^{-1}$)	V (mesopore) ($cc\ g^{-1}$)	V (micropore) ($cc\ g^{-1}$)
CG400	12	0.5	0.04	0
CG600	34	1.3	0.18	0
CG800	128	27.4	0.24	0.04
CG800-6	681	167.4	0.32	0.3

Source: Reproduced with permission from Lee et al. [82]; © 2012, Elsevier.

surrounded by neighboring carbon atoms, forming a more stable pentaerythritol. Hence, the amounts of pyridinic-N and pyrrolic-N reduce obviously, particularly for the carbonization temperatures above 600 °C. However, the amount of quaternary-N remains nearly the same. Thirdly, beneficial from abundant carboxyl groups in the precursor, the carbonized material retains oxygen-containing functional groups. When the carbonization temperature rises, the oxygen content reduces rapidly.

In addition to N doping, the crystallinity of carbon graphite is also one of the important factors affecting material properties. The graphitization degree of collagen-derived carbon materials was studied by Raman spectroscopy. The Raman spectrum of carbon materials has two main peaks, namely D and G bands around 1355 and 1580 cm^{-1}, respectively. As shown in Figure 7.3b, for collagen-derived carbons, the existence of D band indicates that there are disordered graphitic structures and edge defects in these N-doped porous carbons. The D and G bands of CG400 have large half-peak widths and low peaks, indicating that when the carbonization temperature of collagen is lower than 400 °C, graphite crystallinity is very poor. However, these two peaks become sharper when the carbonization temperature rises. The increase of the G-band strength shows an increase in the graphite structure and the removal of oxygen-containing functional groups on the surface. However, more graphitized structures formed in small domains will increase the graphite defects, which form at the edges of these graphite domains, thereby significantly increasing the intensity of the D band. Therefore, these two results demonstrate that the overall graphite crystallinity of collagen-derived carbon is improved at a higher carbonization temperature. All in all, higher carbonization temperature produces better porosity and facilitates the formation of quaternary ammonium salts and graphite structures in the carbon matrix, thereby increasing the crystallinity of these N-doped porous carbons. It can be seen from the material analysis that different degrees of porosity, N doping, and graphite crystallinity are able to be simply controlled via changing the carbonization temperature and time of collagen.

Yang and coworkers synthesized N-doped porous carbon materials via carbonizing a mixture of pigskin and $CaCO_3$ at different temperatures without further activation [83]. Pigskin consists of approximately 47.6 wt% carbon, 31.7 wt% oxygen, 14.1 wt% nitrogen, 6.38 wt % hydrogen, and 0.16 wt% sulfur. Thus, the carbon material prepared by using pigskin as a precursor can have a good doping structure.

Table 7.2 shows the element distribution of the material at different carbonization temperatures.

Carbon content of collagen derived porous carbon (CPC) increases continuously from 64.7 to 91.9 wt%, but the contents of nitrogen and oxygen reduce (N: 10.4–1.6 wt%, and O: 23.3–4.5 wt%) as the carbonization temperature increases. The nitrogen content drops remarkably to 6.2 wt% at 700 °C and retains 1.6 wt% at 900 °C. And oxygen content reduces to 8.2 wt% at 700 °C. CPC possesses a great N/O retention rate at 600 °C. XPS (Figure 7.27) measurement is employed to study

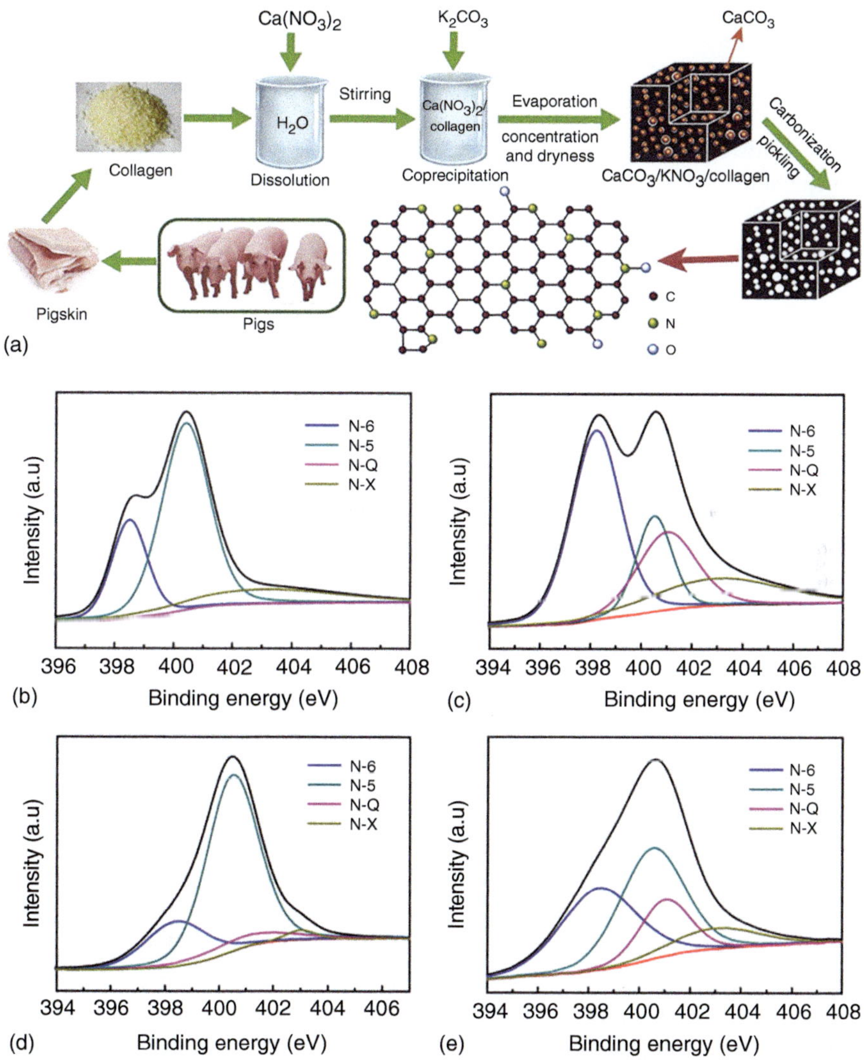

Figure 7.27 (a) The schematic diagram of preparation of pigskin-derived activated carbon and 2N 1s core level spectra of the sample carbonized at 500 °C and the corresponding different temperature activated samples, respectively. (b) CPC-600, (c) CPC-700, (d) CPC-800, (e) CPC-900. Source: Gao et al. [83]/with permission of © Elsevier Ltd.

Table 7.7 Elementary composition evaluated from elemental analysis and relative contents of functional groups in N 1s peaks from XPS spectra of CPCs [83].

	Elemental analysis					XPS (nitrogen)				
	N	C	H	S	O	N	N-6	N-5	N-Q	N-X
Samples	wt%	wt%	wt%	wt%	wt%	at%	at%	at%	at%	at%
CPC-600	10.4	64.7	1.5	0.13	23.25	7.06	1.68	4.32	0	1.06
CPC-700	6.2	84.8	0.5	0.27	8.21	1.05	0.47	0.17	0.23	0.18
CPC-800	2.6	86.8	1.8	0.94	7.90	1.84	0.38	1.29	0.14	0.03
CPC-900	1.6	91.9	1.2	0.87	4.45	0.93	0.32	0.37	0.15	0.09

Source: Reproduced with permission from Gao et al. [83]; © 2017, Elsevier.

the type of nitrogen doping. Through the peak splitting of N 1s XPS, four different nitrogen chemical bonds can be obtained, which represent pyridine N (N-6), pyrrole or pyridine N (N-5), quaternary ammonium N (NQ), and oxide N (NX). Through integral calculation, the nitrogen content of pyridine and pyrrole of CPC-600 is relatively high, and the nitrogen content on the surface of CPC decreases greatly as the carbonization temperature increases (Table 7.7).

Apart from the in situ doping of N heteroatoms from collagen precursors, collagen-derived carbons can be doped by S and P heteroatoms using additional dopants. The carbon network doped with multiple elements can produce a synergistic coupling effect between heteroatoms to improve physical or chemical properties.

It was reported that coordinated effect of N,S-co-doping, coexisting in adjacent carbon hexagonal rings, is able to increase the doping level. N doping enhances conductivity, and S expands the interlayer distance and offers active electrochemical points for pseudocapacitance. Through density function theory (DFT) research, Zhou et al. found that carbon co-doped with N and S exhibited greater performance with LiSH molecules than single N and S doping because of higher binding energy (E_b) [84]. As compared to undoped graphene, the E_b at thionic S site increases and thiophenic S site reduces (Figure 7.28a), while E_b of LiSH at pyridinic N and pyrrolic N site increases with strong interaction (Figure 7.28b). As shown in Figure 7.28c, in N,S-co-doped graphene, LiSH molecules have greater binding energy with nearby thio-S and pyridine or pyrroline N. Therefore, since the electronegativity of N and S atoms is stronger than that of C atoms, they act as electron attractors and cause the polarization of adjacent C atoms and C–O groups to make the surface hydrophilic. Consequently, the double-doped graphene surpasses the single-doped graphene, showing the synergistic effect proved by the DFT calculation. Figure 7.28d shows the possible locations for heteroatoms of O, N, and S incorporation into the carbon skeleton. As an example, Figure 7.28e illustrates the synthesis of O, N, S-doped microporous carbon nanosheets (MPCNSs) with the assistant of KOH from fish skin and its mechanism of energy storage and mass transfer in MPCNS electrode.

Aquatic animals contain a lot of nitrogen and sulfur, elements, which are ideal co-doping precursors. For instance, fish scales containing lecithin and protein as

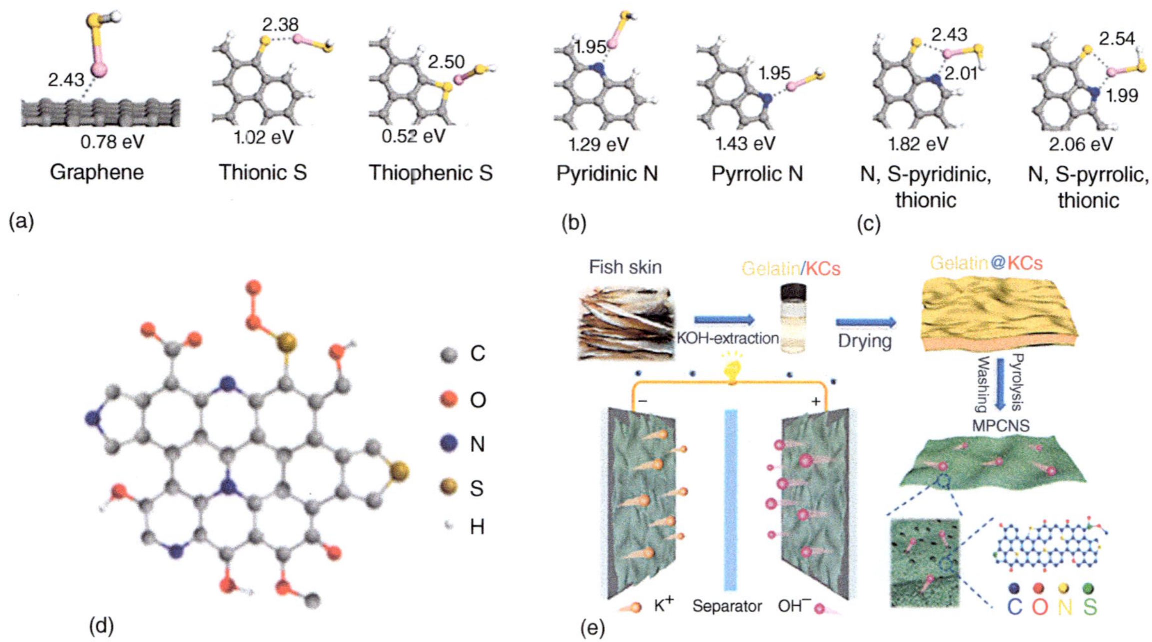

Figure 7.28 (a) Model of binding energy of undoped and S-doped graphene, (b) N-doped graphene, (c) N,S-co-doped graphene with LiSH interaction; (d) possible locations for heteroatoms of O, N, and S incorporation into the carbon skeleton; (e) schematic illustration of the synthesis of O,N,S-doped microporous carbon nanosheets (MPCNSs) with the assistance of KOH from fish skin and its mechanism of energy storage and mass transfer in MPCNS electrode. Source: Gopalakrishnan and Badhulika [75]/with permission of © Elsevier Ltd.

sources of N and P were applied to synthesize porous carbon with a lamellar structure co-doped with N (6.88 at%) and P (1.12 at%) [85]. The P doping generates a protective layer at the carbon surface, which avoids the formation of unstable surface oxygen groups. Apart from scales, fish skins containing collagen (abundant in C, O, N, and S) are applied to synthesize N,S-co-doped (N: 8.18% and S: 2.25%) porous carbon [64]. Swimming bladder containing amino acids (e.g. glycine, proline, and hydroxyproline) is a by-product of the fish industry, which are applied to develop N,S-co-doped porous carbon [86].

Qu et al. prepared highly ordered mesoporous carbon with abundant N atoms (HNMC) by employing SBA-15 as the template and gelatin as the precursor (Figure 7.29) [87]. HNMC was mixed with sublimed sulfur. After that, the mixture was subjected to heat treatment to 155 °C for 20 hours in an argon atmosphere to obtain the S-HNMC composite material. The mapping analysis shows that sulfur was evenly distributed in the mesopores of HNMC. HNMC showed a rod-like morphological structure. They were stacked together in an orderly manner, similar to ordered mesoporous silica. The special highly ordered mesoporous structure of HNMC can promote the even dispersion of sulfur in the carbon base and is conducive to the transmission of electrolyte ions.

The doping of P element can provide a pair of lone electrons, which improves the electronegativity of carbon and enhances its hydrophilic property. Theoretical studies point out that compared with other heteroatom doping, the defects produced

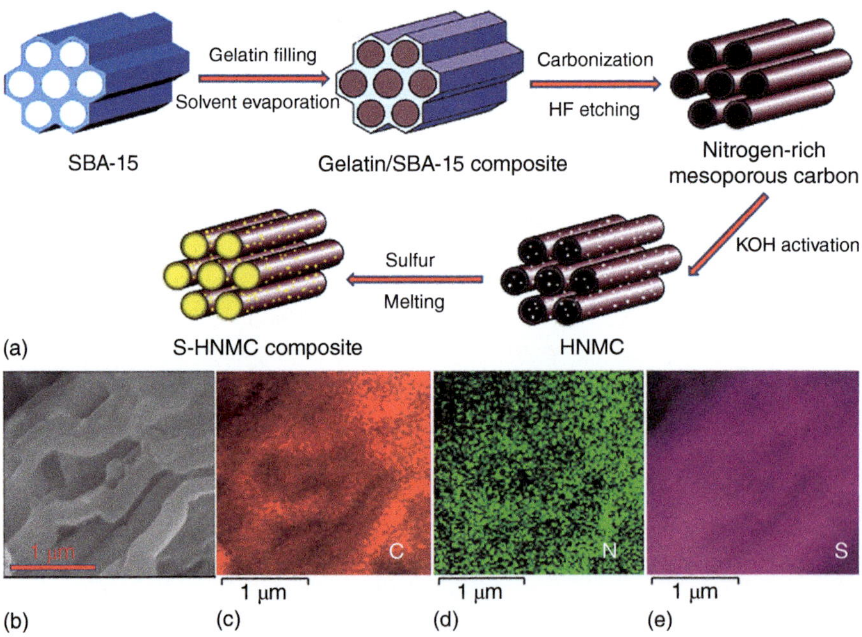

Figure 7.29 Schematic illustration of the synthesis procedure of the S-HNMC composite. SEM image of a selected region of the S-HNMC (53.3 wt% sulfur) composite (b), and the corresponding elemental maps of (c) carbon, (d) nitrogen, and (e) sulfur, respectively. Source: Qu et al. [87]/with permission of © Elsevier Ltd.

by P–N co-doping cause a highly localized state that is intriguing to electrochemical redox reactions.

The N- and P-co-doped 3D hierarchically porous carbon electrocatalysts with N-doped hierarchical porous carbon (NHPC) extracted from bovine bone were developed by Zan et al. via carbonization of phytic acid (PA) and dicyandiamide (DCDA) [88]. The mixture of PA, DCDA, and NHPC is co-pyrolyzed to obtain a 3D hierarchical porous carbon network doped with N and P, wherein NHPC from bovine bone is used as the carbon support, and DCDA and PA are used as the precursors of N and P, respectively. The FE–SEM and TEM images of the synthesized N,P-HPC electrocatalyst show that it maintains the 3D layered porous network structure of NHPC with interconnected mesopores and macropores. (Figure 7.30a–c,e). In addition, compared with NHPC, N,P-HPC shows a slightly larger pore size and more surface defects, which may be induced by the etching effect of volatile products produced by the decomposition of PA and DCDA. It is worth noting that with the decrease of Nextra/P molar ratio (e.g. the content of PA increases or DCDA decreases), the microstructure gradually deviates from NHPC, and more distinct pore collapse appears, indicating that PA has caused a more serious etching effect. The TEM elemental mapping images indicate that the N and P heteroatoms of N,P-HPC are evenly dispersed on the carbon network. Compared with NHPC, the intensity of N is quite high (Figure 7.30d,f), indicating abundant active sites. The dots may be a relatively uniform distribution formed via the co-catalysis of PA, DCDA, and NHPC.

The XRD patterns of four samples (Figure 7.31a) are almost similar, with a wide peak at about 22.3°–25.6°, corresponding to the (002) diffraction surface of graphite, showing that the carbon in the obtained sample is partially crystalline. The Raman spectra are divided into four peaks, namely I peak (1180 cm^{-1}), D peak (1350 cm^{-1}), D'' peak (1470 cm^{-1}), and G peak (1560 cm^{-1}). Among them, I peak can be ascribed to the heteroatom defects. D peak (1350 cm^{-1}) and D'' peak (1470 cm^{-1}) correspond to disordered edges of the graphene layer and stacking defects, respectively. G peak results from sp^2-hybridized carbon vibration (Figure 7.31b). The P-HPC and N,P-HPC exhibit the higher peak intensity ratio of the D peak to the G peak (I_D/I_G) than N-HPC and NHPC (Figure. 7.31b). This means that the incorporation of P may promote to form edge defects on the porous carbon network.

The XPS spectrum (Figure 7.31c) of N,P-HPC displays C (76.0 at%), N (3.2 at%), P (4.0 at%), and O (16.8 at%) species, confirming the existence of N and P heteroatoms within the porous carbon networks. Four types of N can be obtained by deconvoluting the high-resolution N 1s spectrum of N,P-HPC (Figure 7.31d), namely pyridinic N (399.0 eV), pyrrolic N (400.3 eV), graphitic N (401.1 eV), and oxidized pyridinic N (402.4 eV). It is reported that by changing the chemical/electronic environment of adjacent carbon atoms, pyridyl nitrogen makes a great contribution to oxygen reduction reaction (ORR) activity [19]. By calculation, the pyridyl nitrogen content of N,P-HPC (1.8 at%, 55.7% of total nitrogen) is higher than that of NHPC (1.1 at%, 43.7% of total nitrogen), which may be due to the addition of additional nitrogen precursor (DCDA). The high-resolution P 2p spectrum of the N,P-HPC is deconvoluted into two peaks at 132.9 and 134.1 eV, corresponding to P–C and P–O,

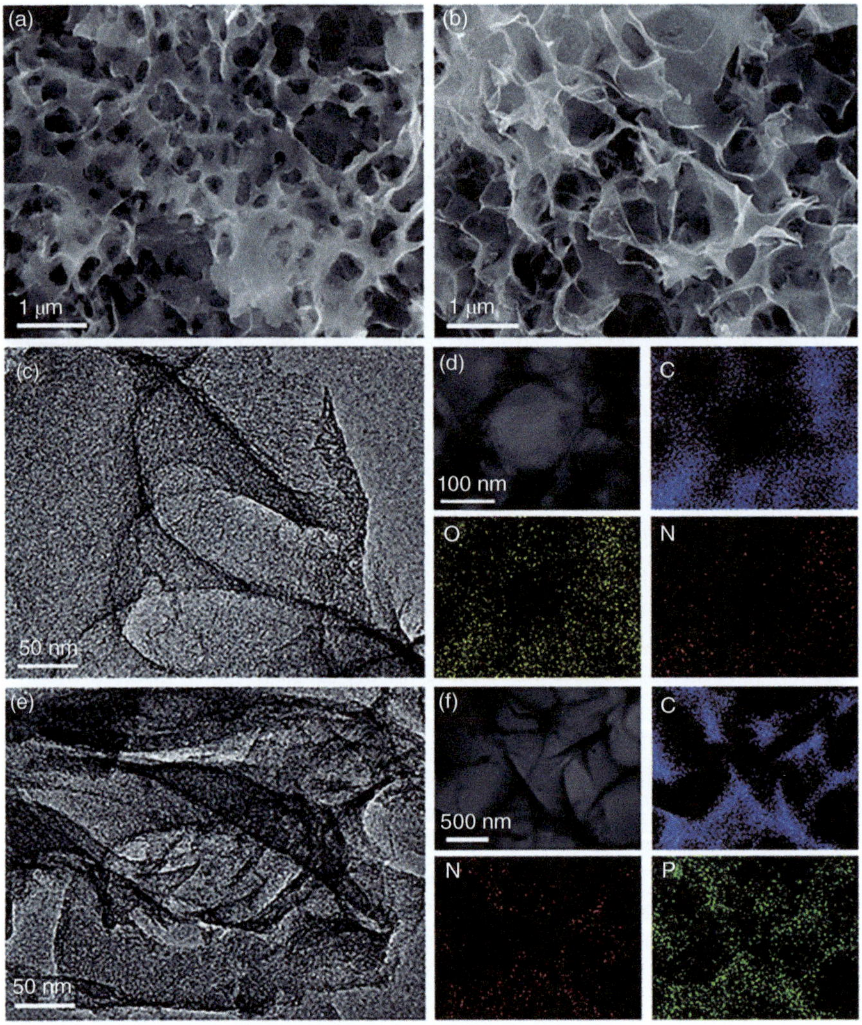

Figure 7.30 Representative SEM and TEM images of (a, c) NHPC and (b, e) N,P-HPC. TEM and corresponding elemental mapping images of (d) NHPC and (f) N,P-HPC. Source: Zan et al. [88]/with permission of © The Royal Society of Chemistry.

respectively (Figure 7.31e). Since the electronegativity of P (2.19) is lower than that of N (3.04), the P—C and the N—C chemical bonds have different polarities, which can become the catalytically active site of ORR. The dominantly well-defined P–C peak (66.88%) for N,P-HPC indicates that almost all P atoms have been successfully doped into the porous carbon. The existence of P–C species is further confirmed by the high-resolution C 1s spectrum (Figure 7.31f) with the characteristic peak of the C–P bond (285.46 eV).

In addition to dual doping of N, S or N, P, the ternary doping of N, P, and S can improve the pseudo-capacitance characteristics while enhancing the electronic conductivity, introduce an n-type dopant with a polarized surface for the pseudo

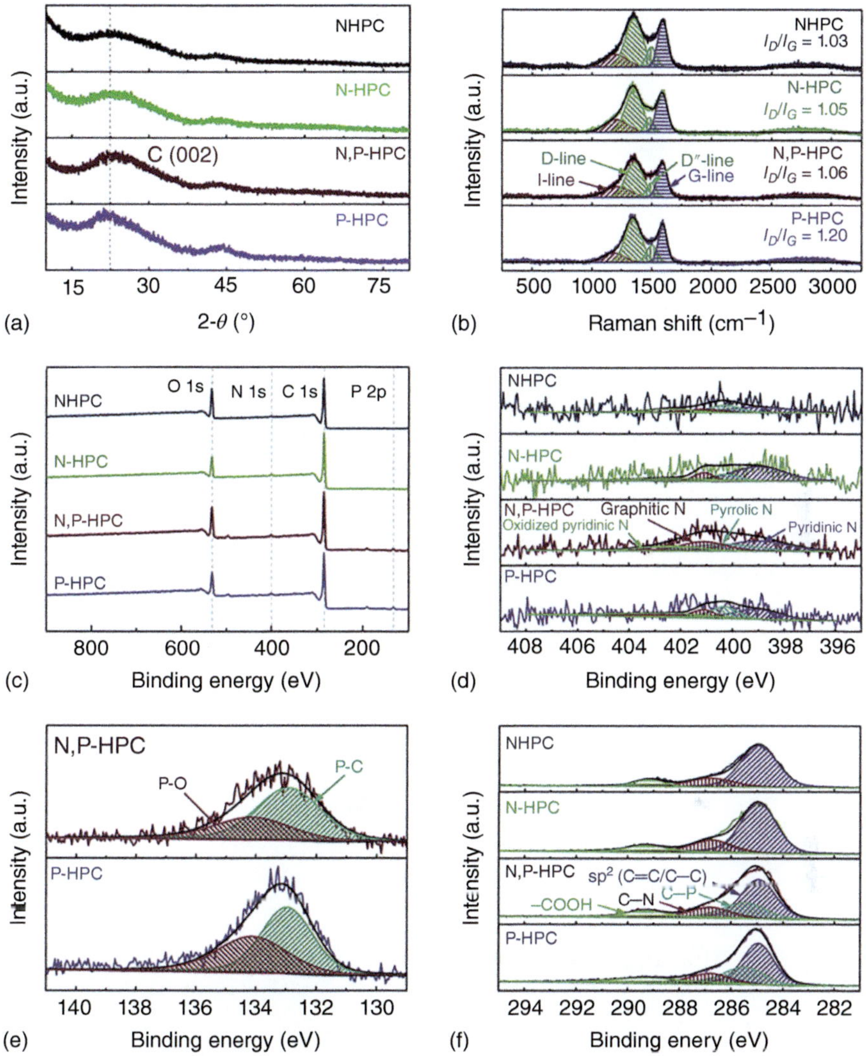

Figure 7.31 (a) XRD patterns, (b) Raman spectra of NHPC, N-HPC, N,P-HPC, and P-HPC. (c) XPS survey spectra of NHPC, N-HPC, N,P-HPC, and P-HPC. High-resolution XPS spectra of (d) N 1s, (e) P 2p, and (f) C 1s for the above samples. Source: Reproduced with permission from Zan et al. [88]; © 2017, The Royal Society of Chemistry.

base point, and stabilize the oxygen function via increasing the stability of the reaction.

Ai et al. used tuna fish bone as precursor to synthesize N/S/O-doped highly porous carbon (FBPC) materials for Li–S batteries (Figure 7.32a) through a low-cost and environment-friendly approach [60]. Nano-hydroxyapatite (n-HA) is evenly dispersed in the mineralized collagen and used as an in situ template to create a porous structure in the carbonization process. Rich and diverse special amino acids serve as C, N, O, and S precursors.

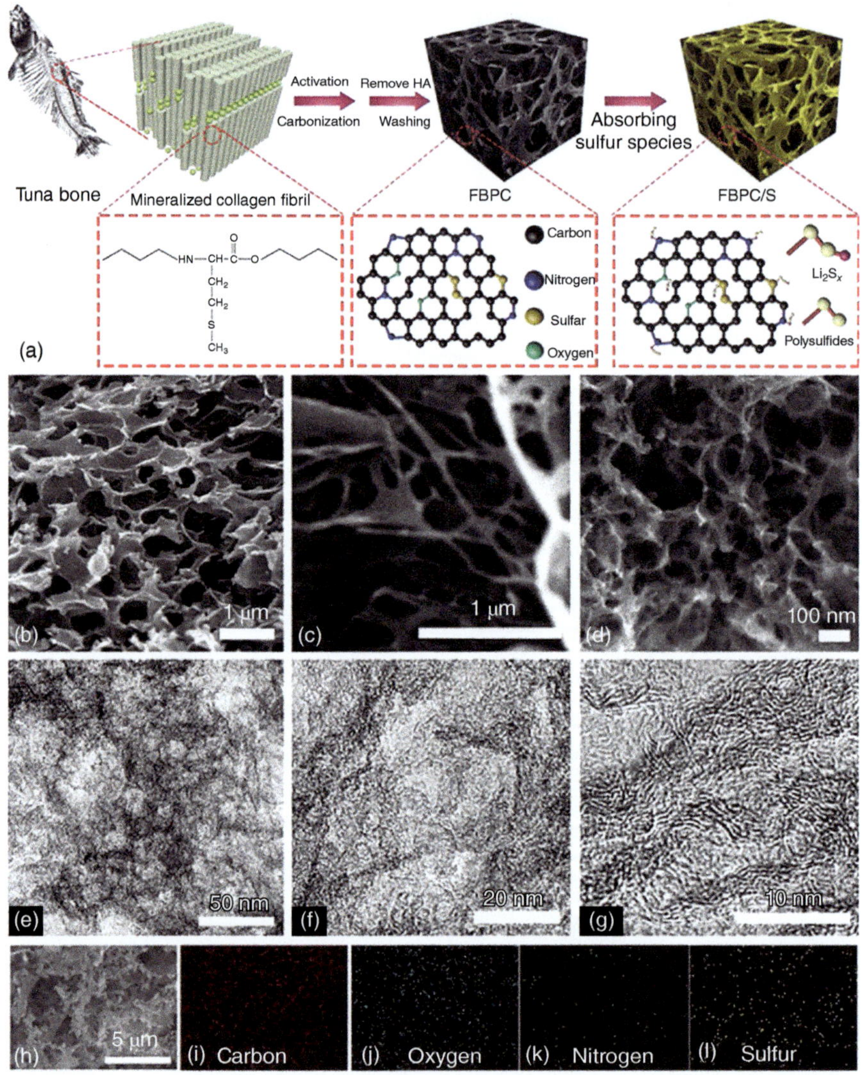

Figure 7.32 (a) Schematic diagram for the preparation procedure of FBPC. Morphology of FBPC. (b–d) SEM and (e–g) TEM images of FBPC at different magnifications. (h–l) Energy dispersive spectroscopy (EDS) measurement of (i) carbon, (j) oxygen, (k) nitrogen, and (l) sulfur, which were derived from the whole area exhibited in (h). Source: Ai et al. [60]/with permission of © Elsevier.

After further hydrochloric acid treatment, the HA particles were completely removed, and a porous surface was obtained (Figure 7.32b). Owing to the partial agglomeration of the HA material at 900 °C, abundant pores with more than 100 nm in diameter are produced (Figure 7.32c). TEM images (Figure 7.32d–f) indicate that FBPC has developed mesopores and micropores sizes with diameters of about 50 and 1–2 nm, respectively. These mesopores can not only accommodate the dissolved LiPSs and most of the electrolyte [25], but can also accommodate the huge volume expansion/contraction in the charge/discharge process and promote the transmission of ions. In addition, the micropores can block LiPSs and increase the active surface. In addition, Figure 7.32g reflects the partial graphitization formed in the carbonization process at 900 °C. The elemental mapping of FBPC (Figure 7.32h–l) demonstrates that there are N, O, and S heteroatoms, and they are uniformly distributed onto the surface, suggesting that they have been maintained in the carbonization process.

XRD and Raman spectra are employed to investigate the crystallographic structure of FBPC (Figure 7.33a,b). FBPC displays a wide diffraction peak at about 25° and a peak ranging from 40° to 50°, corresponding to the (002) and (100) crystal planes of graphite, respectively. The results demonstrate that FBPC has an amorphous carbon structure, which is consistent with the TEM. The Raman spectrum of FBPC exhibits two distinct peaks at ~1350 cm^{-1} (D band) and ~1590 cm^{-1} (G band), assigned to the disordered carbon with sp^3 defects and graphitized carbon reflecting sp^2-hybridized carbon, respectively. The ratio of I_D to I_G is calculated to be 1.01, indicating that FBPC has partially graphitized during the carbonization process. To further analysis the composition of FBPC, XPS was adopted (Figure 7.33c,d), confirming the N and S doping within FBPC.

In summary, the introduction of heteroatoms into the sp^2-conjugated carbon network can enhance the conductivity, surface wettability, as well as electrostatic

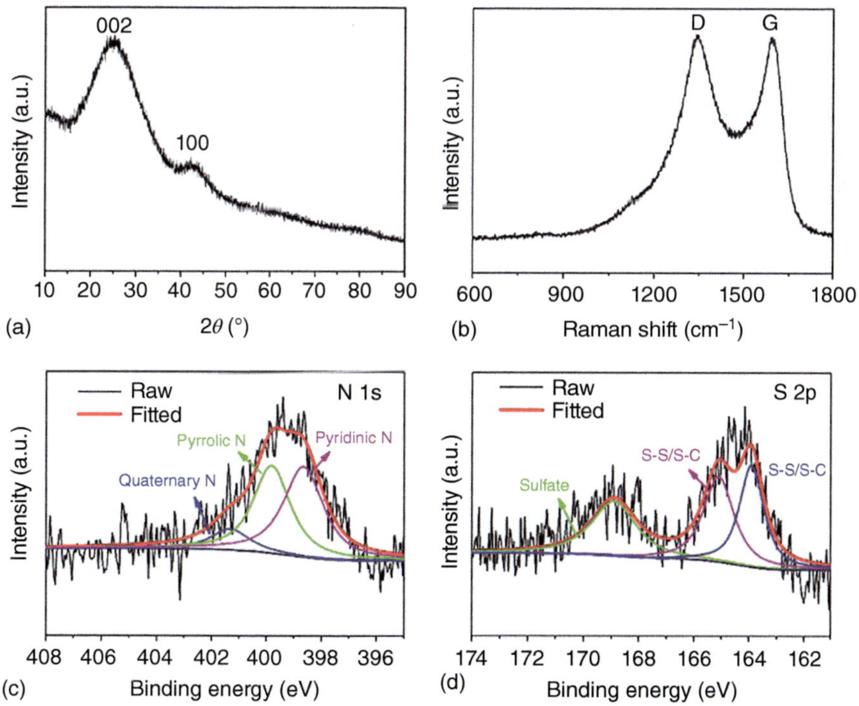

Figure 7.33 Characterization of FBPC. (a) XRD pattern, (b) Raman spectrum, (c, d) XPS spectra of FBPC. (c) N1s spectra, (d) S2p spectra of FBPC. Source: Reproduced with permission from Zan et al. [88]; © 2017 Elsevier Ltd.

adsorption capacity of the material. At the same time, this is an effective approach to adjust the electronic structure. As the charge/spin distribution of the carbon atoms is altered, the active points for electrocatalytic energy conversion are generated. However, it must be noted that the precise regulation of heteroatom types, configurations, and contents is still hard to achieve in experiment [89–91]. In addition, the synergistic effect of the co-doping of various heteroatoms is another key factor. On this basis, more active sites can be created on adjacent carbon atoms for electrocatalytic reactions. Though there are more and more computational researches to investigate the impact of heteroatom doping, the bottom-up construction of carbon-rich materials with well-defined doping profiles remains necessary to solve and verify the relationship of structure and property [74, 92–95].

References

1 De, S., Acharya, S., Sahoo, S., and Chandra Nayak, G. (2020). *Nanostructured, Functional, and Flexible Materials for Energy Conversion and Storage Systems*. Amsterdam: Elsevier.

2 Deng, D., Liao, X., and Shi, B. (2008). Synthesis of porous carbon fibers from collagen fiber. *ChemSusChem* 1: 298–301.

3 Yang, N., Shao, R., Zhang, Z. et al. (2021). Gelatin-derived carbon: carbonization mechanism and application in K-ion storage. *Carbon* 178: 775–782.

4 Yuan, X., Zhu, B., Feng, J. et al. (2021). Biomass bone-derived, N/P-doped hierarchical hard carbon for high-energy potassium-ion batteries. *Materials Research Bulletin* 139: 111282.

5 Yu, Z., Tetard, L., Zhai, L., and Thomas, J. (2015). Supercapacitor electrode materials: nanostructures from 0 to 3 dimensions. *Energy & Environmental Science* 8: 702–730.

6 Li, B., Zheng, J., Zhang, H. et al. (2018). Electrode materials, electrolytes, and challenges in nonaqueous lithium-ion capacitors. *Advanced Materials* 30, 1705670.

7 Yang, X.-F., Wang, A., Qiao, B. et al. (2013). Single-atom catalysts: a new frontier in heterogeneous catalysis. *Accounts of Chemical Research* 46: 1740–1748.

8 Liu, J. (2017). Catalysis by supported single metal atoms. *ACS Catalysis* 7: 34–59.

9 Chen, W., Zhang, H., Huang, Y., and Wang, W. (2010). A fish scale based hierarchical lamellar porous carbon material obtained using a natural template for high performance electrochemical capacitors. *Journal of Materials Chemistry* 20: 4773–4775.

10 Cuong, D.V., Matsagar, B.M., Lee, M. et al. (2021). A critical review on biochar-based engineered hierarchical porous carbon for capacitive charge storage. *Renewable and Sustainable Energy Reviews* 145, 111029.

11 Dou, M., He, D., Shao, W. et al. (2016). Pyrolysis of animal bones with vitamin B12: a facile route to efficient transition metal–nitrogen–carbon (TM–N–C) electrocatalysts for oxygen reduction. *Chemistry – A European Journal* 22: 2896–2901.

12 Dubey, P., Shrivastav, V., Maheshwari, P.H., and Sundriyal, S. (2020). Recent advances in biomass derived activated carbon electrodes for hybrid electrochemical capacitor applications: challenges and opportunities. *Carbon* 170: 1–29.

13 Niu, J., Shao, R., Liu, M. et al. (2019). Porous carbons derived from collagen-enriched biomass tailored design, synthesis, and application in electrochemical energy storage and conversion. *Advanced Functional Materials*, 1905095 29.

14 Shoulders, M.D. and Raines, R.T. (2009). Collagen structure and stability. *Annual Review of Biochemistry* 78: 929–958.

15 Gong, J., Chen, X., and Tang, T. (2019). Recent progress in controlled carbonization of (waste) polymers. *Progress in Polymer Science* 94: 1–32.

16 Qiu, D., Guan, J., Li, M. et al. (2020). Cucurbit[6]uril-derived nitrogen-doped hierarchical porous carbon confined in graphene network for potassium-ion hybrid capacitors. *Advanced Science* 7, 2001681.

17 Zhang, Y., Kang, L., Li, H. et al. (2019). Characterization of moxa floss combustion by TG/DSC, TG-FTIR and IR. *Bioresource Technology* 288: 121516.

18 Zhu, H.M., Yan, J.H., Jiang, X.G. et al. (2008). Study on pyrolysis of typical medical waste materials by using TG-FTIR analysis. *Journal of Hazardous Materials* 153: 670–676.

19 Muyonga, J.H., Cole, C.G.B., and Duodu, K.G. (2004). Fourier transform infrared (FTIR) spectroscopic study of acid soluble collagen and gelatin from skins and bones of young and adult Nile perch (*Lates niloticus*). *Food Chemistry* 86: 325–332.

20 Aliev, A.E. (2005). Solid-state NMR studies of collagen-based parchments and gelatin. *Biopolymers* 77: 230–245.

21 Kuroki, S., Hosaka, Y., and Yamauchi, C. (2013). A solid-state NMR study of the carbonization of polyaniline. *Carbon* 55: 160–167.

22 Niu, J., Shao, R., Liang, J. et al. (2017). Biomass-derived mesopore-dominant porous carbons with large specific surface area and high defect density as high performance electrode materials for Li-ion batteries and supercapacitors. *Nano Energy* 36: 322–330.

23 Shao, R., Niu, J., Liang, J. et al. (2017). Mesopore- and macropore-dominant nitrogen-doped hierarchically porous carbons for high-energy and ultrafast supercapacitors in non-aqueous electrolytes. *ACS Applied Materials & Interfaces* 9: 42797–42805.

24 Shao, R., Niu, J., Zhu, F. et al. (2019). A facile and versatile strategy towards high-performance Si anodes for Li-ion capacitors: concomitant conductive network construction and dual-interfacial engineering. *Nano Energy* 63: 103824.

25 Jorge, A.B., Jervis, R., Periasamy, A.P. et al. (2020). 3D carbon materials for efficient oxygen and hydrogen electrocatalysis. *Advanced Energy Materials* 10, 1902494.

26 Jiang, L., Sheng, L., and Fan, Z. (2018). Biomass-derived carbon materials with structural diversities and their applications in energy storage. *Science China Materials* 61: 133–158.

27 Marichi, R.B., Sahu, V., Sharma, R.K., and Singh, G. (2017). *Handbook of Ecomaterials*. Switzerland: Springer International Publishing.

28 Xu, X., Ray, R., Gu, Y. et al. (2004). Electrophoretic analysis and purification of fluorescent single-walled carbon nanotube fragments. *Journal of the American Chemical Society* 126: 12736–12737.

29 Sun, Y.-P., Zhou, B., Lin, Y. et al. (2006). Quantum-sized carbon dots for bright and colorful photoluminescence. *Journal of the American Chemical Society* 128: 7756–7757.

30 Wang, X., Cao, L., Yang, S.-T. et al. (2010). Bandgap-like strong fluorescence in functionalized carbon nanoparticles. *Angewandte Chemie International Edition* 49: 5310–5314.

31 Liu, H., Ye, T., and Mao, C. (2007). Fluorescent carbon nanoparticles derived from candle soot. *Angewandte Chemie International Edition* 46: 6473–6475.

32 Das, R., Bandyopadhyay, R., and Pramanik, P. (2018). Carbon quantum dots from natural resource: a review. *Materials Today Chemistry* 8: 96–109.

33 Peng, H. and Travas-Sejdic, J. (2009). Simple aqueous solution route to luminescent carbogenic dots from carbohydrates. *Chemistry of Materials* 21: 5563–5565.

34 Li, F., Li, Y., Yang, X. et al. (2018). Highly fluorescent chiral N-S-doped carbon dots from cysteine: affecting cellular energy metabolism. *Angewandte Chemie International Edition* 57: 2377–2382.

35 Lou, Y., Hao, X., Liao, L. et al. (2021). Recent advances of biomass carbon dots on syntheses, characterization, luminescence mechanism, and sensing applications. *Nano Select* 2: 1117–1145.

36 Zhou, J., Booker, C., Li, R. et al. (2007). An electrochemical avenue to blue luminescent nanocrystals from multiwalled carbon nanotubes (MWCNTs). *Journal of the American Chemical Society* 129: 744–745.

37 Wang, C., Wu, X., Li, X. et al. (2012). Upconversion fluorescent carbon nanodots enriched with nitrogen for light harvesting. *Journal of Materials Chemistry* 22: 15522–15525.

38 Bourlinos, A.B., Stassinopoulos, A., Anglos, D. et al. (2008). Surface functionalized carbogenic quantum dots. *Small* 4: 455–458.

39 Dehghani, A., Ardekani, S.M., Hassan, M., and Gomes, V.G. (2018). Collagen derived carbon quantum dots for cell imaging in 3D scaffolds via two-photon spectroscopy. *Carbon* 131: 238–245.

40 Liang, Q., Ma, W., Shi, Y. et al. (2013). Easy synthesis of highly fluorescent carbon quantum dots from gelatin and their luminescent properties and applications. *Carbon* 60: 421–428.

41 Mukherjee, B. (2021). *Carbon nanofiber for hydrogen storage*. In: *Carbon Nanofibers* (ed. M. Sharon), 175–209. Salem: Scrivener Publishing LLC.

42 Smith, J.W. (1968). Molecular pattern in native collagen. *Nature* 219: 157–158.

43 Friess, W. (1998). Collagen – biomaterial for drug delivery. *European Journal of Pharmaceutics and Biopharmaceutics* 45: 113–136.

44 Liao, X.P. and Shi, B.I.J.E.S. (2005). Adsorption of fluoride on zirconium(IV)-impregnated collagen fiber. *Environmental Science & Technology* 39: 4628–4632.

45 Liao, X., Lu, Z., Du, X. et al. (2004). Collagen fiber immobilized *Myrica rubra* tannin and its adsorption to UO_2^{2+}. *Environmental Science & Technology* 38: 324–328.

46 Covington, A.D. (1997). Modern tanning chemistry. *Chemical Society Reviews* 26: 111–126.

47 Ling, Z., Wang, Z., Zhang, M. et al. (2016). Sustainable synthesis and assembly of biomass-derived B/N Co-doped carbon nanosheets with ultrahigh aspect ratio for high-performance supercapacitors. *Advanced Functional Materials* 26: 111–119.

48 Zan, Y., Zhang, Z., Zhu, B. et al. (2021). Ultrathin two-dimensional phosphorus and nitrogen Co-doped carbon nanosheet as efficient oxygen reduction electrocatalyst. *Carbon* 174: 404–412.

49 Mao, X., Cao, Z., Chen, S. et al. (2019). Facile synthesis of N,P-doped hierarchical porous carbon framework catalysts based on gelatin/phytic acid supermolecules for electrocatalytic oxygen reduction. *International Journal of Hydrogen Energy* 44: 5890–5898.

50 Liu, Y., Zhang, X., Gu, X. et al. (2020). One-step turning leather wastes into heteroatom doped carbon aerogel for performance enhanced capacitive deionization. *Microporous and Mesoporous Materials* 303: 110303.

51 Xie, L., Jin, Z., Dai, Z. et al. (2020). Porous carbons synthesized by templating approach from fluid precursors and their applications in environment and energy storage: a review. *Carbon* 170: 100–118.

52 Sing, K.J.P. and Chemistry, A. (1985). Reporting physisorption data for gas/solid systems with special reference to the determination of surface area and porosity (Recommendations 1984). *Pure and Applied Chemistry* 57: 603–619.

53 Forse, A.C., Griffin, J.M., Merlet, C. et al. (2017). Direct observation of ion dynamics in supercapacitor electrodes using in situ diffusion NMR spectroscopy. *Nature Energy* 2: 16216.

54 Jia, H., Zhang, H., Wan, S. et al. (2019). Preparation of nitrogen-doped porous carbon via adsorption-doping for highly efficient energy storage. *Journal of Power Sources* 433: 226712.

55 Sun, C., Guo, D., Shao, Q., and Chen, J. (2021). Preparation of gelatin-derived nitrogen-doped large pore volume porous carbons as sulfur hosts for lithium-sulfur batteries. *New Carbon Materials* 36: 198–208.

56 Zhang, S.-X., Xu, L., Chen, Z.-H. et al. Hierarchical porous carbon derived from green cyclodextrin metal-organic framework and its application in microwave absorption. *Journal of Applied Polymer Science* 138, 50849.

57 Liu, Z., Zhang, F., Liu, T. et al. (2016). Removal of azo dye by a highly graphitized and heteroatom doped carbon derived from fish waste: adsorption equilibrium and kinetics. *Journal of Environmental Management* 182: 446–454.

58 Huang, W., Zhang, H., Huang, Y. et al. (2011). Hierarchical porous carbon obtained from animal bone and evaluation in electric double-layer capacitors. *Carbon* 49: 838–843.

59 Song, H., Li, H., Wang, H. et al. (2014). Chicken bone-derived N-doped porous carbon materials as an oxygen reduction electrocatalyst. *Electrochimica Acta* 147: 520–526.

60 Ai, F., Liu, N., Wang, W. et al. (2017). Heteroatoms-doped porous carbon derived from tuna bone for high performance Li–S batteries. *Electrochimica Acta* 258: 80–89.
61 Liu, M., Niu, J., Zhang, Z. et al. (2019). Porous carbons with tailored heteroatom doping and well-defined porosity as high-performance electrodes for robust Na-ion capacitors. *Journal of Power Sources* 414: 68–75.
62 Niu, J., Liang, J., Shao, R. et al. (2017). Tremella-like N,O-codoped hierarchically porous carbon nanosheets as high-performance anode materials for high energy and ultrafast Na-ion capacitors. *Nano Energy* 41: 285–292.
63 Liu, M., Niu, J., Zhang, Z. et al. (2018). Potassium compound-assistant synthesis of multi-heteroatom doped ultrathin porous carbon nanosheets for high performance supercapacitors. *Nano Energy* 51: 366–372.
64 Niu, J., Liu, M., Xu, F. et al. (2018). Synchronously boosting gravimetric and volumetric performance: biomass-derived ternary-doped microporous carbon nanosheet electrodes for supercapacitors. *Carbon* 140: 664–672.
65 Wang, R., Wang, K., Wang, Z. et al. (2015). Pig bones derived N-doped carbon with multi-level pores as electrocatalyst for oxygen reduction. *Journal of Power Sources* 297: 295–301.
66 Xu, B., Hou, S., Cao, G. et al. (2012). Sustainable nitrogen-doped porous carbon with high surface areas prepared from gelatin for supercapacitors. *Journal of Materials Chemistry* 22: 19088–19093.
67 Zhou, X., Jia, Z., Feng, A. et al. (2019). Synthesis of fish skin-derived 3D carbon foams with broadened bandwidth and excellent electromagnetic wave absorption performance. *Carbon* 152: 827–836.
68 He, D., Niu, J., Dou, M. et al. (2017). Nitrogen and oxygen co-doped carbon networks with a mesopore-dominant hierarchical porosity for high energy and power density supercapacitors. *Electrochimica Acta* 238: 310–318.
69 Niu, L., Shen, C., Yan, L. et al. (2019). Waste bones derived nitrogen–doped carbon with high micropore ratio towards supercapacitor applications. *Journal of Colloid and Interface Science* 547: 92–101.
70 Shan, B., Cui, Y., Liu, W. et al. (2018). Fibrous bio-carbon foams: a new material for lithium-ion hybrid supercapacitors with ultrahigh integrated energy/power density and ultralong cycle life. *ACS Sustainable Chemistry & Engineering* 6: 14989–15000.
71 Salim, N.V., Jin, X., Mateti, S. et al. (2019). Porous carbon fibers made from collagen derived from an animal by-product. *Materials Today Advances* 1, 100005.
72 Selvamani, V., Ravikumar, R., Suryanarayanan, V. et al. (2015). Fish scale derived nitrogen doped hierarchical porous carbon – a high rate performing anode for lithium ion cell. *Electrochimica Acta* 182: 1–10.
73 Jiang, J., Zhang, Q., Zhan, X., and Chen, F. (2019). A multifunctional gelatin-based aerogel with superior pollutants adsorption, oil/water separation and photocatalytic properties. *Chemical Engineering Journal* 358: 1539–1551.
74 Kong, D., Gao, Y., Xiao, Z. et al. (2019). Rational design of carbon-rich materials for energy storage and conversion. *Advanced Materials* 31, 1804973.

75 Gopalakrishnan, A. and Badhulika, S. (2020). Effect of self-doped heteroatoms on the performance of biomass-derived carbon for supercapacitor applications. *Journal of Power Sources* 480: 228830.

76 Uchaker, E. and Cao, G. (2015). The role of intentionally introduced defects on electrode materials for alkali-ion batteries. *Chemistry – An Asian Journal* 10: 1608–1617.

77 Wang, X.H., Li, J.G., Kamiyama, H. et al. (2005). Pyrogenic iron(III)-doped TiO_2 nanopowders synthesized in RF thermal plasma: phase formation, defect structure, band gap, and magnetic properties. *Journal of the American Chemical Society* 127: 10982–10990.

78 Huang, Z., Liu, H., Hu, R. et al. (2020). Structures, properties and application of 2D monoelemental materials (Xenes) as graphene analogues under defect engineering. *Nano Today* 35: 100906.

79 Gopalakrishnan, A. and Badhulika, S. (2020). Sulfonated porous carbon nanosheets derived from oak nutshell based high-performance supercapacitor for powering electronic devices. *Renewable Energy* 161: 173–183.

80 Zhao, X., Zhang, Q., Chen, C.-M. et al. (2012). Aromatic sulfide, sulfoxide, and sulfone mediated mesoporous carbon monolith for use in supercapacitor. *Nano Energy* 1: 624–630.

81 Denis, P.A. (2010). Band gap opening of monolayer and bilayer graphene doped with aluminium, silicon, phosphorus, and sulfur. *Chemical Physics Letters* 492: 251–257.

82 Lee, Y.-H., Li, F., Chang, K.-H. et al. (2012). Novel synthesis of N-doped porous carbons from collagen for electrocatalytic production of H_2O_2. *Applied Catalysis B: Environmental* 126: 208–214.

83 Gao, A., Guo, N., Yan, M. et al. (2018). Hierarchical porous carbon activated by $CaCO_3$ from pigskin collagen for CO_2 and H_2 adsorption. *Microporous and Mesoporous Materials* 260: 172–179.

84 Zhou, G., Paek, E., Hwang, G.S., and Manthiram, A. (2015). Long-life Li/polysulphide batteries with high sulphur loading enabled by lightweight three-dimensional nitrogen/sulphur-codoped graphene sponge. *Nature Communications* 6: 7760.

85 Wang, J., Shen, L., Xu, Y. et al. (2015). Lamellar-structured biomass-derived phosphorus- and nitrogen-co-doped porous carbon for high-performance supercapacitors. *New Journal of Chemistry* 39: 9497–9503.

86 Hu, L., Hou, J., Ma, Y. et al. (2016). Multi-heteroatom self-doped porous carbon derived from swim bladders for large capacitance supercapacitors. *Journal of Materials Chemistry A* 4: 15006–15014.

87 Qu, Y., Zhang, Z., Zhang, X. et al. (2015). Highly ordered nitrogen-rich mesoporous carbon derived from biomass waste for high-performance lithium–sulfur batteries. *Carbon* 84: 399–408.

88 Zan, Y., Zhang, Z., Liu, H. et al. (2017). Nitrogen and phosphorus co-doped hierarchically porous carbons derived from cattle bones as efficient metal-free electrocatalysts for the oxygen reduction reaction. *Journal of Materials Chemistry A* 5: 24329–24334.

89 Liu, F., Wang, Z., Zhang, H. et al. (2019). Nitrogen, oxygen and sulfur co-doped hierarchical porous carbons toward high-performance supercapacitors by direct pyrolysis of kraft lignin. *Carbon* 149: 105–116.
90 Zhou, J. (2020). Theories of pedagogic codes and knowledge structures and their development. *Open Journal of Social Sciences* 08: 105–110.
91 Lei, J., Guo, Q., Yao, W. et al. (2018). Bioconcentration of organic dyes: via fungal hyphae and their derived carbon fibers for supercapacitors. *Journal of Materials Chemistry A* 6: 10710–10717.
92 Cazetta, A.L., Zhang, T., Silva, T.L. et al. (2018). Bone char-derived metal-free N- and S-co-doped nanoporous carbon and its efficient electrocatalytic activity for hydrazine oxidation. *Applied Catalysis B: Environmental* 225: 30–39.
93 Zhao, G., Li, Y., Zhu, G. et al. (2019). Biomass-based N, P, and S self-doped porous carbon for high-performance supercapacitors. *ACS Sustainable Chemistry & Engineering* 7: 12052–12060.
94 Razmjooei, F., Singh, K., Kang, T.H. et al. (2017). Urine to highly porous heteroatom-doped carbons for supercapacitor: a value added journey for human waste. *Scientific Reports* 7: 10910.
95 Zhang, H., Zhou, W., Huang, D. et al. (2021). Functionalized hierarchical porous carbon with sulfur/nitrogen/oxygen tri-doped as high quality sulfur hosts for lithium–sulfur batteries. *Journal of Alloys and Compounds* 858: 157647.

8

Synthesis of Collagen-Derived Carbons

8.1 Conventional Preparation Methods

Conventional methods to prepare collagen-derived carbons include activation method, templating method, in situ-doping method, and post-doping method [1].

8.1.1 Templating Method

8.1.1.1 Hard-Templating Method

Templating method is an attractive method to prepare porous carbon materials with well-defined pore structures. Many hard templates have rigid structures, such as nanoparticles, nanofibers, and nanofilms. The templating method includes the following steps: (i) synthesize hard templates, (ii) mix hard templates with carbon precursors [2], (iii) heat or chemical treatments for the transformation from the precursors to carbon materials, and (iv) remove the hard templates.

Numerous porous carbons can be synthesized by using magnesium oxide (MgO) as hard template. The carbon precursors can also react with chemicals. These bifunctional chemicals will generate gases during the carbonization process. Zn metal is also a hard template, which can be removed through evaporation under high temperatures [3, 4]. Therefore, if the temperature is higher than the boiling point of zinc (907 °C), the template need not be removed [5]. Metallic zinc can act as a template to synthesize porous carbons, in which many biomass materials can be used as carbon precursors.

Carbide is another template to fabricate porous carbon materials. Carbide-derived carbons are produced by removing metal carbides using chlorine gas. Through controlling the synthesis parameters, the carbide-derived carbons with higher specific surface area (SSA) and narrower pore-size distribution can be obtained [5, 6]. By using hard templates, porous carbons with controllable porosities can be easily prepared. However, the high cost of the template and the time-consuming preparation process seriously hinder their large-scale applications. Another problem is the ability to control the pore-wall thickness, which is crucial to adjust the pore sizes [7]. Moreover, the templates also need to be removed [8].

Collagen-Derived Materials: Synthesis and Applications in Electrochemical Energy Storage and Conversion, First Edition. Feng Wang, Yaqin Huang, and Jin Niu.
© 2022 WILEY-VCH GmbH. Published 2022 by WILEY-VCH GmbH.

8.1.1.2 Soft-Templating Method

Soft templates are usually organic molecules with functional groups. The functional groups can afford strong interactions in the solvent, such as electrostatic interactions [9], hydrogen bonding [10], or hydrophilic/hydrophobic interactions [11]. The soft templates turn into micelles immediately when the solvent is added. The micelles combine with carbon precursors to form heterogeneous composites, which are further used to prepare carbon materials with typical porous structures during pyrolysis process [12]. Typically, the carbon pore size can be controlled by adjusting the ratio of micelle to solvent.

By carbonizing the mixture of the carbon precursor and soft template, the carbon materials with mesoporous structure can be obtained. However, the preparation process of soft templates is stringent and complex. Based on the Coulomb interaction between positively charged surfactant and negatively charged phenolic resin, the synthesis of mesoporous carbon from cetyltrimethylammonium bromide was attempted for the first time [13]. However, due to the micelles collapse at 200 °C, porous structure was not obtained. A successful soft template should be able to assemble into nanostructures and sustain the porogens before the carbon-skeleton formation [14]. Based on these requirements, the soft templates such as the poly(ethylene oxide)-*b*-poly(propylene oxide)-*b*-poly(ethylene oxide) (PEO-PPO-PEO) (P123 [15] 151, F108 [16], and F127, [15, 17, 18]), polystyrene-*b*-poly(4-vinylpyridine) (PS-P4VP) [19], and some surfactants have been extensively studied.

In a recent study, cationic, anionic, and nonionic surfactants have been used as soft templates [20]. In particular, ionic surfactants produce small mesopores (<4 nm) or micropores, while nonionic surfactants generate mesopores with pore size over 10 nm [21]. It should be noted that assembly is usually achieved in hydrothermal processes [22]. A further heat or chemical treatment helps remove the residual solvent and templates to prepare mesoporous carbon materials with ordered structure.

8.1.2 Activation Method

There are many kinds of activation methods, such as physical activation method, chemical activation method, or physical–chemical activation method. Physical activation is achieved by the reaction of gaseous etchants (such as H_2O, NH_3, or CO_2) with carbon precursors [23], while chemical activation is achieved by the reaction between activating agents (such as KOH, $ZnCl_2$, $NaHCO_3$, K_2CO_3, H_2SO_4, or H_3PO_4) and carbon precursors [24]. In this approach, some cost-effective and renewable biomass materials are particularly preferred as carbon precursors. It makes it easy to produce hierarchically porous carbons with high SSAs and tailored porous structures.

8.1.2.1 Physical Activation

Physical activation usually takes place in a tube furnace with inert gases from 600 to 1200 °C. It is noted that H_2O and CO_2 are more suitable for practical application due to their lower corrosive effects with the production facilities [25]. The pores are

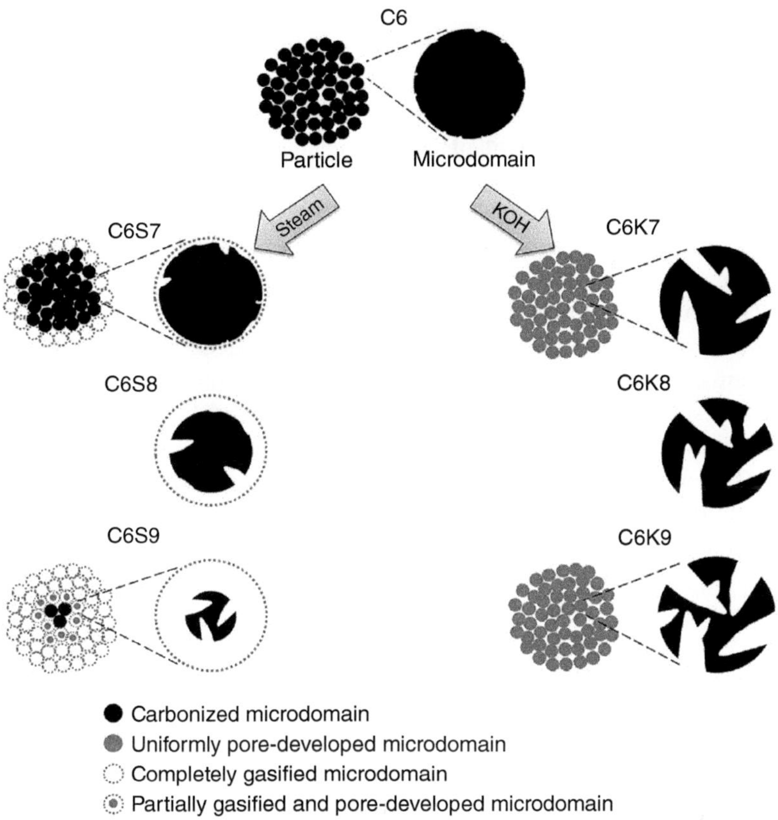

Figure 8.1 Structural mechanism model of pore formation in the carbon material using steam and KOH as activation agents. Source: Reproduced with permission from Kim et al. [27]; © 2017, Elsevier.

seldom formed by the formation and evolution of gas. The pores in porous carbon are mainly formed by the oxidation reaction of oxidizing atmosphere (such as H_2O and CO_2) in the physical activation process (Eqs. (8.1) and (8.2)) [26].

As compared with chemical activation, physical activation produces more micropores and the pore-size distribution is more accurate and narrow. Moreover, physical activation reduces both the microdomain and particle sizes in the activated carbon materials. However, physical activation usually needs longer activation time and higher activation temperatures than chemical activation. In addition, it results in smaller pore sizes, lower tap density, and lower SSA (Figure 8.1) [27]. However, for large-scale production, physical activation method exhibits lower corrosion toward the reactor than chemical activation.

$$C + H_2O \rightarrow CO + H_2 \quad (8.1)$$

$$C + CO_2 \rightarrow 2CO \quad (8.2)$$

8.1.2.2 Chemical Activation

Chemical activation has the merits of low-temperature processes, high carbon yield, and high mesopore ratios. Various studies have shown that porous carbon has been synthesized by chemical activation with catalysts, such as basic metal carbonate, basic metal hydroxide, and phosphoric acid. In addition, researchers can utilize the unique characteristics of carbon precursors to synthesize various porous carbons. Traditionally, $ZnCl_2$ [28], NaOH [29], KOH [30], H_3PO_4 [31], K_2CO_3 [32], and Na_2CO_3 [33] have been used as activation agents.

However, the activation mechanism of the chemical activation agents is still not clear. For KOH activation, pores formed below 500 °C are due to the volatile evaporation. As the chemical activation starts, the pores are further developed combined with the carbon consumption (Eq. (8.3)). Although the reaction occurs at low temperatures, the resulting K further reacts with KOH to form K_2O. K_2O reacts with CO_2 pyrolysis gas at about 400 °C to form K_2CO_3 (Eq. (8.4)). KOH is completely consumed at 600 °C, and the main etching agents are K_2CO_3 and K_2O (Eqs. (8.5) and (8.6)) at about 700 °C. At the same time, the activation products of CO_2 may also participate in the physical activation process to generate pores.

$$6KOH + 2C \rightarrow 2K + 3H_2 + 2K_2CO_3 \tag{8.3}$$

$$K_2O + CO_2 \rightarrow 2K + 3CO \tag{8.4}$$

$$K_2CO_3 + 2C \rightarrow 2K + 3CO \tag{8.5}$$

$$K_2O + C \rightarrow 2K + CO \tag{8.6}$$

High SSA porous carbons can be easily obtained by chemical activation. However, traditional chemical-activation methods have many disadvantages, such as severe corrosivity, environmental pollution, and pore-structure collapse after high-temperature activation. In recent years, great efforts have been made to explore efficient chemical activation agents.

Molten salts are effective agents for chemical activation. For example, $CuCl_2$ does less damage to the natural structure of biomass precursor (Figure 8.1) [34]. The mechanisms of $CuCl_2$ are described as follows [35]:

$$3CuCl_2 + C + 2H_2O \rightarrow Cu + 2CuCl + CO_2 + 4HCl \tag{8.7}$$

$$4CuCl + C + 2H_2O \rightarrow 4Cu + CO_2 + 4HCl \tag{8.8}$$

The spherical pore carbon prepared by $CuCl_2$ activation has a high SSA and the pore-size distribution is less than 2 nm. The $CuCl_2$ activation mechanism is demonstrated by the formation of Cu and CuCl intermediate species. $FeCl_3$ can also be used as an activation agent. The $FeCl_3$ activation mechanism was studied [36]. It is found that the dehydroxylation reactions occurred at 200–330 °C (Eq. (8.9)). As shown in Eqs. (8.10) and (8.11), the generation of Fe_2O_3 and CO_2 promotes the formation of micropores at 330–700 °C. In this process, the formation of Lewis acid under the synergistic action of Fe^{3+} and Cl^- further develops the microporous structure of porous carbon. Besides, the reaction products of iron oxides also help form micropores. The reduction of iron oxide further modifies the microporosity of the resulting carbon (Eqs. (8.12)–(8.14)).

$$FeCl_3 + 2H_2O \rightarrow FeOOH + 3HCl \tag{8.9}$$

$$2FeOOH \rightarrow Fe_2O_3 + H_2O \tag{8.10}$$

$$6Fe_2O_3 + C \rightarrow 4Fe_3O_4 + CO_2 \tag{8.11}$$

$$2Fe_2O_3 + 3C \rightarrow 4Fe + 3CO_2 \tag{8.12}$$

$$Fe_3O_4 + 2C \rightarrow 3Fe + 2CO_2 \tag{8.13}$$

$$Fe_3O_4 + 4C \rightarrow 3Fe + 4CO \tag{8.14}$$

As a kind of dehydration agent, zinc salts adsorb water molecules in organic carbon source and are beneficial to the carbon formation [37]. In the presence of Zn component, aromatic hydrocarbons are more prone to condensation reaction [3]. Therefore, the zinc component can be used as a cycloaddition pore-forming agent. Polyacrylonitrile/asphalt/lignin-based carbon nanofibers (CNFs), containing zinc oxide (PPL-Zn), were prepared by electrospinning using polyacrylonitrile (PAN), asphalt, lignin, and zinc acetate as raw materials. In the high-temperature annealing process, the decomposition product zinc is removed. CNF exhibited an SSA of 1194 $m^2\ g^{-1}$ (Figure 8.2A) [38]. The SSA of porous carbon prepared with decomposed zinc acetate was almost twice as high as that of CNF prepared without zinc acetate.

Li et al. prepared a cross-linked macroporous and mesoporous carbon material using $NaNO_3$ as a pore-forming agent in gelatin biopolymer aerogels (Figure 8.2B) [39]. The porous carbon had a high SSA of 2872.2 $cm^2\ g^{-1}$. It is found that the decomposition of $NaNO_3$ starts above 600 °C to produce N_2, O_2, and NO, which contribute to the formation of mesoporous structures in the resulting porous carbon products.

Antonietti and coworkers investigated the ability of a series of oxygen-containing anionic salts to produce pores in a molten LiCl/KCl [40]. Porous carbon with SSA up to 3200 $m^2\ g^{-1}$ was prepared by a series of oxides in LiCl/KCl liquid medium [41, 42]. The possible pore-forming mechanisms of oxide anions are described as follows:

$$OH^-:\ C + 2OH^- \rightarrow CO + H_2\uparrow + O^{2-} \tag{8.15}$$

$$CO_3{}^{2-}:\ C + CO_3{}^{2-} \rightarrow 2CO + O^{2-} \tag{8.16}$$

$$NO_3{}^-:\ 5C + 4NO_3{}^- \rightarrow 5CO_2 + 2N_2\uparrow + O_2\uparrow \tag{8.17}$$

$$PO_4{}^{3-}:\ a:\ 2C + PO_4{}^{3-} \rightarrow 2CO + PO_2{}^{3-} \tag{8.18}$$

$$b:\ C + 2PO_2{}^{3-} \rightarrow CO + 2P\uparrow + 3O^{2-} \tag{8.19}$$

$$SO_4{}^{2-}:\ a:\ C + SO_4{}^{2-} \rightarrow CO + SO_3{}^{2-} \tag{8.20}$$

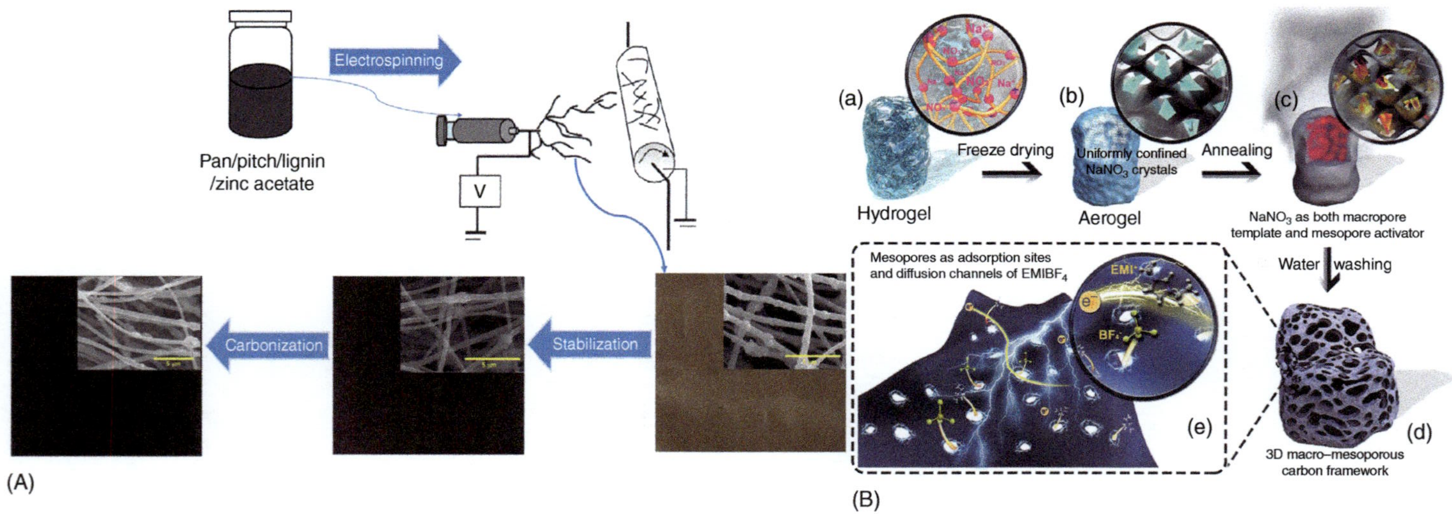

Figure 8.2 (A) Preparation of PPL–Zn composite (above). Source: Yun et al. [38]/with permission of © Elsevier Ltd. (B) Synthesis process of 3D cross-coupled macro- and mesoporous carbon. Source: Li et al. [39]/with permission of © John Wiley & Sons.

$$b:\ 2C + SO_3^{2-} \rightarrow 2CO + S\uparrow + O^{2-} \tag{8.21}$$

$$ClO_3^- :\ 3C + 2ClO_3^- \rightarrow 3CO_2 + 2Cl^- \tag{8.22}$$

Although the exact mechanisms still need to be further confirmed by different physical/chemical methods and advanced characterization techniques, oxygenated salts are promising activators for the synthesis of porous carbons [43].

The pore engineering of carbon materials can also be realized by using oxidation activators. Porous carbon was produced by oxidizing carbon with HNO_3, [44] $KMnO_4$, [45] and $Mn(NO_3)_2$ [46] under room-temperature or high-temperature annealing conditions. When $KMnO_4$ was used as an activator, the derived carbon materials had more macropores and mesopores than the KOH-activated carbon materials [47]. Numerous mesopores and macropores were generated through the reactions described in Eqs. (8.23)–(8.25). $KMnO_4$ is decomposed into K_2MnO_4, MnO_2, and O_2. These products play important roles in etching carbon skeleton and forming mesoporous structure. When the temperature rises to 700 °C, the decomposition of K_2CO_3 has a significant effect on the microporous structure of carbon materials.

$$2KMnO_4 \rightarrow K_2MnO_4 + MnO_2 + O_2 \tag{8.23}$$

$$K_2MnO_4 + C \rightarrow K_2CO_3 + MnO \tag{8.24}$$

$$2MnO_2 + C \rightarrow 2MnO + CO_2 \tag{8.25}$$

Chemical activation method is the most widely used method to prepare porous carbon materials in lab-scale research. However, owing to the high corrosive properties of the activators, chemical activation is not suitable for large-scale production. Moreover, chemical activation also results in the environment pollution. Thus, green and environment-friendly synthesis methods need to be developed.

8.1.3 Post-doping Method

Zan et al. proposed an effective strategy to synthesize N- and P-co-doped 3-D hierarchically porous carbon (N,P-HPC) by pyrolyzing dicyandiamide (DCDA) and phytic acid (PA) with N-doped hierarchically porous carbon (NHPC) derived from cattle bones [48].

Figure 8.3 shows the synthesis of the N,P-HPC. After co-pyrolysis, the mixture of PA, DCDA, and NHPC is transformed into N- and P-co-doped 3-D HPC, while DCDA and PA are precursors of N and P, respectively. The resultant N,P-HPC retains the 3-D hierarchically porous network characteristic of NHPC with interconnected mesopores and macropores (Figure 8.4a,b). Moreover, compared with NHPC, N,P-HPC showed larger pore size and more surface defects, which may be due to the corrosive effect of volatile products produced by the decomposition of PA and DCDA.

N,P-HPC has a hierarchically microporous/mesoporous structure (Figure 8.4c). The Brunauer–Emmett–Teller (BET) surface areas of NHPC, N-HPC, N,P-HPC, and

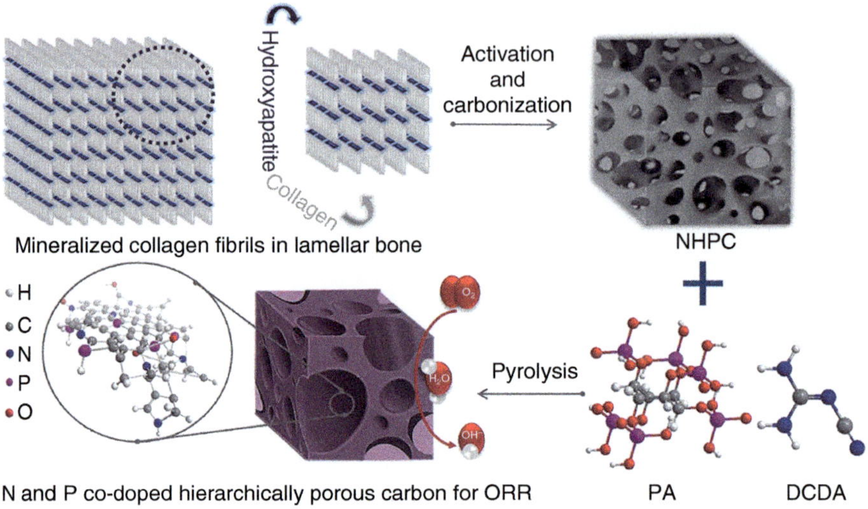

Figure 8.3 Synthesis process of the N,P-HPC [48]. Source: Reproduced with permission from Zan et al. [48]; © 2017, The Royal Society of Chemistry.

P-HPC are 2165, 1747, 1516, and 1389 $m^2\ g^{-1}$, respectively. Compared with NHPC, the specific surface areas of N and P-HPC decreased significantly, which was due to the pore blockage caused by pyrolysis products of PA and DCDA. In addition, with the decrease of N extra/P molar ratio, the specific surface area decreased more obviously, especially for P-HPC, which may be due to the higher PA content and the more serious corrosion effect. N,P-HPC exhibits the content of N (3.2 at%), P (4.0 at%), C (76.0 at%), and O (16.8 at%), indicating a successful co-doping of N and P heteroatoms into the porous carbon networks. The pyridinic N content of N,P-HPC is 1.8 at%, higher than that of NHPC, which may be owing to the addition of additional N precursor (DCDA) (Figure 8.4d). P–C and P–O species can be observed in N,P-HPC according to the high-resolution P 2p spectrum (Figure 8.4e) [20]. Owing to the electronegativity of P (2.19) being lower than that of N (3.04), the polarity of P–C species is different from that of N–C species. The P–C peak (66.88%) of N,P-HPC indicates that almost all P atoms have been introduced into the porous carbons. The presence of P–C species is further confirmed by the high-resolution C 1s spectrum (Figure 8.4f), and the characteristic peak of C—P bond is observed at 285.46 eV.

After that, Zan et al. also presented an effective preparation method of a high-performance S, N, and P ternary-doped 3-D hierarchically porous carbon (S,N,P-HPC) through co-pyrolysis of thiourea (THU) with N,P-HPC (Figure 8.5) [49]. N,P-HPC was obtained by heat treatment of a mixture of HPC, PA, and DCDA.

TG analysis shows that the co-pyrolysis process of N,P-HPC and Thu is divided into two stages (Figure 8.6a). The first stage occurs in the range of 200–320 °C with a weight loss of 63.13%, possibly due to the release of organic material (Figure 8.6b). The TG–MS spectra show two peaks at m/z of 17 and 16, confirming the existence of NH_3 (Figure 8.6c) [50]. According to TG–Fourier transform infrared spectroscopy (FTIR) and TG–MS analyses, the second carbonization stage occurs at 300 °C, with

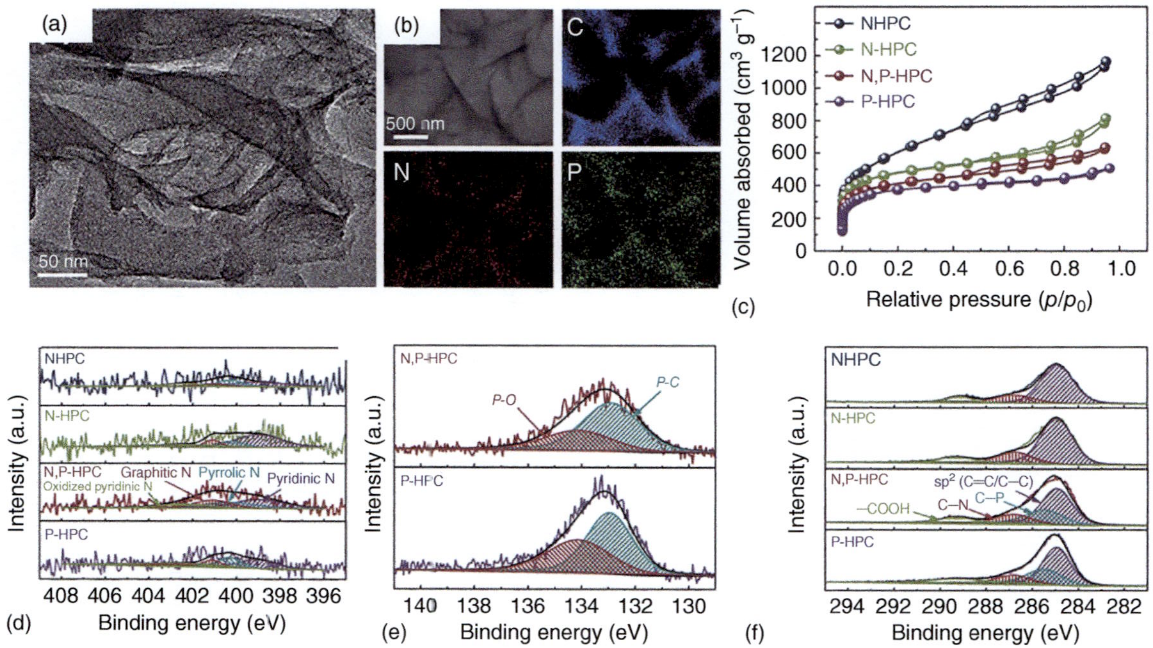

Figure 8.4 (a, b) TEM images of N,P-HPC. (c) Nitrogen adsorption–desorption isotherms of NHPC; N-HPC; N,P-HPC; and P-HPC. High-resolution XPS spectra of (d) N 1s, (e) P 2p, and (f) C 1s for the above samples. Source: Zan et al. [48]/with permission of © The Royal Society of Chemistry.

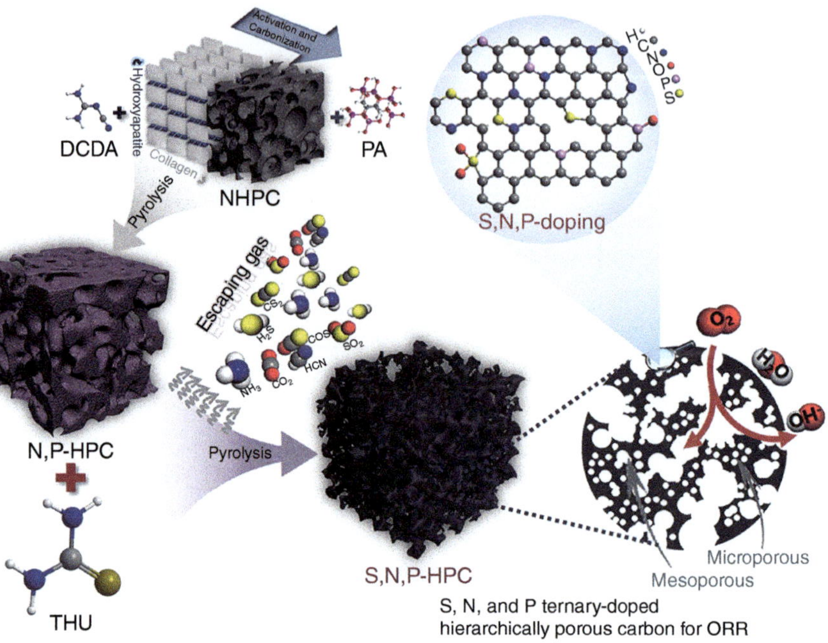

Figure 8.5 Synthesis process of S,N,P-HPC. Source: Reproduced with permission from Zan et al. [49]; © 2017, The Royal Society of Chemistry.

a weight loss of about 36.08%. During the carbonization process, NH_3, CO_2, HCN, and H_2S gases are produced. These newly emerging N and S species should not only be conducive to heteroatom doping, but may also produce rich pores through the activation of small-molecule gases (e.g. CS_2, NH_3, and H_2S) [50].

The FE–scanning electron microscopy (SEM) and transmission electron microscopy (TEM) images show that the S,N,P-HPC inherited the hierarchically porous structures of N,P-HPC, and the mesopores and macropores are interconnected (Figure 8.6d,e). With the increase of THU content, the microstructure of S,N,P-HPC gradually deviated from that of N,P-HPC, and the porous structure gradually collapsed, indicating that the gas generated in the pyrolysis process caused the destruction of carbon matrix.

The TEM elemental mapping diagrams show that the heteroatoms (S,N,P-HPC-1) are distributed homogeneously on the carbon skeleton of S,N,P-HPC-1 (Figure 8.6f). S,N,P-HPC-1 exhibits an SSA of 1533 m^2 g^{-1}. The X-ray photoelectron spectroscopy (XPS) spectra of S,N,P-HPC-1 showed C (80.35 at%), N (4.35 at%), O (11.31 at%), S (1.29 at%), and P (2.96 at%) on the carbon network. It is further confirmed that the heteroatoms S, N, and P are successfully doped into the carbon matrix.

Zan et al. also proposed an effective dual-template method to prepare a P and N co-doped ultrathin two-dimensional carbon nanosheet (P-NCNS) using gelatin as the carbon and nitrogen precursor and red phosphorus (red P) as phosphorus source [51]. During the pyrolysis process, the sublimated red P could generate P vapor, which can form abundant pore space. The obtained nanosheet shows a large

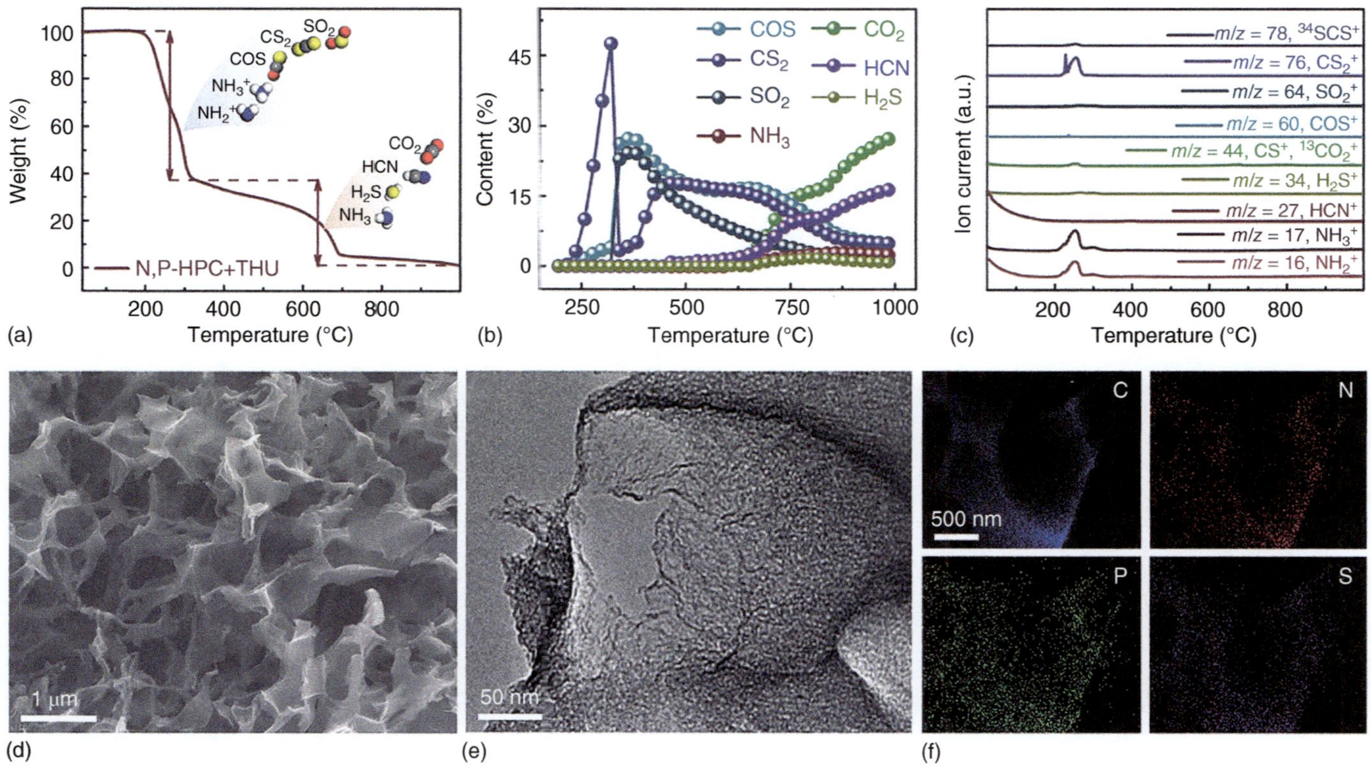

Figure 8.6 (a) TG curve of the mixture of N,P-HPC and THU. (b) The decomposition product contents. (c) Single-ion current curves of co-pyrolysis of N,P-HPC with THU according to TG–MS. Representative (d) SEM and (e) TEM images of S,N,P-HPC-1. (f) The elemental mapping images of C, N, P, and S in S,N,P-HPC-1. Source: Zan et al. [49]/with permission of © The Royal Society of Chemistry.

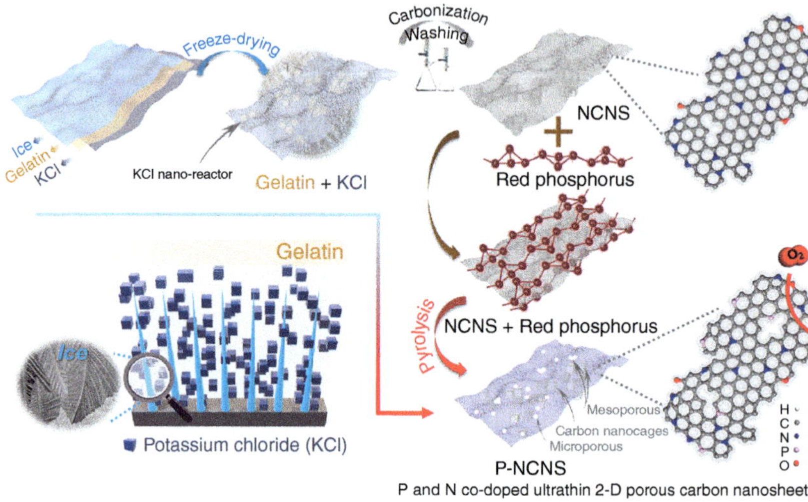

Figure 8.7 The synthesis process of P-NCNS. Source: Reproduced with permission from Zan et al. [51]; © 2021, Elsevier.

SSA (499.5 m^2 g^{-1}) with abundant N,P-doping sites. The thickness of the carbon nanosheet is 1–2 nm.

Figure 8.7 shows the process of synthesizing ultrathin two-dimensional P-NCNS using an effective method. KCl was used as an independent sealed nanoreactor to conduct closed pyrolysis. Thus, the N-doped 2D porous carbon nanosheets (NCNS) exhibit high N content (10.38 at%), Then, the obtained NCNS was pyrolyzed with red P for a second time to transform into N and P co-doped 2-D carbon nanosheets (P-NCNS). TG analysis shows that the co-pyrolysis stage of NCNS containing red P appeared in the range of 360–540 °C (Figure 8.8a) with a significant weight loss of 80.1%. As shown in TG–MS and TG–FTIR spectra (Figure 8.8b,c), P-based species (P—O—C, P—O/P=O, and P—OH) and CO_2 were generated. Then, at temperatures above 800 °C, the weightlessness persisted at 4.0% due to the separation of heteroatoms in the form of P-based species, NH_3, and CO_2 [52]. When the P-containing species flow along the carbon nanosheets, they are not only beneficial to heteroatomic doping, but also may be used as pore-forming agents to produce abundant porosity [53].

As shown in Figure 8.9, the P weight percentage in P-NCNS was estimated to be 5 wt%. The high-resolution N 1s spectrum of P-NCNS can be decomposed into oxidized pyridinic N (402.2 V), graphitic N (400.9 V), pyrrolic N (399.9 V), and pyridinic N (399.0 V). The N content of P-NCNS (5.00 at%) is much lower than that of NCNS (10.38 at%), which is owing to the elimination of numerous N atoms. However, the N content of P-NCNS is still higher than that of H-NCNS (4.01 at%). This phenomenon indicates that the co-pyrolysis results in the decrease of N content, and the introduction of P may fix N [54]. In addition, P-NCNS has a higher content of pyridinic N and graphitic N than H-NCNS, which may benefit from the introduction of red P [55]. These results show that the N content of NCNS (10.38 at%) is much higher than NC

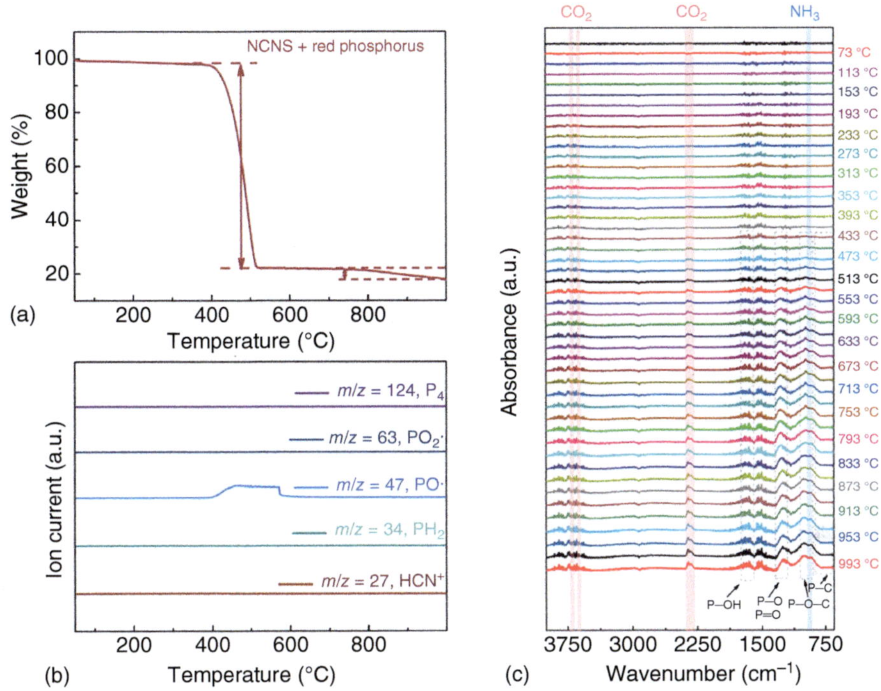

Figure 8.8 (a) TG curve. (b) Single-ion current curves of co-pyrolysis of NCNS with red P based on TG–MS. (c) FTIR spectra during the co-pyrolysis process. Source: Reproduced with permission from Zan et al. [51]; © 2021, Elsevier.

(3.51 at%), while similar to NC (10.11%), indicating that the samples without KCl suffered a great deal of loss of active N species. The deconvoluted XPS spectrum of P 2p shows the presence of P–C and P–O moieties, indicating that P atoms have been successfully doped into the resulting carbon.

X-ray adsorption near-edge structure (XANES) measurements are also conducted to analyze the binding states of P and N heteroatoms in the carbons. The K-edge XANES spectra of N have three broad peaks centered at 403.0, 401.5, and 400.0 eV, representing graphitic N, pyrrolic N, and pyridinic N species, respectively, which are consistent with the results obtained by XPS. The P K-edge XANES spectrum has a strong peak at 2153.0 eV, which is attributed to the P 1s/s transition. What's more, a 2.0 eV shoulder peak is detected at lower energy, which is attributed to P—C bond formation.

8.2 Nature-Inspired Methods

Although carbon materials with different porosities and heteroatoms (including collagen-derived carbons) can be prepared by traditional methods, the complexity of templates removal and high cost make these methods unsuitable for large-scale preparation [29]. Owing to the closed structure of biomass precursors preventing

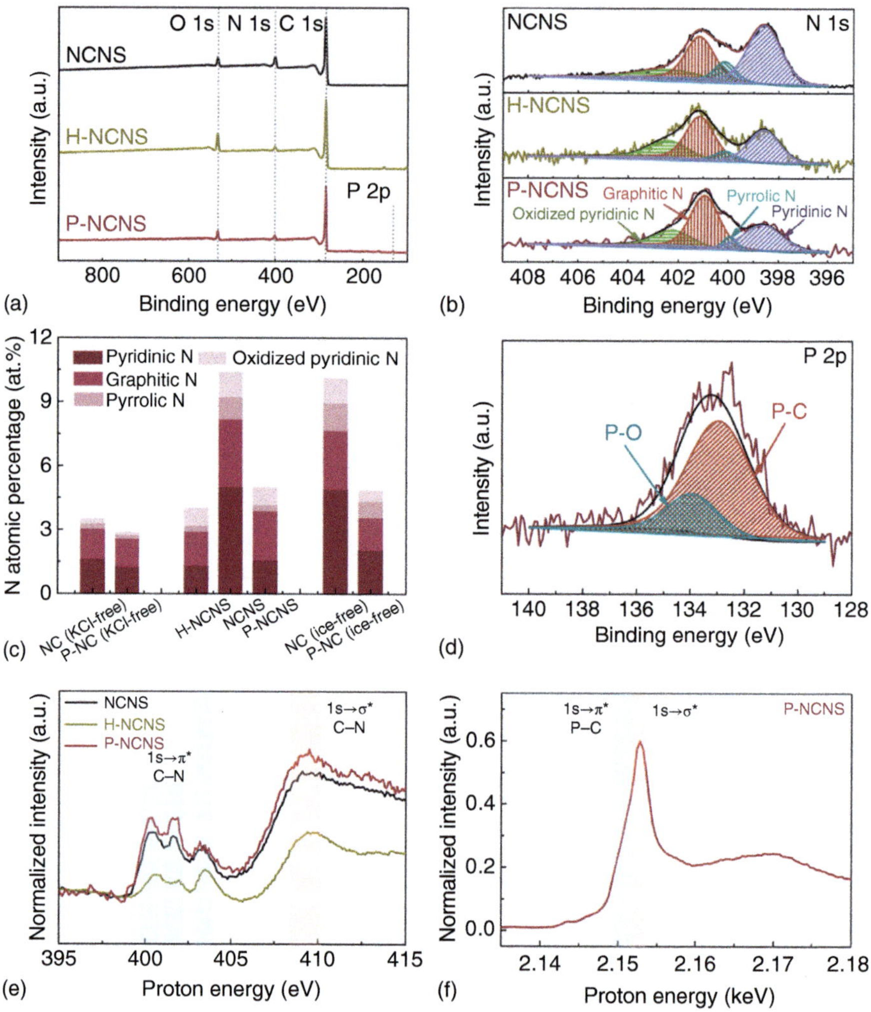

Figure 8.9 (a) XPS survey and high-resolution XPS spectra of (b) N 1s for NCNS and P-NCNS. (c) Summary of N atomic percentage for NCNS, P-NCNS, NC (KCl-free), P-NC (KCl-free), NC (ice-free), and P-NC (ice-free). (d) P 2p for P-NCNS. (e, f) N and P K-edge XANES spectra of NCNS and P-NCNS. Source: Reproduced with permission from Zan et al. [51]; © 2021, Elsevier.

the activators from entering the biomass precursors during activation, most activators only produce pores on the surface or near the surface of the biomass precursors [56]. Moreover, to ensure large pore volume and high SSA, high pyrolysis temperature is usually needed, which easily leads to the elimination of heteroatoms and uncontrollable morphology [57].

Collagen-enriched biomass materials have many merits to prepare carbon materials with different porosities and heteroatoms. Firstly, hydroxyapatite nanocrystals are uniformly dispersed in fish scales and bones. These can act as template to directly prepare collagen-derived carbons without additional templates and activators. Secondly, the synergistic effect of internal hydroxyapatite template and external activators can lead to novel morphologies and porosities. Thirdly, in hot water and acid or alkaline solutions, the collagen can be changed to water-soluble gelatin and amino acids. Gelatin and amino acids can mix with activators and water-soluble templates uniformly, which is conducive to prepare collagen-derived carbons with controlled morphology and porosity. Finally, numerous heteroatoms in collagen can help to produce in situ heteroatom-doped carbons. To synthesize collagen-derived carbons with heteroatoms and open porous structures, many effective methods have been proposed.

8.2.1 Hydroxyapatite-Induced Self-Activation Method

In 2010, Huang and coworkers proposed a facile strategy to prepare collagen-derived carbons using animal bone as the carbon precursor for the first time [58]. The pig bone was directly carbonized at 850 °C and then pickled, and the obtained collagen-derived carbon showed a hierarchical porous structure with high SSA of 861 $m^2\ g^{-1}$. They claimed that the formation of mesopores and micropores is attributed to the template effect of hydroxyapatite and the pyrolysis of organic species, respectively. Two years later, Redepenning et al. prepared a collagen-derived carbon with a higher SSA of 1383 $m^2\ g^{-1}$ by directly pyrolyzing cattle bone at 1000 °C. They also proposed that the hierarchical porous structure was caused by hydroxyapatite crystals.

Recently, Wang and coworkers proposed a hydroxyapatite-induced self-activation mechanism to explain the hierarchical porosity of the above bone-derived carbons [59]. To investigate the pore-forming mechanism, they prepared several collagen-derived carbons by directly pyrolyzing cattle bone at different temperatures (600, 850, 900, 1000, 1100, 1200 °C). The product carbonized at T °C was denoted as CP-T. As the temperature increased, hydroxyapatite shrunk from acicular to irregular, especially above 900 °C, which can be seen in the high resolution transmission electron microscopy (HRTEM) images of CPs. After removing hydroxyapatite with dilute hydrochloric acid, two kinds of carbon (nonporous carbon [NPC-T] and porous carbon [PC-T]) were obtained according to temperature conditions. All the carbon materials maintained the fiber bundle-like structure after pickling. As the temperature rises, the surfaces of carbons change from nonporous to porous.

The HRTEM image in Figure 8.10a shows the specific structure of carbon. The carbon layer prepared at 600 °C showed a disordered accumulation structure and no obvious pores. When the temperature rises to 850 °C, there are some micropores. Although only 50 °C was added, elliptical mesopores with a size of about 4 nm appeared in the samples obtained at 900 °C. With the continuous increase of temperature, numerous mesopores were generated at 1000 °C. HRTEM images of PC-1100 and PC-1200 show that mesopores are closely connected with ultrathin pore walls, and some of them even coalesced.

The PC-900, PC-1000, PC-1100, and PC-1200 exhibited much larger hysteresis loops than that of PC-850 (Figure 8.10b, inset), suggesting that the mesopores were

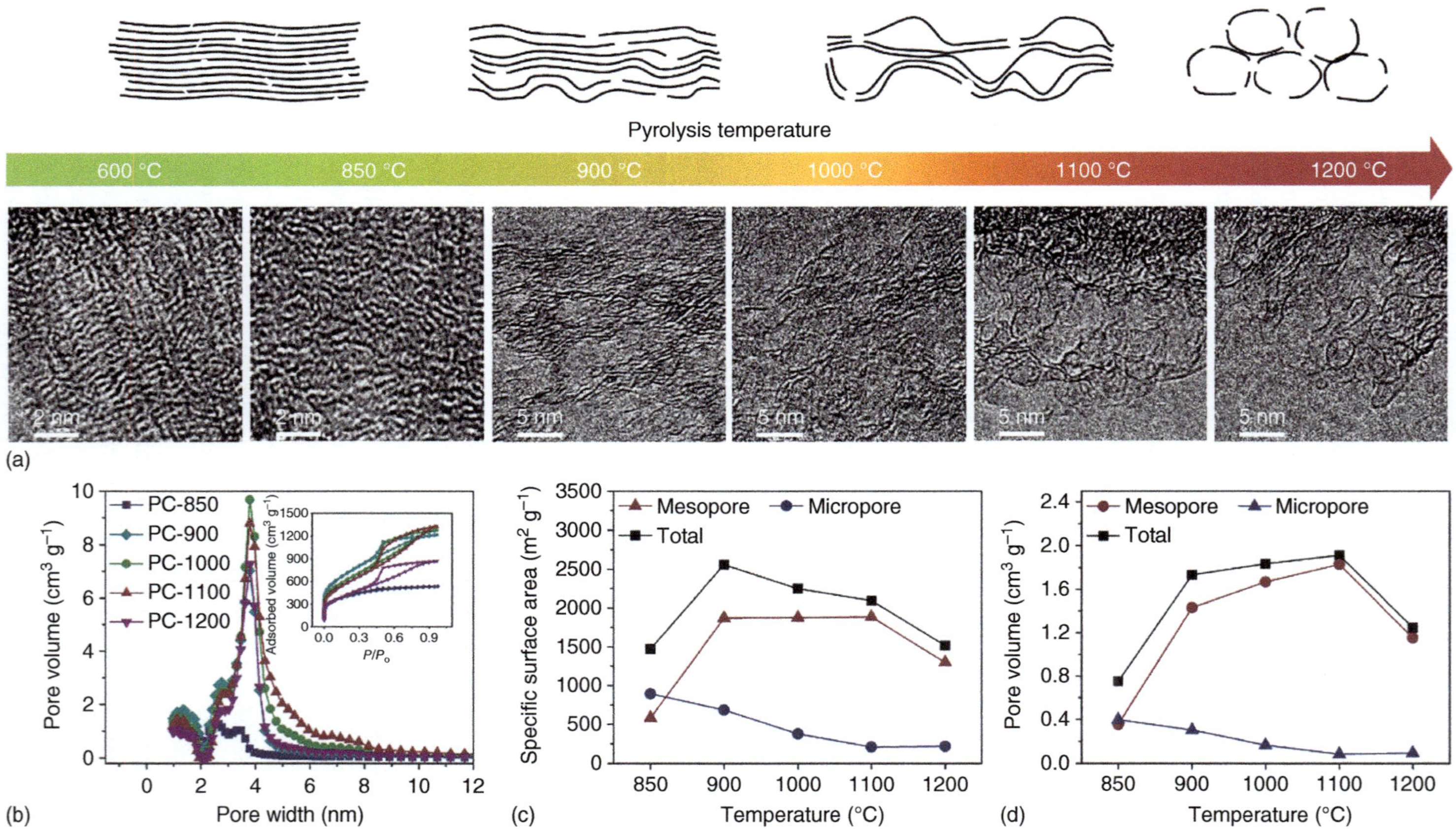

Figure 8.10 (a) TEM images of CP-T. (b) Nitrogen adsorption–desorption isotherms (inset) and pore-size distributions of PCs. (c) SSA and (d) pore volume of PCs. Source: Niu et al. [58]/with permission of © Elsevier Ltd.

mainly formed above 900 °C. The pore size of PC-850 was from 1 to 4 nm. The pore size of PC-900 was increased, and the pore-size diameter (PSD) is centered at 4 nm. Besides, its mesoporous SSA increased significantly, while microporous SSA decreased (Figure 8.10c). It is worth noting that the 4 nm-centered PSD remains constant when the temperature rises from 900 to 1200 °C. At 1100 °C, the SSA of mesopores and micropores reached maximum and minimum values of 1884.9 and 210.7 $m^2 g^{-1}$, respectively. The PC-1100 also shows a high mesoporosity with the highest mesopore volume of 1.829 $cm^3 g^{-1}$. At 1200 °C, the mesopore SSA and pore volume all decreased (Figure 8.10d).

The formation process of PCS was determined by thermogravimetric analyzer/differential scanning calorimetry (TGA/DSC) method. In an Ar atmosphere (Figure 8.11a), two exothermic peaks at 380 and 460 °C were clearly observed, which are due to the primary carbonization of collagen. A small exothermic peak was observed around 750 °C. Different from the first two peaks, this peak did not change even in an air atmosphere, suggesting that the change in bone at 750 °C was not related to carbon but hydroxyapatite. This phenomenon was further studied by TG–IR. As shown in Figure 8.11b, when the temperature increases from 730 to 770 °C, the characteristic band of H_2O at 3400–4000 cm^{-1} is enhanced [60]. This indicates that the exothermic peak at 750 °C in the DSC curve is the result of the thermal decomposition of hydroxyapatite, and H_2O was generated. Moreover, the intensity of CO in the characteristic band of 2030–2240 cm^{-1} increases. According to Eq. (8.1), the increase of CO content may be due to the steam activation process. XRD pattern of CPs without acid treatment is shown in Figure 8.11c. With the increase of temperature, the characteristic peak of hydroxyapatite becomes intense and narrow, which is due to the increase of crystallinity caused by the regrowth of hydroxyapatite [54]. Moreover, the PC-1200 has a peak at 37.4°. This peak is characteristic of CaO, which is the product of further decomposition of hydroxyapatite. It is also shown that CaO was basically formed at 900 °C and gradually generated with H_2O due to the further decomposition of hydroxyapatite at 1200 °C. FTIR patterns of the cattle bones carbonized in an Ar atmosphere (Figure 8.11d) exhibit the bands at 1418 and 1456 cm^{-1}, which correspond to B-type carbonate groups [55]. This indicated the existence of a large number of carbonate groups in the hydroxyapatite. These two bands disappear when the temperature was raised to 900 °C, indicating that CO_2 is generated by CaO [54, 55]. According to Eq. (8.2), CO_2 can also be used as an activator. The use of CO_2–vapor mixtures produced large pore sizes and high specific surface areas due to combined effects [61].

Based on the above analysis, it could be inferred that hydroxyapatite decomposes to H_2O at ~750 °C, and the pore size was less than 4 nm after steam activation at 850 °C. At 900 °C, hydroxyapatite was further decomposed to generate CO_2, and the pore size increases due to steam–CO_2 activation. When the pore diameter reaches the thickness of hydroxyapatite (~4 nm), the hydroxyapatite crystals detach from the carbon-coating structure and regrow. The remaining porous carbon retained its porosity owing to the hydroxyapatite-induced activation abortion. Thus, even if the temperature continued to rise, the PSD center was still at ~4 nm. The micropores were gradually expanded to mesopores, resulting in the decreased SSA of micropore.

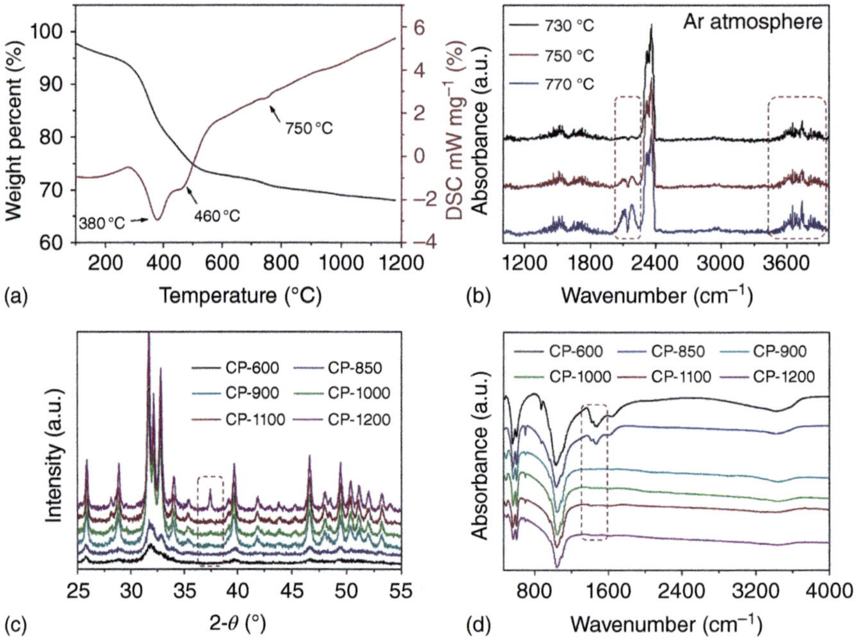

Figure 8.11 (a) TGA/DSC curves and (b) TG–IR patterns of cattle bone tested in an Ar atmosphere. (c) XRD patterns and (d) FTIR patterns of CP-T in an Ar atmosphere. Source: Reproduced with permission from Niu et al. [58]; © 2017, Elsevier.

At high temperatures of 1000 and 1100 °C, due to strong chemical activation, some pores of 5–8 nm size were produced, resulting in pore coalescence and ultrathin pore walls. After steam–CO_2 activation at 1200 °C, some mesopores were broken. The pore volume and SSA of porous carbon decreased accordingly.

Raman spectroscopy and XRD were used to analyze the defects and structural changes of the carbon layer. The intensity of D peak was much higher than that of G peak and increased with the increase of temperature (Figure 8.12a). This result indicated that the activation process produced more defects on the carbon layer. The content of (D′+ D″) reached the maximum value at 850 °C but decreased with the increase of temperature, suggesting that the stacked structure was formed by steam activation at 850 °C, and the structure was etched away by the enhanced CO_2–steam activation above 900 °C. XRD analysis also confirmed the structural changes (Figure 8.12b). For PCs, the peak (002) widens and weakens with the increase of temperature and almost disappears in other PCs. The G-peak content also had the same trend, which led to the decrease of resistivity. Although PC-850 had the most defects, a large number of oxygen-containing functional groups, which were not beneficial for electron transport, were removed at 850 °C. As a result, its electroconductivity was higher than that of NPC-600 (Figure 8.12c). Owing to the PC-1200 structure being partly broken by the enhanced activation, the resistivity of PC-1200 was increased compared with PC-1100. Due to the integrated fiber bundle-like structure, low O content, and relatively high content of

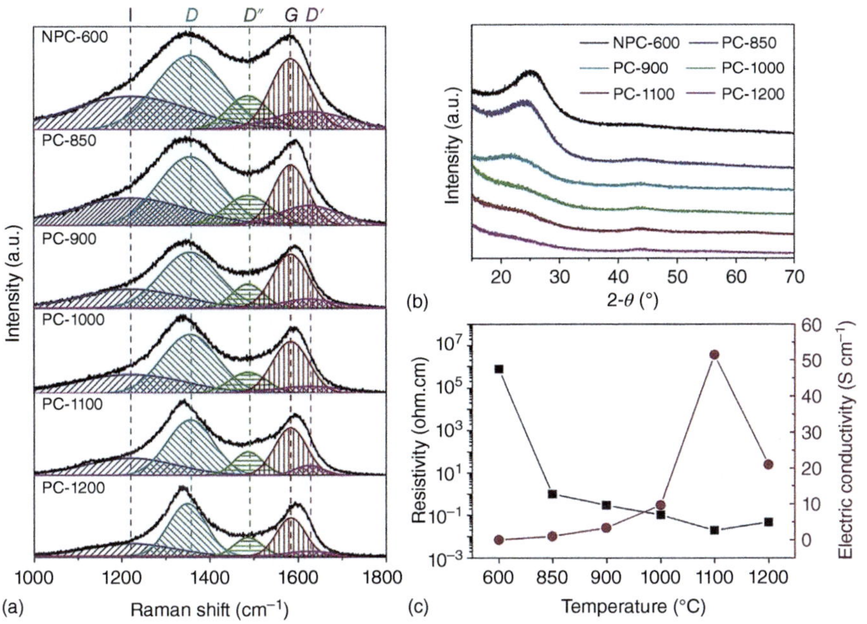

Figure 8.12 (a) Raman spectra, (b) XRD patterns, and (c) resistivities and electrical conductivities of all the carbons. Source: Reproduced with permission from Niu et al. [58]; © 2017, Elsevier.

sp^2-boned graphitic carbons, PC-1100 exhibited the highest electrical conductivity of 5141 S m^{-1}.

To eliminate the influence of precursors on porosity, collagen-derived carbon was also prepared by pyrolysis of sheep bone at 1100 °C [62]. The morphology and porosity of the resulting material (HDMPC) were similar to PC-1100. The HDMPC was obtained by directly carbonizing sheep bone powder at 1100 °C followed by acid pickling. The carbonized product retains the fibrous bundle shape formed by the originally mineralized collagen fibrous bundle structure in the sheep bone [55], and the hydroxyapatite after heat treatment recrystallized from acicular to massive. The integrity of HDMPC remains unchanged even after pickling with dilute hydrochloric acid (Figure 8.13a). High-magnification SEM (Figure 8.13b) shows a rough surface of HDMPC. TEM images show a large number of pores inside the HDMPC (Figure 8.13c). As shown in Figure 8.13d, HDMPC was mainly composed of small mesopores with interconnected structure.

Figure 8.14a (inset) is the adsorption–desorption isotherm of HDMPC. The curve shows a type I/IV adsorption–desorption isotherm. When the relative pressure is greater than 0.4, the adsorption–desorption curve exhibited an obvious hysteresis loop, indicating the presence of micropores and mesopores. The total pore volume was 1.99 cm^3 and specific surface area was 2192 m^2 g^{-1}. The pore-size distribution showed that the porosity of HDMPC was mainly mesoporous, and the pore size is mainly about 4 nm (Figure 8.14a). The mesopore volume was calculated to be 1.95 cm^3 g^{-1} (93.9% of the total pore volume) and the specific surface area mesopore

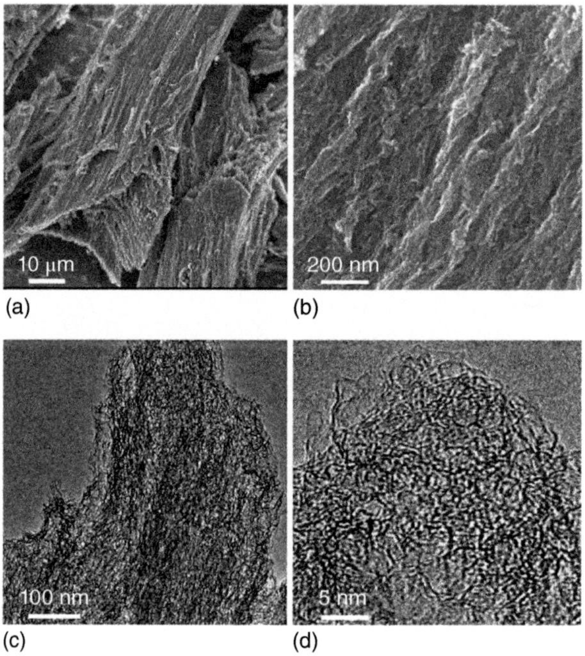

Figure 8.13 (a, b) SEM images, (c) TEM image, and (d) HRTEM image of HDMPC. Source: Niu et al. [62]/with permission of © Elsevier Ltd.

was 2058 $m^2 g^{-1}$ (98.0% of the total specific surface area). The porous structure of HDMPC was significantly different from that of previous porous carbon prepared by activator or template [63]. TEM image of the carbonized product without acid pickling showed that the hydroxyapatite size was more than 20 nm. However, there were only a few pores larger than 20 nm in HDMPC. This indicated that hydroxyapatite did not play a template role in the formation of mesoporous pores.

The pore-formation mechanism of HDMPC was analyzed by TGA/MS. The TG and DTG curves are shown in Figure 8.14b. The significant weight loss between 200 and 400 °C was because of the thermal decomposition and primary carbonization of collagen [59]. When the DTG value increases from 400 to 600 °C, the corresponding weightlessness rate decreases. However, the DTG value decreased after 600 °C and lost weight rapidly above 950 °C, which was different from other carbon precursors that lost weight slowly at high temperature. This result suggested that a self-activation process might occur after 600 °C. The evolution of CO, CO_2, H_2O, and H_2 gases during carbonization was detected by MS to explore the mechanism of self-activation. The evolution value of H_2O increased significantly after 600 °C (Figure 8.14c), which might be the result of the gradual generation of H_2O after hydroxyapatite decomposition [64]. Moreover, according to Eq. (8.1), the increased evolution values of H_2 and CO were caused by the enhancement of steam activation. With the increase of temperature, the evolution values of CO, H_2O, and H_2 increased continuously. However, the generation rate of H_2 was much lower than that of CO, suggesting that another reaction has taken place to generate CO. Figure 8.14d shows the FTIR spectra of the raw bone powder and the carbonized material. After carbonization process, the B-type carbonate group peaks in hydroxyapatite

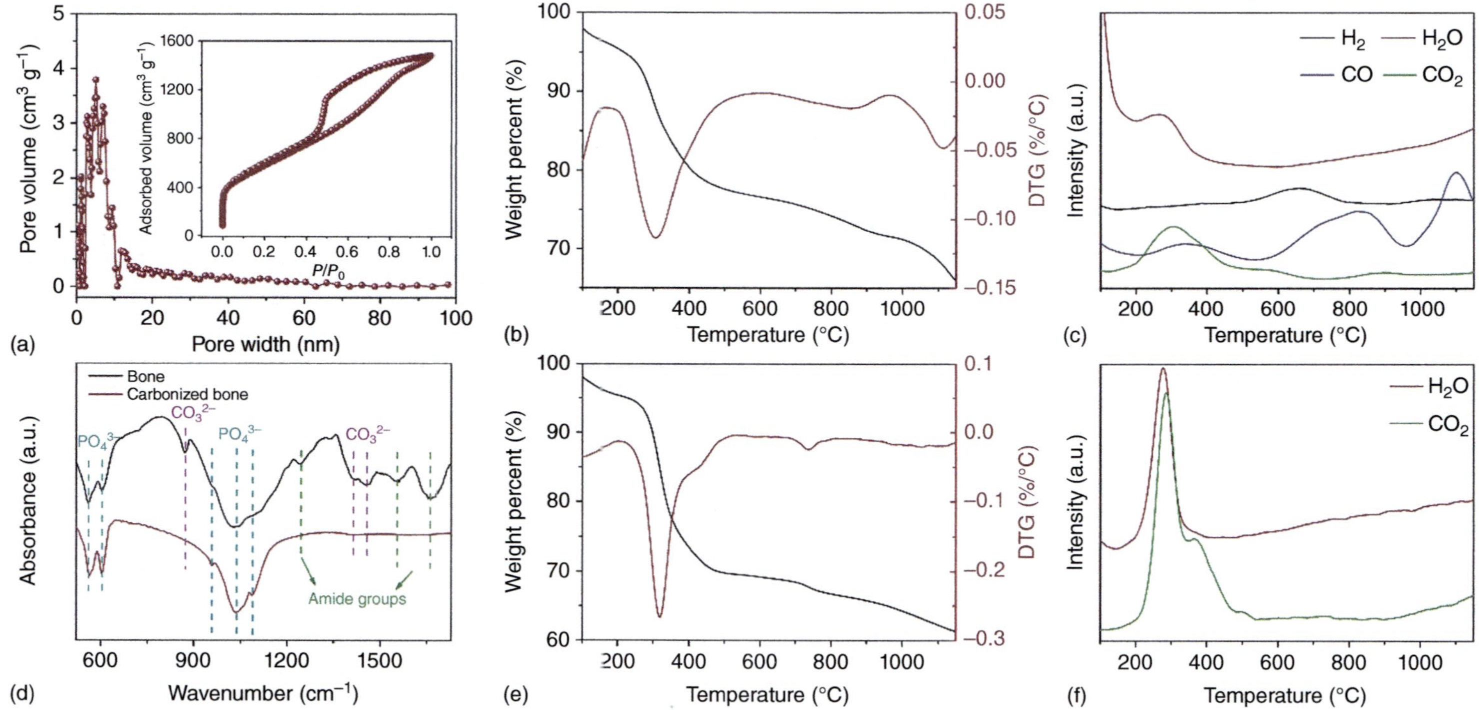

Figure 8.14 (a) Nitrogen adsorption–desorption isotherm (inset) and pore-size distribution of HDMPC. (b) TGA and (c) corresponding MS profiles of sheep bone pyrolyzed in the Ar atmosphere. (d) FTIR patterns of sheep bone and carbonized product without acid pickling. (e) TGA and (f) corresponding MS profiles of sheep bone pyrolyzed in the air atmosphere. Source: Reproduced with permission Niu et al. [62]; © 2018, Elsevier.

(876, 1417, and 1456 cm^{-1}) [56] disappeared, indicating that CO_2 was generated from hydroxyapatite decomposition over 950 °C. According to Eq. (8.2), CO_2 could act as an activator, and the evolution value of CO is larger after 950 °C. Notably, at such high temperatures (1100 °C), the in situ co-activation of steam and CO_2 was much stronger than conventional physical activation, resulting in mesopores rather than micropores [61].

TGA measurements were also conducted to further confirm the inference. The weight of sheep bone decreased continuously above 670 °C (Figure 8.14e), suggesting that hydroxyapatite was decomposed. As shown in Figure 8.14f, the corresponding MS curves further confirmed that hydroxyapatite decomposed at high temperature and generated CO_2 and H_2O. Because of the in situ hydroxyapatite-induced activation, the mesopore-dominant porous carbon was obtained from sheep bone.

8.2.2 Hydroxyapatite–KOH Synergistic Method

Even though the self-activation method induced by hydroxyapatite can be used to prepare collagen-derived carbons without additional activators and templates, hydroxyapatite cannot be used as a template to create pores. In addition, hydroxyapatite-induced self-activation of collagen-derived carbons commonly maintains the bulk structures of the carbon precursors. To solve these problems, KOH was also used to regulate the porous structures, morphologies, and heteroatoms of collagen-derived carbons.

In the subsequent work, Wang and coworkers proposed a hydroxyapatite/KOH co-activation mechanism [65]. It was found that when both hydroxyapatite and KOH were presented, carbon networks (NHPC–HA/KOH) could be obtained using cattle bone as precursor. When hydroxyapatite or KOH existed, only porous carbon bulk could be obtained. They proposed that hydroxyapatite-induced pores provide many channels for KOH to penetrate the precursors, enhancing activation inside the material, resulting in major mesoporous/macroporous pores for NHPC–HA/KOH. In further research, they reported a HA/KOH synergistic mechanism for the carbon formation.

He et al. reported a simple method to synthesize N,O co-doped HPC (N,O-HPC) networks using cattle bones as precursors [66]. The cattle bones were pre-carbonized in an Ar atmosphere at low temperature to maintain a considerable amount of heteroatom-doped carbon network. Then, N,O-HPC was obtained by pyrolyzing the pre-carbonized powders with KOH. It should be noted that N,O-HPC carbonized at 850 °C (N,O-HPC-850) has a unique mesoporous-dominated hierarchical porosity (Figure 8.15a).

Figure 8.15(b–e) shows the SEM images of N,O-HPCs to characterize their morphologies and structures. All the samples had unique 3D porous networks with numerous mesoporous/macroporous structure, which resulted in a sheet-like pore wall. Nitrogen adsorption/desorption measurements were further carried out to determine the porous structures of N,O-HPCs (Figure 8.16a) [50]. The pore volume and BET surface area of N,O-HPC-850 were 3.18 cm^3 g^{-1} and 2520 m^2 g^{-1},

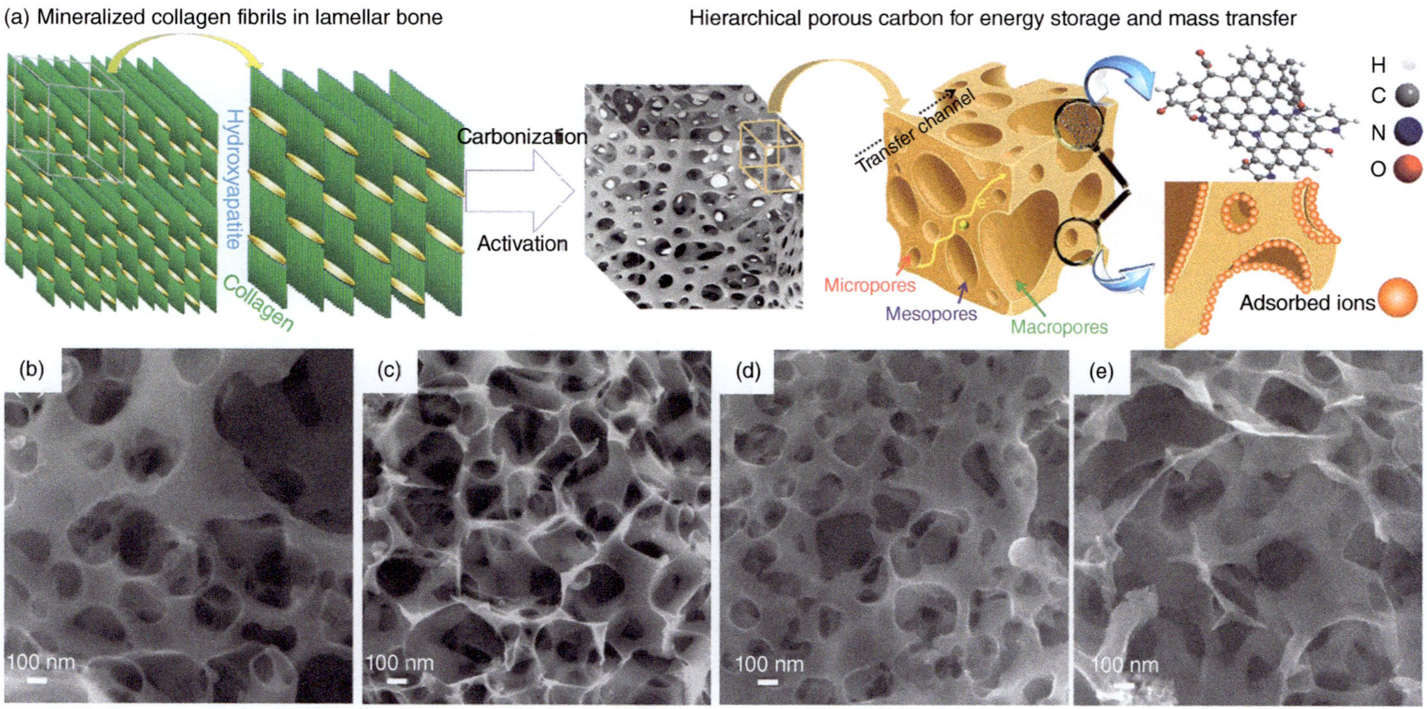

Figure 8.15 (a) Schematic illustration of the preparation of N,O-HPC. SEM images of (b) N,O-HPC-650; (c) N,O-HPC-750; (d) N,O-HPC-850; and (e) N,O-HPC-950. Source: He et al. [66]/with permission of © Elsevier Ltd.

respectively, which were greatly larger than those of other N,O-HPCs. Figure 8.16b shows the pore-size distribution plots for all N,O-HPCs. N,O-HPC-850 exhibited more mesopores than other N,O-HPCs, suggesting a mesopore-dominant porosity. Apparently, N,O-HPC-850 had the highest mesoporous volume and mesoporous specific surface area in N,O-HPC samples. The higher intensity of D peaks in Raman spectra (Figure 8.16c) also confirmed the disordered structure of N,O-HPCs. In addition, with the increase of carbonization temperature, the defects produced by KOH activation increased, and the I_D/I_G increased with the increase of carbonization temperature.

High-resolution XPS spectra were analyzed to study the chemical compositions of the N,O-HPC samples (Figure 8.16d–f). The C 1s XPS spectra were deconvoluted into four peaks of C=C, C—C, C—O/C—N, and C=O. With the increase in carbonization temperature, the C=C content decreased, indicating the structures of N,O-HPCs are getting more disordered. It also revealed that the doped N atoms were in the form of graphitic N (401.2 eV), pyrrolic N (400.2 eV), and pyridinic N (399.5 eV) [67]. Pyrrolic N and pyridinic N have been proved to can generate pseudocapacitance because of the following reactions:

$$>C{=}NH_2 + 2OH^- \leftrightarrow\ >C{=}NH + 2e^- + H_2O \tag{8.26}$$

$$>C{-}NH_2 + 2OH^- \leftrightarrow\ >C{-}NHOH + 2e^- + H_2O \tag{8.27}$$

Moreover, quaternary nitrogen may be beneficial to improving electron transfer. The pyrrolic N and pyridinic N contents of N,O-HPCs prepared at low temperature were higher. The O 1s XPS spectra of N,O-HPCs were deconvoluted into four peaks: C=O (531.2 eV), C—O (532.2 eV), O=C—O (533.2 eV), and chemisorbed O (534.3 eV) [68]. Although the conductivities of N,O-HPCs may be reduced by O atoms doped on carbons, the O heteroatoms can improve the surface wettability and contribute pseudocapacitance as follows:

$$>C{-}OH + OH^- \leftrightarrow\ >CO^- + H_2O \tag{8.28}$$

$$>C{=}O + OH^- \leftrightarrow -COOH + e^- \tag{8.29}$$

$$-COOH + OH^- \leftrightarrow -COO^- + H_2O \tag{8.30}$$

Based on the previous work, by using cattle bone as a precursor, Shao et al. prepared a mesopore- and macropore-dominant NHPC with 3D-interconnected porosity by synergistic activation of hydroxyapatite and KOH (NHPC–HA/KOH). For comparison, the preparation of NHPC by hydroxyapatite self-activation (NHPC–HA) and KOH chemical activation (NHPC–KOH) was also made.

SEM and TEM were used to characterize the morphologies of the NHPCs. Figure 8.17a1 shows that the fiber bundle-like structure of NHPC–HA was reserved after self-activation. Moreover, as shown in Figure 8.17a2,a3, the rough surface of NHPC–HA with some mesopores could be observed. When KOH was used as the activator and the hydroxyapatite was removed by acid washing first, NHPC–KOH exhibited irregular bulk structure (Figure 8.17b1). Figure 8.17b2,b3 shows that

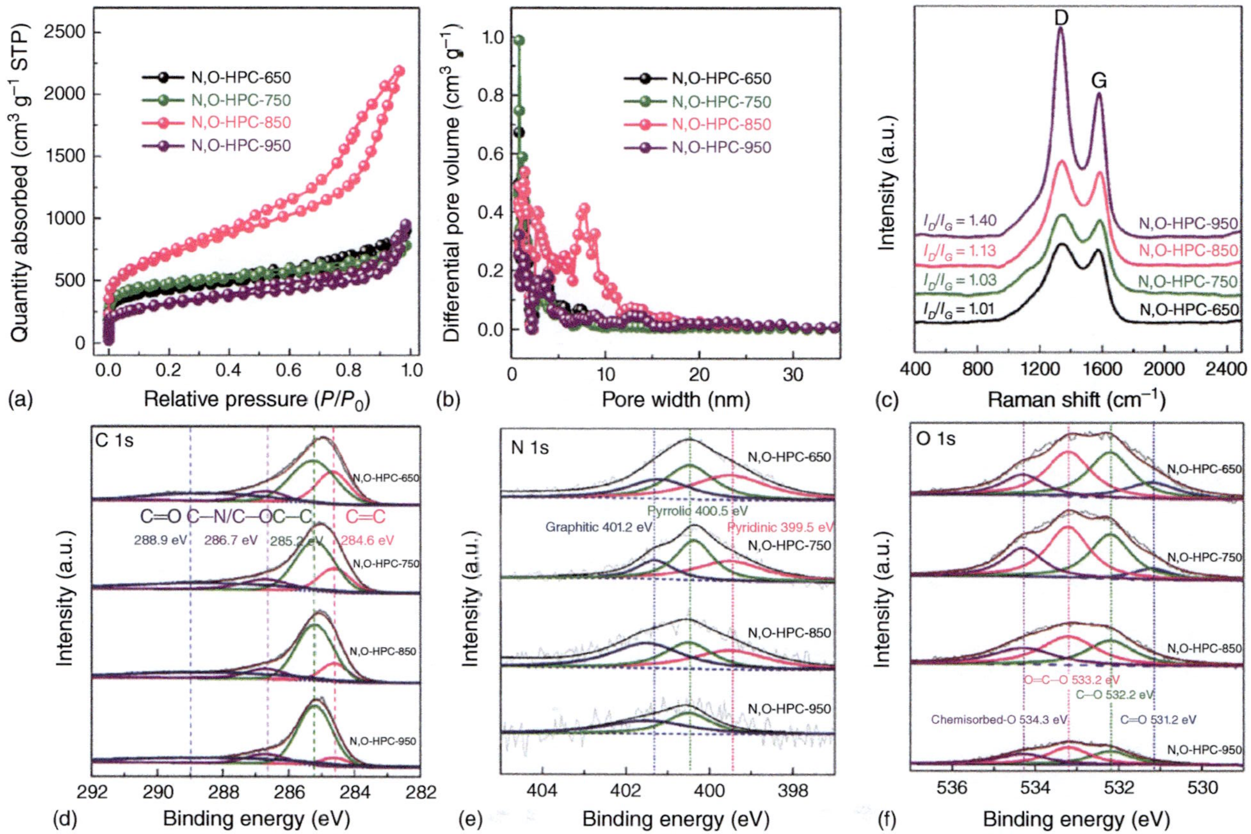

Figure 8.16 (a) N_2 absorption–desorption isotherms, (b) pore-size distributions, (c) Raman spectra. High-resolution (d) C 1s, (e) N 1s, and (f) O 1s spectra of N,O-HPCs. Source: Reproduced with permission from He et al. [66]; © 2017, Elsevier.

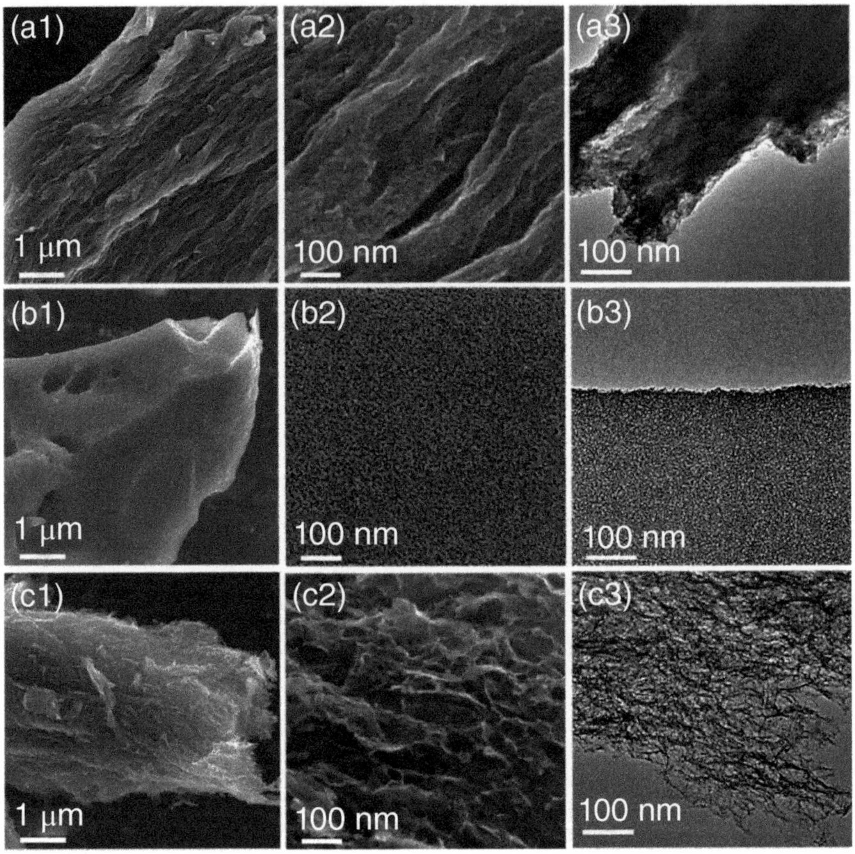

Figure 8.17 SEM images of (a1, a2) NHPC–HA, (b1, b2) NHPC–KOH, and (c1, c2) NHPC–HA/KOH; HRTEM images of (a3) NHPC–HA, (b3) NHPC–KOH, and (c3) NHPC–HA/KOH. Source: Shao et al. [69]/with permission of © American Chemical Society.

KOH activation produced a number of small-sized mesopores on the surface of NHPC–KOH.

Due to the coexistence of hydroxyapatite and KOH, there were highly interconnected mesopores and macropores in NHPC–HA/KOH, which were the result of co-activation of hydroxyapatite and KOH in the internal and external parts of the cattle bone (Figure 8.17c1–c3) [69]. It is important to note that the pores formed by the activation of KOH could provide numerous channels for KOH to penetrate into the bone, which enhanced the internal activation, resulting in highly interconnected pores of NHPC–HA/KOH.

N_2 adsorption–desorption measurements were carried out to characterize the porous structures of all the NHPCs. At high relative pressure, all the NHPCs showed type I and IV adsorption–desorption isotherms, which have obvious hysteresis loops (Figure 8.18a), indicating the presence of micropores and mesopores. Notably, the hysteresis loop of NHPC–HA/KOH was much larger than that of NHPC–KOH and NHPC–HA. Moreover, when the relative pressure was greater than 0.7, the hysteresis loop of NHPC–HA/KOH was obviously absorbed, indicating that macroporous structure existed in NHPC–HA/KOH. Figure 8.18b shows that NHPC–HA and NHPC–KOH had a large number of micropores and small mesopores with pore size below 8 nm, while NHPC–HA/KOH consisted of micropores, mesopores, and macropores with the PSD range of 0.5–200 nm. These results were consistent with those of SEM and TEM. The SSA of NHPC–KOH was as high as 3236 $m^2 g^{-1}$, which was mainly derived from micropores. The mesoporous volume was 0.96 $cm^3 g^{-1}$ (50% of the total), and the mesoporous SSA was about 1027 $m^2 g^{-1}$ (31% of the total). NHPC–HA exhibited the lowest mesoporous volume (0.74 $cm^3 g^{-1}$) and mesoporous SSA (801 $cm^3 g^{-1}$), which had similar porosity to NHPC–KOH. Unlike NHPC–HA and NHPC–KOH, NHPC–HA/KOH had ultrahigh mesoporous and macroporous SSA of 2203 $m^2 g^{-1}$. The mesopore and macropore pore volumes were 3.00 $cm^3 g^{-1}$, accounting for 96% of the total pore volumes, suggesting that the porosity of NHPC–HA/KOH was dominated by mesopores and macropores.

As shown in Figure 8.18c, Raman spectra for NHPCs were fitted by five peaks of I peak, D peak, D″ peak, G peak, and D′ peak, which centered at 1220, 1350, 1490, 1580, and 1620 cm^{-1}. The G peak represents the sp^2-boned graphitic carbons, and D peak arises from the sp^3 defects on the graphene layers. The D′ and D″ peaks are caused by the graphene-layer stacking fault and lattice disorder, and I peak is attributed to the doped nitrogen. Figure 8.18d shows that the G-peak content for all the NHPCs was less than 20%, suggesting high defect density. In addition, NHPC–KOH and NHPC–HA/KOH had lower (D′+D″) content and higher D peak content than NHPC–HA. The results are in good agreement with the XRD analysis. In addition to inherent defects, NHPCs also have heteroatom-induced defects. The *I*-peak content of all NHPCs was more than 20%.

XPS measurements were carried out to characterize the surface heteroatoms of NHPCs. From Figure 8.18e, we can see that the C 1s spectra of all the NHPCs were fitted into four peaks: C=C (284.6 eV), C—C (285.2 eV), C—N or C—O (286.7 eV), and C=O (288.5 eV) [53] The high-resolution N 1s spectra (Figure 8.18f) showed that the N atoms possessed three types of nitrogen: pyridine nitrogen, pyrrolic nitrogen, and quaternary nitrogen [36], which centered at 398.4, 399.9, and 401.0 eV, respectively. Quaternary nitrogen could facilitate the electron transfer, while pyrrolic N and pyridinic N could provide additional active sites [70] The electrical conductivities of NHPC–HA (782 $S m^{-1}$), NHPC–KOH (134 $S m^{-1}$), and NHPC–HA/KOH (1144 $S m^{-1}$) were higher than that of commercial-activated carbon [39]. NHPC–HA/KOH exhibited the highest conductivity because it had the lowest O content among the three NHPCs and had a complete fibrous bundle structure.

To further modify the morphology of the carbon material, Niu et al. carbonized cattle bone with KOH at a low temperature of 550 °C [59, 71]. The thermal treatment temperature was only 550 °C, which was much lower than the previously reported KOH activation method [56]. To investigate the formation mechanism of HPCNS, two comparative experiments were carried out. Firstly, the pre-carbonized cattle powders were pyrolyzed at 550 °C without KOH. After washing with HCl and deionized water, carbon bulks were obtained (referred to as non-porous carbon bulk (NPCB)). To understand the effect of hydroxyapatite on the porosity, the

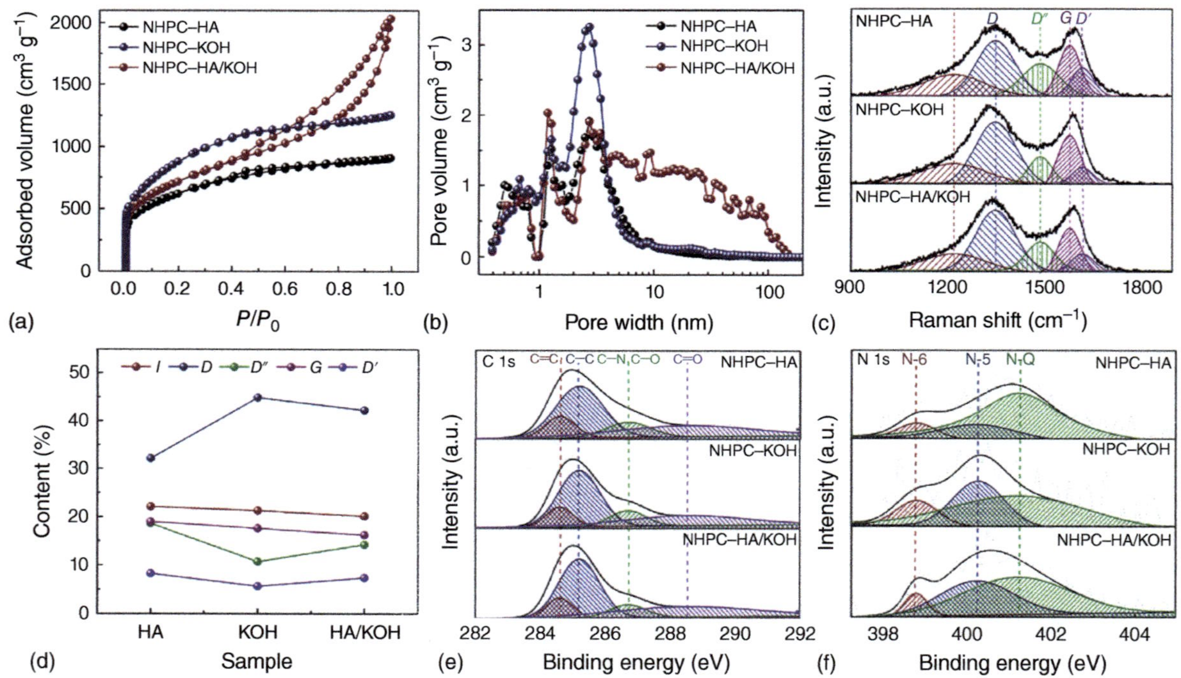

Figure 8.18 (a) N_2 adsorption–desorption isotherms, (b) pore-size distribution plots of NHPCs, (c) Raman spectra for NHPCs, and (d) content of integrated fitting peaks for NHPCs. XPS survey spectra and high-resolution (e) C 1s and (f) N 1s spectra of NHPCs. Source: Reproduced with permission from Shao et al. [69]; © 2018, Elsevier.

pre-carbonized products without hydroxyapatite were pyrolyzed at 550 °C with KOH (referred to as microporous carbon bulk (MPCB)).

With the addition of KOH, hydroxyapatite recrystallized after heat treatment and its size increased. XRD also confirmed the recrystallization behavior of HA. These results suggested that KOH may destroy the carbon coating on the surface of hydroxyapatite and promote the recrystallization of hydroxyapatite. Recrystallized hydroxyapatite crystals could be used as templates to form two-dimensional nanosheets, which provided new space for molten KOH penetration and improved the activation effects. By virtue of the synergistic effect of KOH and hydroxyapatite, the carbon bulks could be fully activated after pickling to form 3D tremella-shaped carbon composed of multilayer porous 2D nanosheets. During the heat treatment, KOH could only result in nanopores on the bulk surface without hydroxyapatite, resulting in abundant micropores and heterogeneous macropores on the surface of MPCB.

Figure 8.19a shows the XRD patterns of all the samples. Three (002) diffraction peaks centered at 25.38°, 24.41°, and 21.77° of HPCNS, MPCB, and NPCB were observed. According to Bragg's law, the corresponding carbon interlayer distances are 0.35, 0.36, and 0.41 nm, respectively. The increase of the distance between the carbon layers may be due to the increase of structural defects caused by the synergistic effect of HA/KOH and many heteroatoms, which is also confirmed by the Raman spectrum as shown in Figure 8.19b. The intensity ratio of D peak to G peak shows the defect level of the porous carbons [72]. The result of I_D/I_G for HPCNS was the highest, indicating that HPCNS had numerous defects.

The co-activation of both KOH and hydroxyapatite not only led to the unique structure and a large number of defects, but also affected their surface chemical state. According to the XPS results (Figure 8.19c), HPCNS had the lowest surface N content (6.4 at%) and the highest surface O content (17.1 at%) among all the samples [56]. Notably, the HPCNS N content was still higher than those of conventional KOH activated carbons [26, 56], which might result in a large number of defects [73]. The high-resolution N 1s and O 1s XPS spectra were analyzed (Figure 8.19d). The doped N atoms could be deconvoluted into pyridinic N (398.5 eV), pyrrolic N (400.1 eV), and quaternary nitrogen (401.2 eV) [54]. Among them, pyrrolic N and pyridinic N have high chemical activity, and quaternary nitrogen helps enhancing electron transfer. Moreover, the deconvolution of O 1s spectrum results in two peaks, which centered at 532.2 and 533.5 eV, indicating the existence of C=O and C—O, respectively [50].

Collagen-derived carbons (HPC-800 and HPC-550) derived from fish scales could be obtained at 800 and 550 °C by adjusting the carbonization temperature and KOH content (Figure 8.20) [74]. The morphology of HPC sample is different from that of traditional KOH-activated carbon with irregular block structure, showing honeycomb-like structures. The SEM images HPC-550 and HPC-800 (Figure 8.21a,e) show honeycomb-like structure. As shown in Figure 8.21b,f, TEM images indicate the presence of a large number of interconnected mesopores. HRTEM images (Figure 8.21c,g) show that the HPCs were in amorphous structure. According to the line-scan profiles of lattice fringes (Figure 8.21c,g, inset), it can be determined that the average carbon interlayer distances of HPCs were about

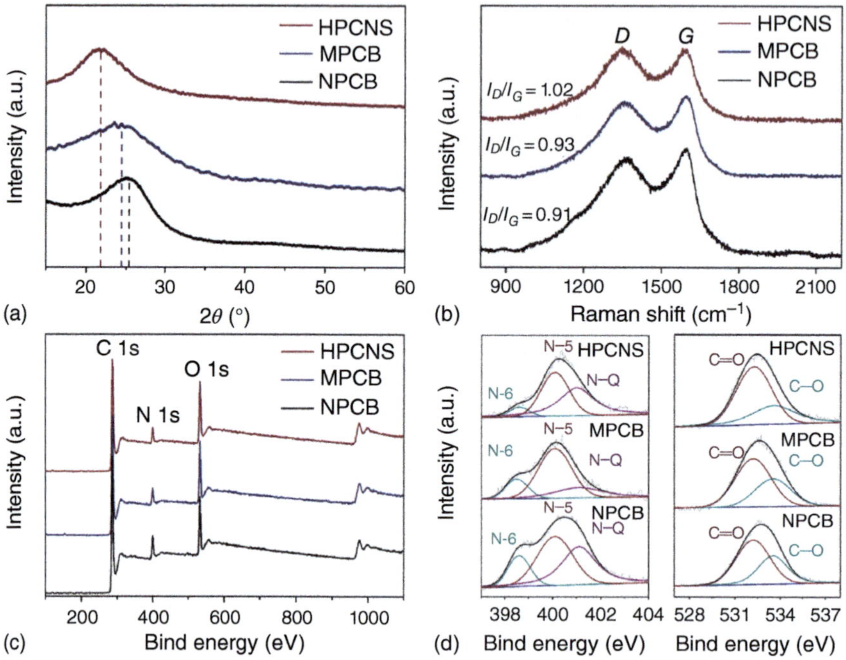

Figure 8.19 (a) Nitrogen adsorption–desorption isotherms and (b) plots of the pore-size distribution. (c) XRD patterns, (d) Raman spectra of HPCNS, MPCB, and NPCB. Source: Reproduced with permission from Refs. [59, 71]; © 2018, Elsevier.

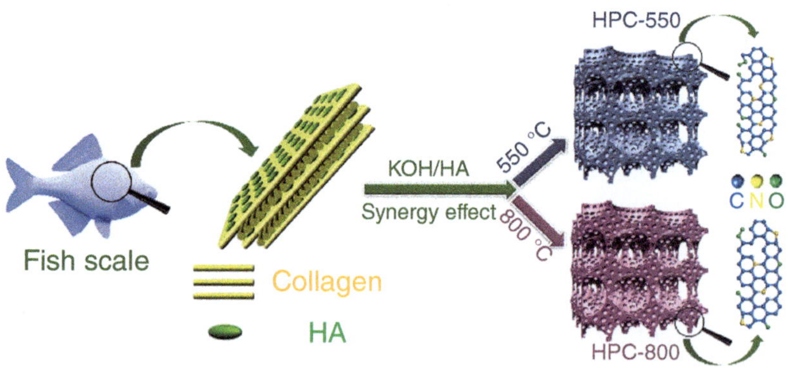

Figure 8.20 Schematic illustration for the preparation of hierarchically porous carbons (HPCs) with N/O heteroatom doping. Source: Reproduced with permission from Liu et al. [74]; © 2018, Elsevier.

0.38 and 0.40 nm, respectively. The energy dispersive X-Ray spectroscopy (EDX) mapping (Figure 8.21d,h) images of HPC-550 and HPC-800 indicated the uniform dispersion of rich N and O atoms in HPC.

The porosity of HPC samples was studied by nitrogen adsorption/desorption measurements. Both HPC samples showed the integration characteristics of type I and

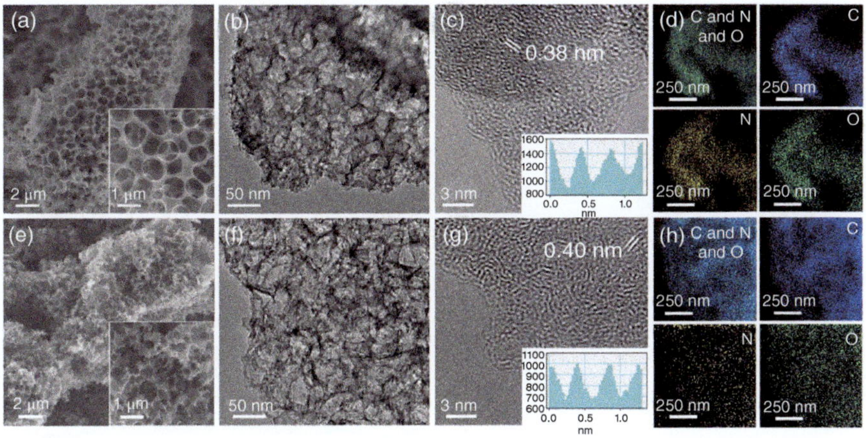

Figure 8.21 (a, e) SEM images, (b, f) TEM images, (c, g) HRTEM images (inset: line-scan profile across the lattice fringes), (d, h) EDX elemental mapping images of HPC-550 and HPC-800, respectively. Source: Liu et al. [74]/with permission of © Elsevier Ltd.

type IV, with obvious adsorption at relatively low pressure and obvious time-lag loop at relatively high pressure, indicating that there were a large number of micropores and mesopores in the HPC samples (Figure 8.22a). In addition, when the relative pressure is greater than 0.9, the hysteresis loop shows obvious absorption, confirming the existence of macropores. The porosity-distribution curve of the HPC sample (Figure 8.22b) also confirms the good stratified porosity. Because of more severe activation at higher carbonization temperature, the pore volume and SSA of HPC-800 were 2.82 $cm^3\ g^{-1}$ and 3285 $m^2\ g^{-1}$, which were much larger than those of HPC-550 (0.38 $cm^3\ g^{-1}$ and 594 $m^2\ g^{-1}$). It was worth noting that even at lower pyrolysis temperatures, mesopores and macropores could be generated, which were different from the traditional micropores-dominated porosity.

The results show that at 550 °C, the mesoporous carbon (PC-HA-550) with very low SSA value was obtained from the carbonization of fish scales without the use of KOH (only containing hydroxyapatite), which indicates that hydroxyapatite maintains the original structure and only acts as hard template to generate mesoporous particles with a particle size of about 4 nm. After the addition of KOH, the hydroxyapatite in the fish scales recrystallized into large particles (>4 nm) during the carbonization process, resulting in the interconnection of mesopores and macropores of HPCs.

To further determine the synergistic effect of KOH/hydroxyapatite on the porous structure of HPCs, hydroxyapatite was first removed with HCl, and then further carbonization was conducted at 550 °C with KOH. The obtained carbon (denoised as PC-KOH-550) has a few lumpy structures with irregular macropores, and the macropores are unevenly dispersed on the surface of carbon bulks. The nitrogen adsorption/desorption isotherm and pore-size distribution diagram were used to further characterize the microporous-dominant porosity of PC-KOH-550. The results

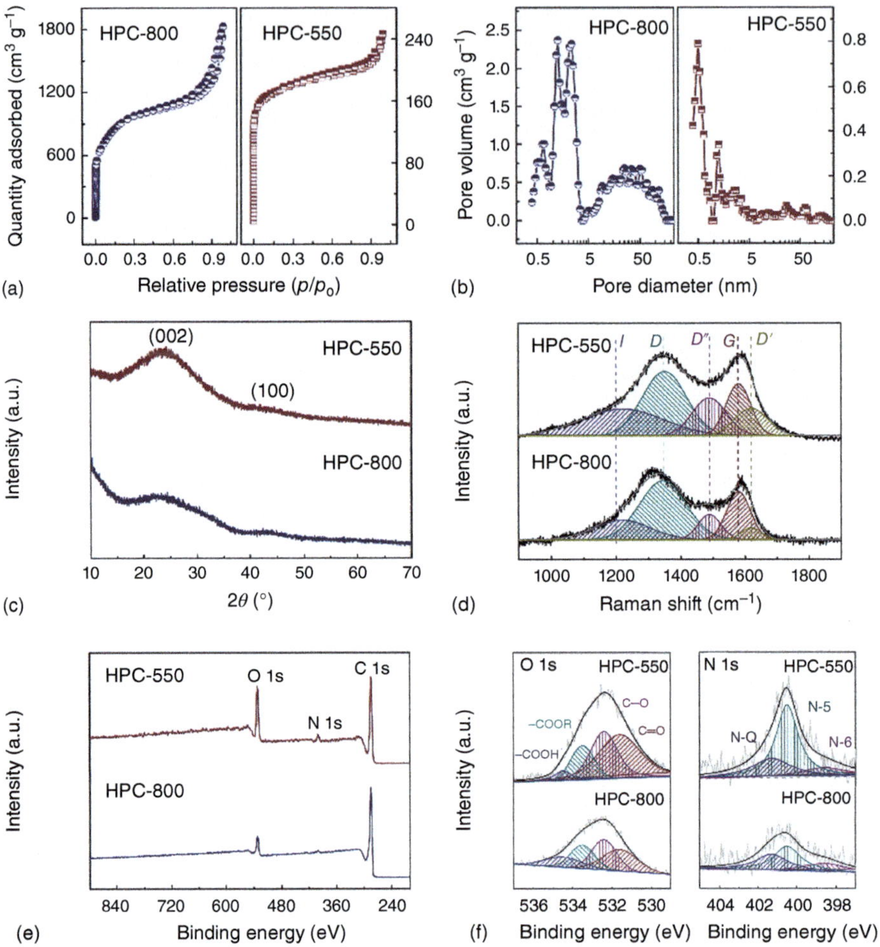

Figure 8.22 (a) Nitrogen adsorption/desorption isotherms, (b) plots of the pore-size distribution, (c) XRD patterns, (d) Raman spectra, (e) XPS spectra, and (f) high-resolution O 1s and N 1s XPS spectra of HPC-550 and HPC-800, respectively. Source: Reproduced with permission from Liu et al. [74]; © 2018, Elsevier.

showed that KOH only produces micropores but not hydroxyapatite, and the connected stratified pores are formed due to the synergistic action of KOH/HA.

XRD and Raman spectroscopy were conducted to analyze the microstructures of HPC samples. As shown in Figure 8.22c, the (002) peak of HPC-800 was broader and weaker than that of HPC-550, suggesting more defects. It is calculated that the carbon interlayer distances of HPC-800 and HPC-550 were 0.40 and 0.38 nm according to Bragg's law. The increase of interlayer distance is related to the increase of the density of structural defects caused by the activation of internal heteroatom and KOH [59]. The Raman spectra of HPCs (Figure 8.22d) were divided into four peaks [75]. The results showed that the G-peak contents of HPC-550 and HPC-800 were 11% and 19%, respectively, indicating that there are a lot of defects in HPC samples.

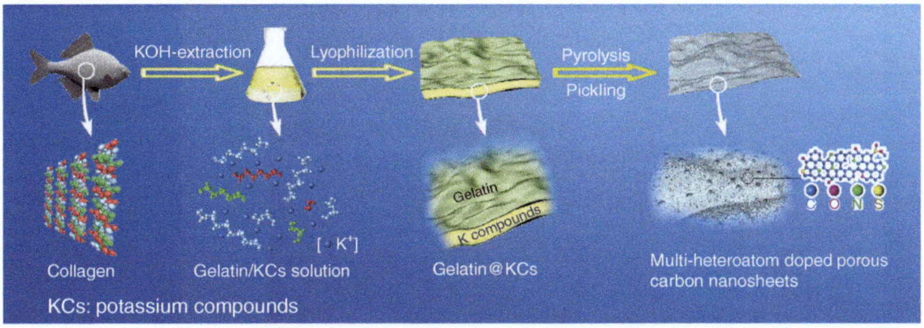

Figure 8.23 Schematic illustration of the synthesis of PCNSs doped by O, N, and S atoms via KCs assistant method from fish scales. Source: Liu et al. [71]/with permission of © Elsevier Ltd.

The doping of heteroatom introduces more inherent defects. The I-peak content of HPC-550 (26%) is higher than that of HPC-800 (19%), suggesting that HPC-550 contains more heteroatoms than HPC-800.

The surface composition and chemical state of PC samples were analyzed by XPS technique. A large number of O (18.8 at%) and N (4.3 at%) heteroatoms were retained in HPC-550 due to the low pyrolysis temperature. High temperature and strong activation kept the N and O contents of HPC-800 at 2.7% and 7.6% (Figure 8.22e). As shown in Figure 8.22f, the high-resolution N 1s spectrum has three peaks, centered at 398.6, 400.5, and 401.3 eV, representing pyridine, pyrrole/pyridone, and quaternary nitrogen, respectively [71]. The pyridinic N and pyrrolic N heteroatoms can be used as active sites, and the quaternary nitrogen heteroatom is beneficial to the conductivity [65].

8.2.3 Metal Hydroxide–Assistant Method

By using fish scale as precursor, Wang et al. further proposed a cost-effective method for the preparation of ultrathin porous carbon nanosheets (PCNSs) with multi-heteroatoms (N, O, S) with the assistance of potassium compounds (KCs) [71]. The PCNSs obtained at 600 °C (PCNS-600) are rich in surface heteroatoms (5.05 at% for N, 14.64 at% for O, and 0.92 at% for S).

PCNSs were prepared as shown in Figure 8.23. KOH solution at 80 °C was used to extract gelatin from collagen within fish scales. The resulting light-yellow solution containing gelatin and KCs was freeze dried. After that, pyrolysis process was carried out in an Ar atmosphere and then pickled with dilute hydrochloric acid and deionized water. According to the pyrolysis temperature, four PCNS samples were obtained (denoted as PCNS-T, $T = 500, 600, 700, 800$ °C).

SEM images clearly showed that after lyophilization, flakes rich in C, O, and K atoms were coated on the KC template (gelatin@KCs) because of the film-forming properties of gelatin (Figure 8.24a). As shown in Figure 8.24b–d, PCNSs were all 2D nanosheet structures after carbonization, suggesting that KCs could also act as chemical activators to produce a large number of nanopores on the gelatin-derived carbon nanosheets.

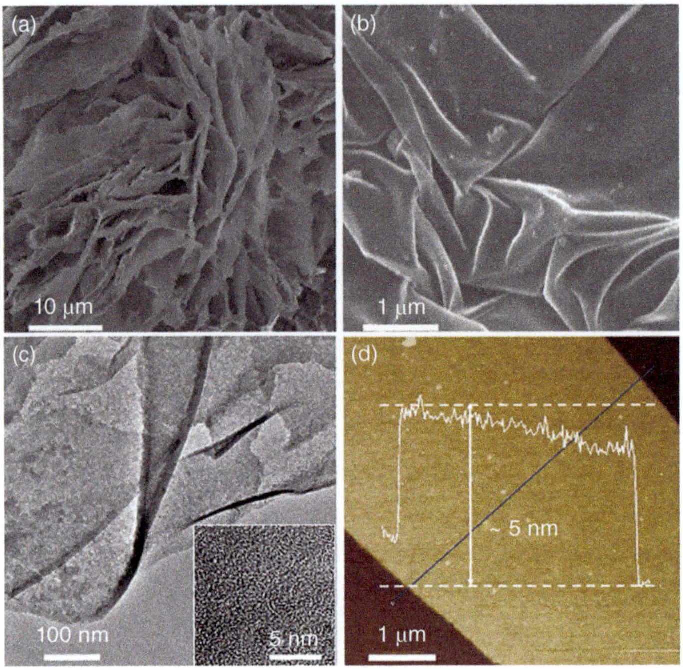

Figure 8.24 (a) SEM images of gelatin@KCs. (b) SEM, (c) TEM, and (d) AFM of PCNS-600. Source: Liu et al. [71]/with permission of © Elsevier Ltd.

Due to the chemical activation of KCs, the nitrogen adsorption–desorption isotherms of all PCNS showed type I isotherm characteristics related to the micropore structure (Figure 8.25a). DFT method was used to identify the micropore characteristics of all PCNS samples through pore-size distribution (as shown in Figure 8.25b).

At high temperature, because of the gradual enhancement of chemical activation, the specific surface area and pore volume of each PCNS sample increased with the rise of pyrolysis temperature. Raman spectroscopy further confirmed the high defect density of PCNS samples, and five peaks were fitted (Figure 8.25c). The content of peak *D* in PCNS samples was higher than that of peak *G*, indicating that there were a large number of sp^3-bonded carbon atoms. In addition to the inherent defects caused by lattice disorder and accumulation of graphene layers (producing D' and D'' peaks), the presence of N, O, and S functional groups in PCNS samples also introduced a large number of defects (producing *I* peaks) into the carbon structure [76]. Owing to the disappearance of heteroatom at high temperature, the content of peak *I* decreases gradually with the increase of pyrolysis temperature.

The chemical composition of each PCNS sample was analyzed by XPS. All the PCNS samples had a large number of sp^3-bonded and heteroatom-bonded C atoms, suggesting that all the PCNS samples have high defect densities. The O 1s spectra for all the PCNS samples could be divided into the following peaks: C=O, C—O, O=C—O—C=O, and O=C—O, which centered at 534.2, 531.7, 532.4, and 533.4 eV

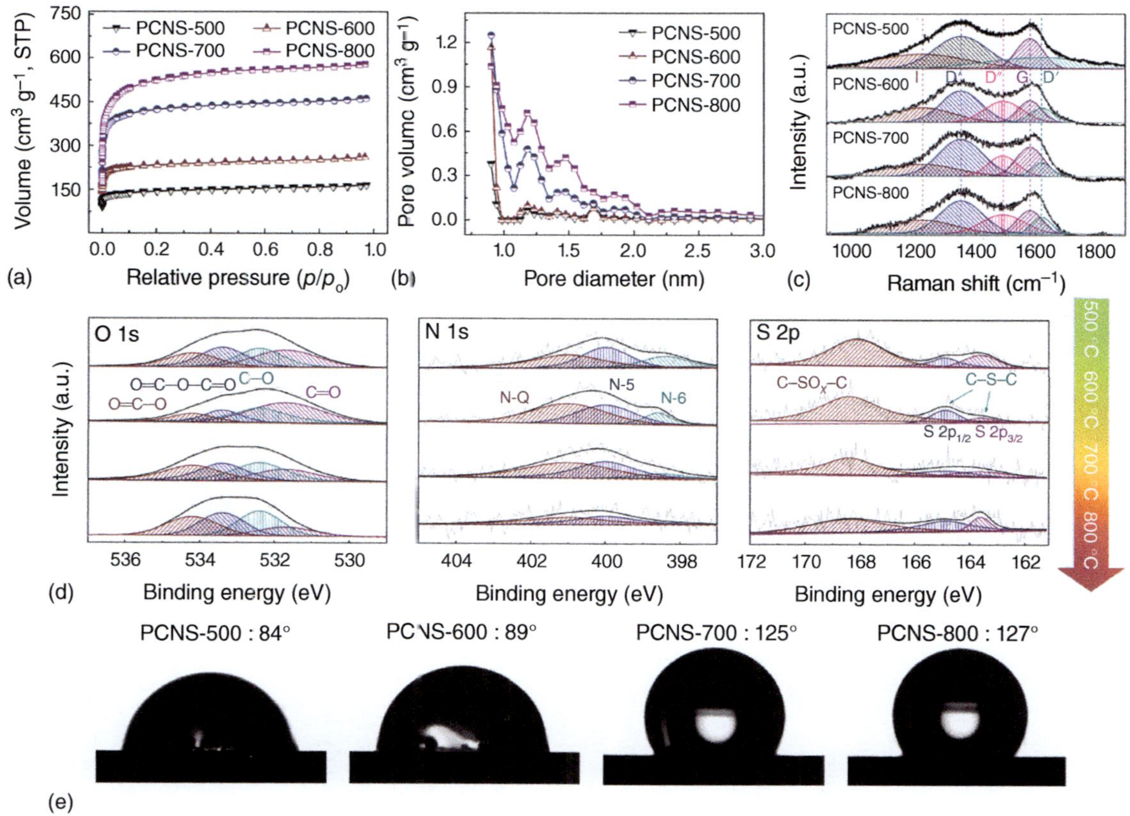

Figure 8.25 (a) Nitrogen adsorption–desorption isotherms, (b) plots of pore-size distribution, (c) Raman spectra, (d) high-resolution O 1s, N 1s, and S 2p XPS spectra, and (e) electrolyte contact-angle measurements for 6 M KOH aqueous electrolyte. Source: Reproduced with permission from Liu et al. [71]; © 2018, Elsevier.

(Figure 8.25d). When the pyrolysis temperature exceeds 700 °C, the content of C=O decreases obviously due to the low thermal stability of C=O, which is the key to electrolyte wettability. In addition to O atoms, N atoms and S atoms of different chemical states also have important effects on improving the wettability of PCNS samples.

Pyridinic-N, pyrrolic-N, and oxidized S (C–SO_x–C) have proved to be good for the electrolyte wettability of heteroatom-doped carbon materials. From the high-resolution spectra for N 1s and S 2p, it can be seen that the contents of pyridinic N, pyrrolic N, and C–SO_x–C decrease significantly when the pyrolysis temperature exceeds 700 °C. PCNS-500 and PCNS-600 electrodes have good hydrophilicity, and contact angles are 84° and 89°, respectively (Figure 8.25e). However, PCNS-700 and PCNS-800 show hydrophobic property with larger contact angles of 125° and 127°, respectively [77]. Combined with XPS analysis, the reduction of surface heteroatoms significantly reduced the wettability of PCNS samples, indicating that high temperature is beneficial to improving SSA, but it can reduce the hydrophilicity of the materials.

Notably, the chemical activation of PNS-800 electrode was enhanced at high temperature, and its O content is higher than that of PNS-600 electrode, but its hydrophilic and hydrophobic performance was completely different from that of PNS-600 electrode. This result further indicated that the hydrophilic and hydrophobic properties of PCNS electrodes were not only related to the number of heteroatoms on the surface but also related to the change of chemical states on the surface. The O atoms in the form of C=O, N atoms in the form of pyridinic N/pyrrolic N, and S atoms in the form of C–SO*x*–C are important factors affecting the hydrophilicity of PCNS electrodes to KOH electrolyte.

Wang and coworkers also proposed a facile and efficient strategy to prepare O, N, and S-doped microporous carbon nanosheets (referred to as microporous carbon nanosheets (MPCNSs)) by pyrolyzing fish skin (Figure 8.26) [78].

Fish skin was well converted into water-soluble gelatin after being treated by KOH solution (~1.0 M) at 80 °C and dried into solid products at 100 °C. The results of FTIR analysis indicated that the collagen in the fish skin was decomposed into

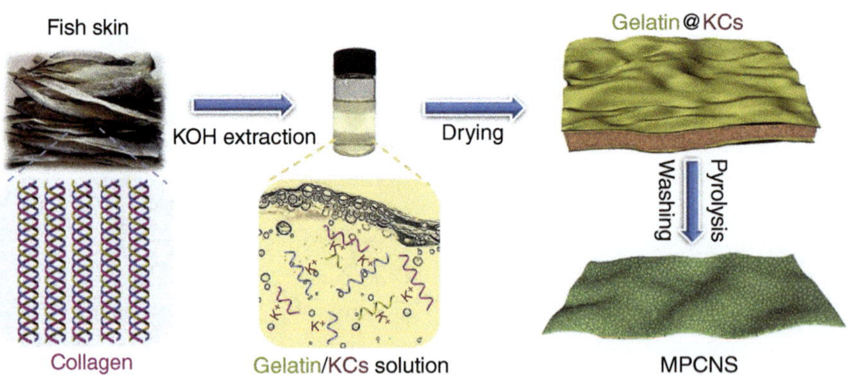

Figure 8.26 Schematic illustration of the preparation of O,N,S-doped MPCNSs with the assistance of KOH from fish skin. Source: Niu et al. [78]/with permission of © Elsevier Ltd.

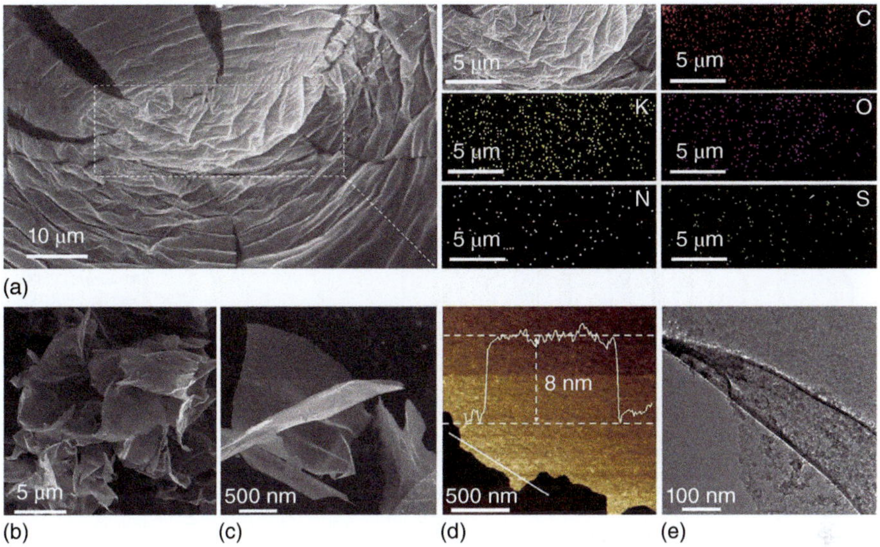

Figure 8.27 (a) SEM image of gelatin@KCs and the corresponding element distribution. (b, c) SEM, (d) AFM, and (e) TEM images of MPCNS. Source: Niu et al. [78]/with permission of © Elsevier Ltd.

gelatin. The gelatin made from fish skin is rich in C, N, O, and S elements and has a wrinkled nanosheet structure (Figure 8.27a). As shown in Figure 8.27b,c, MPCNS showed nanosheet structure, suggesting that KOH-derived KCs might serve as templates. The nanosheet structure was still maintained within MPCNS-800, indicating its good stability. Based on the atomic force microscopy (AFM) image (Figure 8.27d), the MPCNS thickness was about 7 nm. TEM images (Figure 8.27e) showed that MPCNS exhibited a very rough surface.

KOH-derived KCs not only served as templates for the formation of nanosheet structures, but also acted as activators in the process of nanopore generation in MPCNS. Figure 8.28a shows the nitrogen adsorption–desorption isotherms for all the samples. All the carbons showed type I characteristics, indicating that the microporous structure was dominant. The SSA of MPCNS was 1017 $m^2\ g^{-1}$, which was higher than that of MPCB, indicating that MPCNS had better activation effect. After carbonization, the SSA of MPCNS decreased to 913 $m^2\ g^{-1}$. Notably, the pore volumes of all three samples were smaller than 0.422 $cm^3\ g^{-1}$, which were much lower than other porous carbons reported in other works [8].

XRD patterns showed that all samples exhibited two inconspicuous and wide peaks located at about 24° and 44° (Figure 8.28b), corresponding to (002) and (100) peaks for graphite, respectively. The results showed that the samples were mainly amorphous with smaller crystal regions. The Raman spectra of the carbon samples were fitted with five peaks (Figure 8.28c) [69]. The area of *G* peaks only accounted for a small proportion, indicating the high defect densities for all samples. The I_D/I_G value for MPCNS (2.68) was much higher than that for MPCB (2.13), suggesting that the activation degree of MPCNS was higher than that of MPCB. Both the I_D/I_G (2.51)

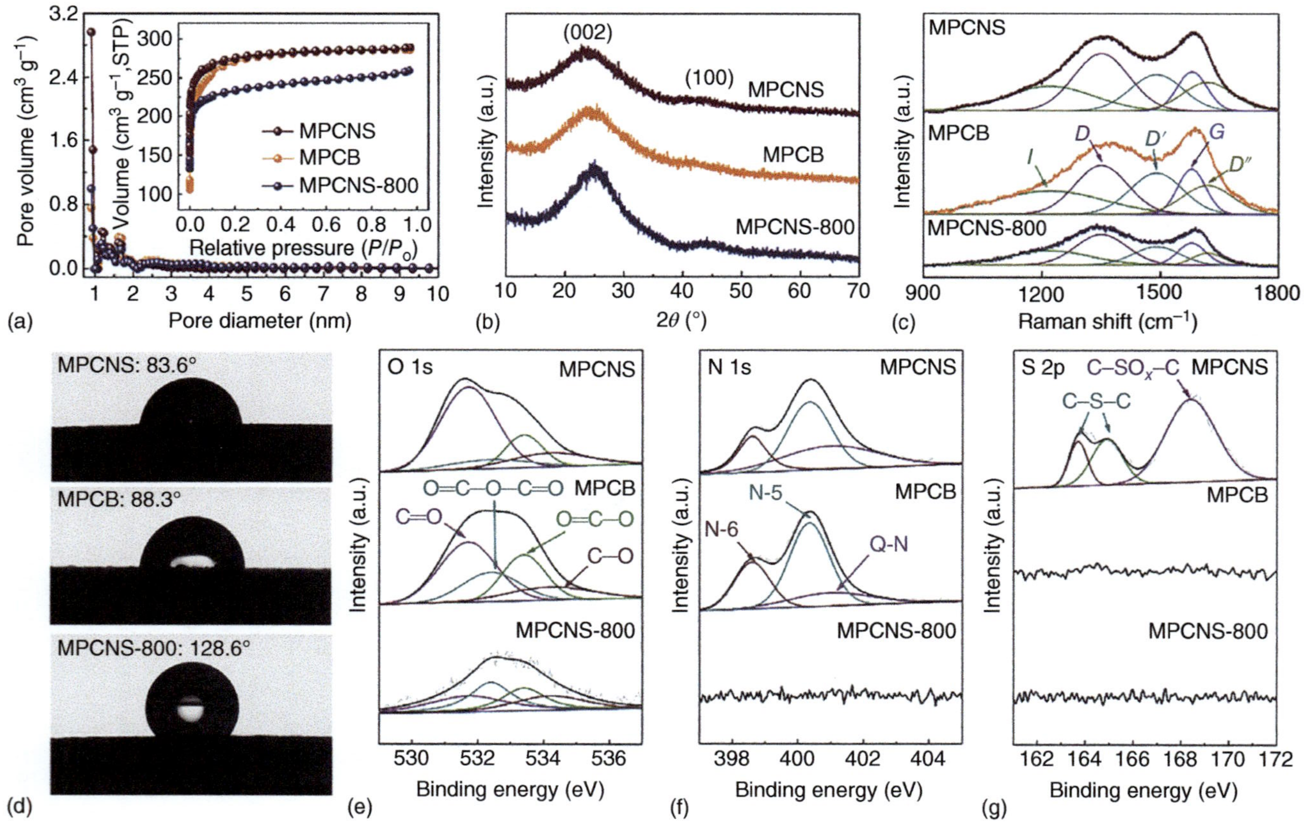

Figure 8.28 (a) Nitrogen adsorption–desorption isotherms (inset) and pore-size distribution plots, (b) XRD patterns, and (c) Raman spectra of MPCNS, MPCB, and MPCNS-80. (d) Electrolyte contact-angle measurements and high-resolution (e) O 1s, (f) N 1s, and (g) S 2p XPS spectra of MPCNS, MPCB, and MPCNS-800. Source: Reproduced with permission from Niu et al. [78]; © 2018, Elsevier.

value and $I_{D'+D''}/I_G$ (1.95) of MPCNS-800 were lower than those of MPCNS (2.83 for $I_{D'+D''}/I_G$), which was owing to the heteroatom elimination at high temperature.

This phenomenon indicated that KOH-derived templates could reserve the surface S atoms because of the confinement effect. After being carbonized, the N and S atoms on the surface of MPCNS disappeared, which may lead to changes in the surface wettability of MPCNS-800 (Figure 8.28d). Notably, MPCNS was more hydrophilic than MPCB although the N/O content of these two samples was similar. To investigate the reasons for the different hydrophilicity, XPS spectra for all the samples were further analyzed to determine their chemical states. All the samples had a number of heteroatom-bonded and sp^3-bonded C atoms. The O 1s spectrum was fitted by four peaks: C=O, C—O, O=C—O—C=O, and O=C—O (Figure 8.28e). Among them, C=O could greatly improve the hydrophilicity of the sample [79]. It was noted that MPCNS possessed more C=O than MPCB, indicating that KOH-derived templates could prevent the C=O elimination. In addition, MPCNS and MPCB had a number of pyrrolic N (399.9 eV) and pyridinic N (398.4 eV), which were also good for wettability (Figure 8.28f) [67]. There was some oxidized S (C–SO_x–C, 168.2 eV) on the surface of MPCNs, which could also be expected to improve electrolyte wettability (Figure 8.28g) [80]. Therefore, MPCNS showed the best hydrophilicity among all the samples.

8.2.4 Metal Chloride–Assistant Method

Wang and coworkers extracted gelatin from fish scales with KCl aqueous solution, lyophilized, carbonized, and washed to prepare N and O co-doped porous carbon nanosheets [81]. With the help of ice/KCl dual templates, the resulting sample has a porous nanosheet structure, about 4 nm, with a large number of N and O heteroatoms (8.2 at% for N, 8.4 at% for O) derived from gelatin.

Figure 8.29a shows the preparation of NOPCNS. Unlike previous studies, the reagents used in this study were both resourceful and non-toxic. Fish scales were treated directly in 80 °C KCl aqueous solution (about 0.6 M) for 12 hours. KCl aqueous solution can convert about 93% of the collagen of fish scales into gelatin. The resulting gelatin/KCl aqueous solution was further lyophilized and pyrolyzed in Ar atmosphere.

It is worth noting that KCl and ice form interconnected porous nanosheet structures as double templates, and gelatin from fish scales acts as precursors of N and O co-doped carbon. Therefore, NOPCNS could be obtained by rinsing with deionized water and vacuum filtration. As a contrast, by using ice or KCl as the only template, carbon blocks (known as NOCB1 and NOCB2) were obtained (Figure 8.29b,c).

SEM images show the morphology and structure of NOPCNS and NOCBs (Figure 8.30a–c). NOPCNS showed interconnected ultrathin nanosheets, while NOCBs possessed a bulk structure. In addition, there were some large square pores

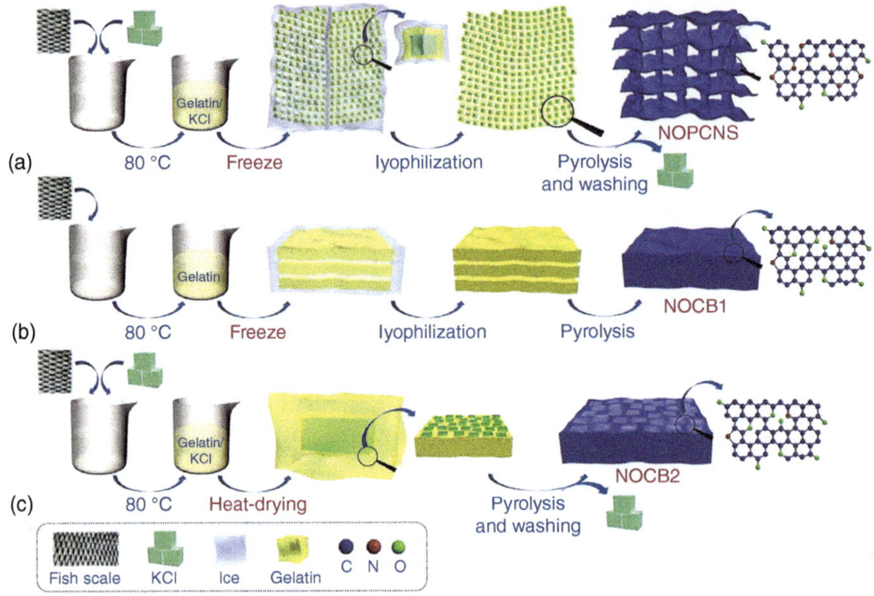

Figure 8.29 Schematic illustration for the formation of (a) NOPCNS, (b) NOCB1, and (c) NOCB2. Source: Reproduced with permission from Liu et al. [81]; © 2019, Elsevier.

on NOCB2 surface, which might have resulted from the removal of KCl templates. TEM images of NOPCNS (Figure 8.30d) further illustrate the ultrathin nanosheet structure. Furthermore, some square pores smaller than 20 nm were found in the ultrathin nanosheet, which might be also generated by KCl crystals. As shown in Figure 8.30e, HRTEM image demonstrated that NOPCNS was almost in amorphous structure with small regions of crystalline. Line-scanning results showed that the carbon layer distance of NOPCNs is about 0.37 nm (Figure 8.30e, inset). Figure 8.30f shows the EDX element mapping diagram, which revealed that N and O atoms are uniformly dispersed in the carbon nanosheets, showing the in situ co-doping of N and O atoms.

The porous structures of NOPCNs and NOCBs were analyzed in detail by N_2 adsorption–desorption measurements (Figure 8.31a). When the relative pressure is greater than 0.45, NOPCNs show a typical IV (H3)-type isotherm with obvious hysteresis loop, demonstrating that their porosity is mainly mesoporous. The mesopores within NOPCNS included square pores and the narrow-slit pores generated by the nanosheet stacking [82]. As shown in Figure 8.31b, the mesopores had a wide pore-size distribution. However, the adsorption and desorption of NOCBs were not obvious with low pore volume. NOPCNS sample exhibited a high SSA of $168\,m^2\,g^{-1}$, which was mainly contributed by mesopores. In contrast, NOCBs had a very low specific surface area. The limited macropores in NOCB2 made little contributions to the porosity of the material.

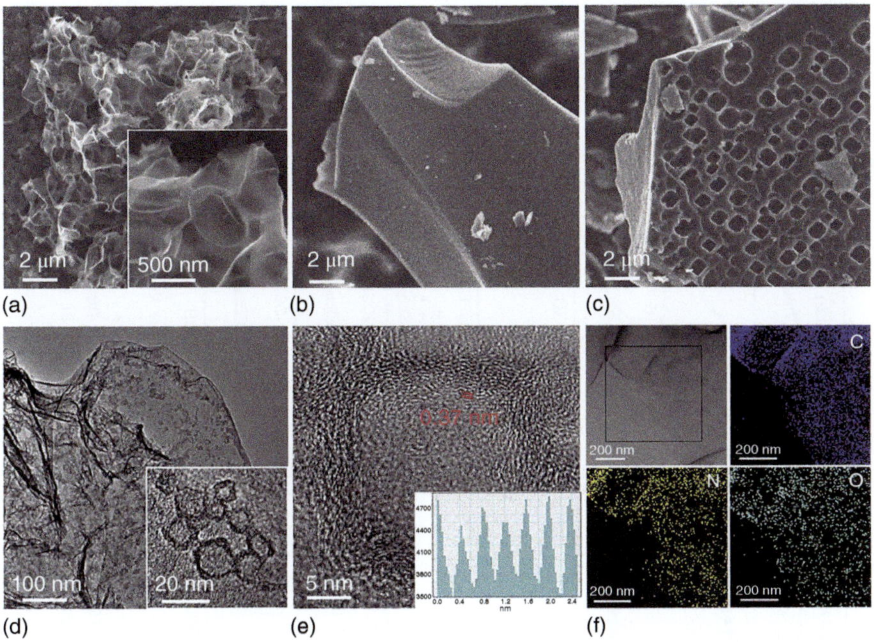

Figure 8.30 SEM images of (a) NOPCNS, (b) NOCB1, and (c) NOCB2. (d) TEM images, (e) HRTEM image (inset: line-scan profile across the lattice fringes), and (f) EDX elemental mapping images of NOPCNS. Source: Liu et al. [81]/with permission of © Elsevier Ltd.

Since the ice or KCl template alone can only produce bulk structures with poor porosity, there should be a synergistic effect between KCl and ice to produce ultra-thin nanosheet structures. To investigate the formation mechanism of NOPCNS and NOCBs, the morphology and structure of the samples before pyrolyzation were characterized by SEM.

In the freeze-dried gelatin without KCl template, the sample showed large-size bulks, demonstrating a serious phase separation occurred between gelatin and ice during the freeze-drying process. The heat-dried gelation/KCl sample also possessed bulk structure when only KCl was used. Notably, a number of cubes in the size of 1–100 mm could be observed (Figure 8.31a,b). This phenomenon suggests that during the heat-drying process, KCl recrystallized and gelatin aggregated into large bulks, resulting in poor dispersion of KCl crystals in gelatin. As a result, ice or KCl alone is not an effective template for the formation of nanostructure. If ice and KCl coexisted, phase separation was significantly reduced. Ice templates created many dendrite-like canals. Gelatin was coated on KCl uniformly along the canals. Carbon nanosheets were successfully obtained by using a unique ice/KCl double template. Moreover, some KCl crystals also served as templates to produce square mesopores.

XRD and Raman measurements were carried out to explore the graphitization and defects of the NOPCNS and NOCBs. Figure 8.31c shows that all the carbon materials presented broad graphite (002) peaks. Their carbon interlayer distances were calculated to be 0.370, 0.363, and 0.351 nm for NOPCNS, NOCB1, and NOCB2

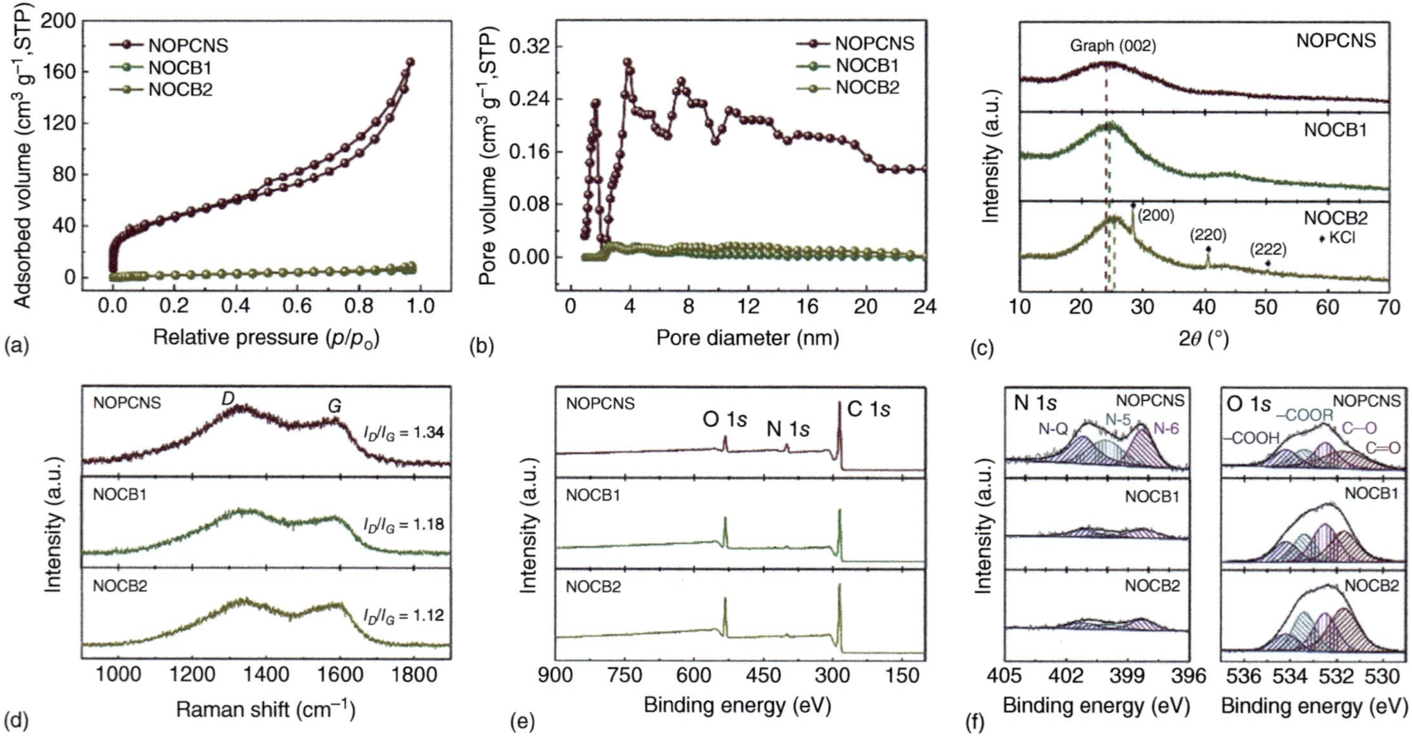

Figure 8.31 (a) Nitrogen adsorption–desorption isotherms, (b) pore-size distribution plots, (c) XRD patterns, (d) Raman spectra, (e) XPS spectra, and (f) high-resolution N 1s/O 1s XPS spectra of NOPCNS, NOCB1, and NOCB2. Source: Reproduced with permission from Liu et al. [81]; © 2019, Elsevier.

according to their specific diffraction peaks, which centered at 24.0°, 24.5°, and 25.3°, respectively. This phenomenon indicated the high defect densities of NOPCNS and NOCBs. It was noted that the KCl peaks were also detected in the XRD pattern of NOCB2. This result indicates that part of the KCl template was trapped in the gelatin-derived carbon, because the closed structure of NOCB2 could not be removed by deionized water. Raman measurements were carried out to further identify the high defect densities of the samples (Figure 8.31d). All the carbon materials obviously possessed D peaks. The I_D/I_G value of NOPCNS (1.34) was much higher than that of NOCB1 (1.18) and NOCB2 (1.12), indicating that NOPCNS exhibited more intrinsic defects and structural distortions. The higher defect density of NOPCNS might be due to the lack of abundant holes in NOCBs. A large amount of structural deformation also leads to the largest carbon interlayer distance of NOPCNS among all samples.

The XPS spectra of NOPCNS and NOCBs confirmed the existence of N and O heteroatoms (Figure 8.31e). Compared with NOCB1 (3.5 at%) and NOCB2 (2.9 at%), NOPCNS (8.2 at%) exhibited higher N content, demonstrating that the ice/KCl dual templates were favorable to the retention of N atoms because of the nano-confining effect [83]. As shown in Figure 8.31f, the high-resolution N 1s spectra for three samples could be divided into pyridinic N, pyrrolic N, and quaternary nitrogen. In addition to N atoms, all carbon materials also contained a large number of O atoms on the surface. Moreover, the high-resolution O 1s XPS spectra showed that the oxygen atoms exist in the form of C=O, C—O, -COOR, and -COOH.

References

1. Roberts, A.D., Li, X., and Zhang, H. (2014). Porous carbon spheres and monoliths: morphology control, pore size tuning and their applications as Li-ion battery anode materials. *Chemical Society Reviews* 43: 4341–4356.
2. Knossalla, J., Jalalpoor, D., and Schüth, F. (2017). Hands-on guide to the synthesis of mesoporous hollow graphitic spheres and core-shell materials. *Chemistry of Materials* 29: 7062–7072.
3. Liu, X., Antonietti, M., and Giordano, C. (2013). Manipulation of phase and microstructure at nanoscale for SiC in molten salt synthesis. *Chemistry of Materials* 25: 2021–2027.
4. Kuhn, P., Forget, A., Su, D. et al. (2008). From microporous regular frameworks to mesoporous materials with ultrahigh surface area: dynamic reorganization of porous polymer networks. *Journal of the American Chemical Society* 130: 13333–13337.
5. He, S., Zhang, C., Du, C. et al. (2019). High rate-performance supercapacitor based on nitrogen-doped hollow hexagonal carbon nanoprism arrays with ultrathin wall thickness in situ fabricated on carbon cloth. *Journal of Power Sources* 434: 226701.
6. Gogotsi, Y., Nikitin, A., Ye, H. et al. (2003). Nanoporous carbide-derived carbon with tunable pore size. *Nature Materials* 2: 591–594.

7 Chuenchom, L., Kraehnert, R., and Smarsly, B.M. (2012). Recent progress in soft-templating of porous carbon materials. *Soft Matter* 8: 10801–10812.

8 Liu, J., Wickramaratne, N.P., Qiao, S.Z., and Jaroniec, M. (2015). Molecular-based design and emerging applications of nanoporous carbon spheres. *Nature Materials* 14: 763–774.

9 Wei, J., Zhou, D., Sun, Z. et al. (2013). A controllable synthesis of rich nitrogen-doped ordered mesoporous carbon for CO_2 capture and supercapacitors. *Advanced Functional Materials* 23: 2322–2328.

10 Chu, W.-C., Bastakoti, B.P., Kaneti, Y.V. et al. (2017). Tailored design of bicontinuous gyroid mesoporous carbon and nitrogen-doped carbon from poly(ethylene oxide-*b*-caprolactone) diblock copolymers. *Chemistry – A European Journal* 23: 13734–13741.

11 Zhang, F., Meng, Y., Gu, D. et al. (2006). An aqueous cooperative assembly route to synthesize ordered mesoporous carbons with controlled structures and morphology. *Chemistry of Materials* 18: 5279–5288.

12 Tian, H., Lin, Z., Xu, F. et al. (2016). Quantitative control of pore size of mesoporous carbon nanospheres through the self-assembly of diblock copolymer micelles in solution. *Small* 12: 3155–3163.

13 Moriguchi, I., Ozono, A., Mikuriya, K. et al. (1999). Micelle-templated mesophases of phenol-formaldehyde polymer. *Chemistry Letters* 28: 1171–1172.

14 Liang, C., Li, Z., and Dai, S. (2008). Mesoporous carbon materials: synthesis and modification. *Angewandte Chemie International Edition* 47: 3696–3717.

15 Wang, J.-G., Liu, H., Sun, H. et al. (2018). One-pot synthesis of nitrogen-doped ordered mesoporous carbon spheres for high-rate and long-cycle life supercapacitors. *Carbon* 127: 85–92.

16 Meng, Y., Gu, D., Zhang, F. et al. (2006). A family of highly ordered mesoporous polymer resin and carbon structures from organic–organic self-assembly. *Chemistry of Materials* 18: 4447–4464.

17 Peng, L., Hung, C.-T., Wang, S. et al. (2019). Versatile nanoemulsion assembly approach to synthesize functional mesoporous carbon nanospheres with tunable pore sizes and architectures. *Journal of the American Chemical Society* 141: 7073–7080.

18 Hasegawa, G., Kanamori, K., Kiyomura, T. et al. (2016). Hierarchically porous carbon monoliths comprising ordered mesoporous nanorod assemblies for high-voltage aqueous supercapacitors. *Chemistry of Materials* 28: 3944–3950.

19 Liang, C., Hong, K., Guiochon, G.A. et al. (2004). Synthesis of a large-scale highly ordered porous carbon film by self-assembly of block copolymers. *Angewandte Chemie International Edition* 43: 5785–5789.

20 Petkovich, N.D. and Stein, A. (2013). Controlling macro- and mesostructures with hierarchical porosity through combined hard and soft templating. *Chemical Society Reviews* 42: 3721–3739.

21 Soler-Illia, G.J.d.A.A., Sanchez, C., Lebeau, B., and Patarin, J. (2002). Chemical strategies to design textured materials: from microporous and mesoporous oxides to nanonetworks and hierarchical structures. *Chemical Reviews* 102: 4093–4138.

22 Soler-Illia, G.J.A.A. and Azzaroni, O. (2011). Multifunctional hybrids by combining ordered mesoporous materials and macromolecular building blocks. *Chemical Society Reviews* 40: 1107–1150.

23 Etacheri, V., Sharon, D., Garsuch, A. et al. (2013). Hierarchical activated carbon microfiber (ACM) electrodes for rechargeable Li-O_2 batteries. *Journal of Materials Chemistry A* 1: 5021–5030.

24 Uçar, S., Erdem, M., Tay, T., and Karagöz, S. (2015). Removal of lead (II) and nickel (II) ions from aqueous solution using activated carbon prepared from rapeseed oil cake by Na_2CO_3 activation. *Clean Technologies and Environmental Policy* 17: 747–756.

25 Yang, K., Peng, J., Xia, H. et al. (2010). Textural characteristics of activated carbon by single step CO_2 activation from coconut shells. *Journal of the Taiwan Institute of Chemical Engineers* 41: 367–372.

26 Navarro, R.M., Peña, M.A., and Fierro, J.L.G. (2007). Hydrogen production reactions from carbon feedstocks: fossil fuels and biomass. *Chemical Reviews* 107: 3952–3991.

27 Kim, D.-W., Kil, H.-S., Nakabayashi, K. et al. (2017). Structural elucidation of physical and chemical activation mechanisms based on the microdomain structure model. *Carbon* 114: 98–105.

28 Sayğılı, H. and Güzel, F. (2016). High surface area mesoporous activated carbon from tomato processing solid waste by zinc chloride activation: process optimization, characterization and dyes adsorption. *Journal of Cleaner Production* 113: 995–1004.

29 Islam, M.A., Ahmed, M.J., Khanday, W.A. et al. (2017). Mesoporous activated coconut shell-derived hydrochar prepared via hydrothermal carbonization-NaOH activation for methylene blue adsorption. *Journal of Environmental Management* 203: 237–244.

30 Lin, G., Ma, R., Zhou, Y. et al. (2018). KOH activation of biomass-derived nitrogen-doped carbons for supercapacitor and electrocatalytic oxygen reduction. *Electrochimica Acta* 261: 49–57.

31 Prahas, D., Kartika, Y., Indraswati, N., and Ismadji, S. (2008). Activated carbon from jackfruit peel waste by H_3PO_4 chemical activation: pore structure and surface chemistry characterization. *Chemical Engineering Journal* 140: 32–42.

32 Zhang, F., Liu, T., Li, M. et al. (2017). Multiscale pore network boosts capacitance of carbon electrodes for ultrafast charging. *Nano Letters* 17: 3097–3104.

33 Yin, H., Lu, B., Xu, Y. et al. (2014). Harvesting capacitive carbon by carbonization of waste biomass in molten salts. *Environmental Science & Technology* 48: 8101–8108.

34 Liu, S., Liang, Y., Zhou, W. et al. (2018). Large-scale synthesis of porous carbon via one-step $CuCl_2$ activation of rape pollen for high-performance supercapacitors. *Journal of Materials Chemistry A* 6: 12046–12055.

35 Tian, Y., Xiao, C., Yin, J. et al. (2019). Hierarchical porous carbon prepared through sustainable $Cucl_2$ activation of rice husk for high-performance supercapacitors. *ChemistrySelect* 4: 2314–2319.

36 Xu, Z., Zhou, Y., Sun, Z. et al. (2020). Understanding reactions and pore-forming mechanisms between waste cotton woven and $FeCl_3$ during the synthesis of magnetic activated carbon. *Chemosphere* 241: 125120.

37 Liu, X., Fechler, N., and Antonietti, M. (2013). Salt melt synthesis of ceramics, semiconductors and carbon nanostructures. *Chemical Society Reviews* 42: 8237–8265.

38 Yun, S.I., Kim, S.H., Kim, D.W. et al. (2019). Facile preparation and capacitive properties of low-cost carbon nanofibers with ZnO derived from lignin and pitch as supercapacitor electrodes. *Carbon* 149: 637–645.

39 Li, J., Wang, N., Tian, J. et al. (2018). Cross-coupled macro-mesoporous carbon network toward record high energy-power density supercapacitor at 4 V. *Advanced Functional Materials* 28, 1806153.

40 Liu, X., Giordano, C., and Antonietti, M. (2014). A facile molten-salt route to graphene synthesis. *Small* 10: 193–200.

41 Liu, X., Fechler, N., Antonietti, M. et al. (2016). Synthesis of novel 2-D carbon materials: sp^2 carbon nanoribbon packing to form well-defined nanosheets. *Materials Horizons* 3: 214–219.

42 Liu, X. and Antonietti, M. (2013). Moderating black powder chemistry for the synthesis of doped and highly porous graphene nanoplatelets and their use in electrocatalysis. *Advanced Materials* 25: 6284–6290.

43 Liu, X. and Antonietti, M. (2014). Molten salt activation for synthesis of porous carbon nanostructures and carbon sheets. *Carbon* 69: 460–466.

44 Wei, Q., Qin, F., Ma, Q., and Shen, W. (2019). Coal tar-and residual oil-derived porous carbon as metal-free catalyst for nitroarene reduction to aminoarene. *Carbon* 141: 542–552.

45 Qiu, D., Kang, C., Gao, A. et al. (2019). Sustainable low-temperature activation to customize pore structure and heteroatoms of biomass-derived carbon enabling unprecedented durable supercapacitors. *ACS Sustainable Chemistry & Engineering* 7: 14629–14638.

46 Demir, M., Kahveci, Z., Aksoy, B. et al. (2015). Graphitic biocarbon from metal-catalyzed hydrothermal carbonization of lignin. *Industrial & Engineering Chemistry Research* 54: 10731–10739.

47 Qiu, D., Guo, N., Gao, A. et al. (2019). Preparation of oxygen-enriched hierarchically porous carbon by KMnO4 one-pot oxidation and activation: mechanism and capacitive energy storage. *Electrochimica Acta* 294: 398–405.

48 Zan, Y., Zhang, Z., Liu, H. et al. (2017). Nitrogen and phosphorus co-doped hierarchically porous carbons derived from cattle bones as efficient metal-free electrocatalysts for the oxygen reduction reaction. *Journal of Materials Chemistry A* 5: 24329–24334.

49 Zan, Y., Zhang, Z., Dou, M., and Wang, F. (2019). Enhancement mechanism of sulfur dopants on the catalytic activity of N and P co-doped three-dimensional hierarchically porous carbon as a metal-free oxygen reduction electrocatalyst. *Catalysis Science & Technology* 9: 5906–5914.

50 Chen, S., Duan, J., Jaroniec, M., and Qiao, S.Z. (2013). Hierarchically porous graphene-based hybrid electrodes with excellent electrochemical performance. *Journal of Materials Chemistry A* 1: 9409–9413.

51 Zan, Y., Zhang, Z., Zhu, B. et al. (2021). Ultrathin two-dimensional phosphorus and nitrogen Co-doped carbon nanosheet as efficient oxygen reduction electrocatalyst. *Carbon* 174: 404–412.

52 Ahamad, T. and Alshehri, S.M. (2012). Thermal degradation and evolved gas analysis of thiourea-formaldehyde resin (TFR) during pyrolysis and combustion. *Journal of Thermal Analysis and Calorimetry* 109: 1039–1047.

53 Jiang, L., Sheng, L., Long, C. et al. (2015). Functional pillared graphene frameworks for ultrahigh volumetric performance supercapacitors. *Advanced Energy Materials* 5, 1500771.

54 Haberko, K., Bućko, M.M., Brzezińska-Miecznik, J. et al. (2006). Natural hydroxyapatite – its behaviour during heat treatment. *Journal of the European Ceramic Society* 26: 537–542.

55 Sofronia, A.M., Baies, R., Anghel, E.M. et al. (2014). Thermal and structural characterization of synthetic and natural nanocrystalline hydroxyapatite. *Materials Science and Engineering: C* 43: 153–163.

56 Wang, J. and Kaskel, S. (2012). KOH activation of carbon-based materials for energy storage. *Journal of Materials Chemistry* 22: 23710–23725.

57 Ding, J., Li, Z., Cui, K. et al. (2016). Heteroatom enhanced sodium ion capacity and rate capability in a hydrogel derived carbon give record performance in a hybrid ion capacitor. *Nano Energy* 23: 129–137.

58 Niu, J., Shao, R., Liang, J. et al. (2017). Biomass-derived mesopore-dominant porous carbons with large specific surface area and high defect density as high performance electrode materials for Li-ion batteries and supercapacitors. *Nano Energy* 36: 322–330.

59 Niu, J., Liang, J., Shao, R. et al. (2017). Tremella-like N,O-codoped hierarchically porous carbon nanosheets as high-performance anode materials for high energy and ultrafast Na-ion capacitors. *Nano Energy* 41: 285–292.

60 Chen, X., Zhuo, J., and Jiao, C. (2012). Thermal degradation characteristics of flame retardant polylactide using TG–IR. *Polymer Degradation and Stability* 97: 2143–2147.

61 Arenas, E. and Chejne, F. (2004). The effect of the activating agent and temperature on the porosity development of physically activated coal chars. *Carbon* 42: 2451–2455.

62 Niu, J., Shao, R., Liu, M. et al. (2018). Porous carbon electrodes with battery-capacitive storage features for high performance Li-ion capacitors. *Energy Storage Materials* 12: 145–152.

63 Deng, Y., Wei, J., Sun, Z., and Zhao, D. (2013). Large-pore ordered mesoporous materials templated from non-pluronic amphiphilic block copolymers. *Chemical Society Reviews* 42: 4054–4070.

64 Cihlář, J., Buchal, A., and Trunec, M. (1999). Kinetics of thermal decomposition of hydroxyapatite bioceramics. *Journal of Materials Science* 34: 6121–6131.

65 Zhang, J., Lv, W., Zheng, D. et al. (2018). The interplay of oxygen functional groups and folded texture in densified graphene electrodes for compact sodium-ion capacitors. *Advanced Energy Materials* 8, 1702395.

66 He, D., Niu, J., Dou, M. et al. (2017). Nitrogen and oxygen co-doped carbon networks with a mesopore-dominant hierarchical porosity for high energy and power density supercapacitors. *Electrochimica Acta* 238: 310–318.

67 Wang, S., Han, C., Wang, J. et al. (2014). Controlled synthesis of ordered mesoporous carbohydrate-derived carbons with flower-like structure and n-doping by self-transformation. *Chemistry of Materials* 26: 6872–6877.

68 Vatamanu, J., Vatamanu, M., and Bedrov, D. (2015). Non-faradaic energy storage by room temperature ionic liquids in nanoporous electrodes. *ACS Nano* 9: 5999–6017.

69 Shao, R., Niu, J., Liang, J. et al. (2017). Mesopore- and macropore-dominant nitrogen-doped hierarchically porous carbons for high-energy and ultrafast supercapacitors in non-aqueous electrolytes. *ACS Applied Materials & Interfaces* 9: 42797–42805.

70 Wu, Q., Yang, L., Wang, X., and Hu, Z. (2020). Mesostructured carbon-based nanocages: an advanced platform for energy chemistry. *SCIENCE CHINA Chemistry* 63: 665–681.

71 Liu, M., Niu, J., Zhang, Z. et al. (2018). Potassium compound-assistant synthesis of multi-heteroatom doped ultrathin porous carbon nanosheets for high performance supercapacitors. *Nano Energy* 51: 366–372.

72 Guo, C., Liao, W., Li, Z. et al. (2015). Easy conversion of protein-rich enoki mushroom biomass to a nitrogen-doped carbon nanomaterial as a promising metal-free catalyst for oxygen reduction reaction. *Nanoscale* 7: 15990–15998.

73 Wen, Y., He, K., Zhu, Y. et al. (2014). Expanded graphite as superior anode for sodium-ion batteries. *Nature Communications* 5: 4033.

74 Liu, M., Niu, J., Zhang, Z. et al. (2019). Porous carbons with tailored heteroatom doping and well-defined porosity as high-performance electrodes for robust Na-ion capacitors. *Journal of Power Sources* 414: 68–75.

75 Zhao, G., Chen, C., Yu, D. et al. (2018). One-step production of O-N-S co-doped three-dimensional hierarchical porous carbons for high-performance supercapacitors. *Nano Energy* 47: 547–555.

76 Martucci, J.F. and Ruseckaite, R.A. (2010). Biodegradable bovine gelatin/Na^+-montmorillonite nanocomposite films. structure, barrier and dynamic mechanical properties. *Polymer-Plastics Technology and Engineering* 49: 581–588.

77 Qie, L., Chen, W., Xu, H. et al. (2013). Synthesis of functionalized 3D hierarchical porous carbon for high-performance supercapacitors. *Energy & Environmental Science* 6: 2497–2504.

78 Niu, J., Liu, M., Xu, F. et al. (2018). Synchronously boosting gravimetric and volumetric performance: biomass-derived ternary-doped microporous carbon nanosheet electrodes for supercapacitors. *Carbon* 140: 664–672.

79 Figueiredo, J.L., Pereira, M.F.R., Freitas, M.M.A., and Órfão, J.J.M. (1999). Modification of the surface chemistry of activated carbons. *Carbon* 37: 1379–1389.

80 Zhao, J., Lai, H., Lyu, Z. et al. (2015). Hydrophilic hierarchical nitrogen-doped carbon nanocages for ultrahigh supercapacitive performance. *Advanced Materials* 27: 3541–3545.

81 Liu, M., Zhang, Z., Dou, M. et al. (2019). Nitrogen and oxygen co-doped porous carbon nanosheets as high-rate and long-lifetime anode materials for high-performance Li-ion capacitors. *Carbon* 151: 28–35.

82 Chen, Y., Shi, L., Guo, S. et al. (2017). A general strategy towards carbon nanosheets from triblock polymers as high-rate anode materials for lithium and sodium ion batteries. *Journal of Materials Chemistry A* 5: 19866–19874.

83 Ding, W., Li, L., Xiong, K. et al. (2015). Shape fixing via salt recrystallization: a morphology-controlled approach to convert nanostructured polymer to carbon nanomaterial as a highly active catalyst for oxygen reduction reaction. *Journal of the American Chemical Society* 137: 5414–5420.

9

Applications of Collagen-Derived Carbons in Electrochemical Energy Storage and Conversion

As sustainable and clean green energy is closely related to the economy, environment, and people's livelihood, it has become one of the hottest topics in the twenty-first century. A wide variety of new electrochemical energy-storage/conversion systems, such as electrochemical capacitors, lithium–sulfur (Li–S) batteries, lithium-ion batteries (LIBs), and fuel cells (FCs), were developed to relieve the inevitable depletion of fossil-fuel resources and growing environmental worries [1–3]. These new electrochemical energy-storage/conversion systems have the characteristics of good mobility, simple operation, little or no environmental pollution, wide application, high power, and energy density. Thanks to the efforts of scientists and engineers, the energy systems have made significant progress in many areas, such as hybrid/pure electric vehicles, smart grids, portable and wearable electronic devices, and other environment-friendly rechargeable electronic devices. It is worth noting that these systems can directly and efficiently convert chemical energy into electrical energy with almost no pollution. However, due to the unsatisfactory performance of some of these systems, which are ultimately related to the challenges of forming well-defined material structures and functions, including poor maneuverability, high cost, and insufficient duration, the wide practical application of these innovative electrochemical energy technologies still has great limitations. Therefore, for these commercially viable energy technologies, rational design and careful exploration of new materials with well-defined structures and functions are necessary to improve efficiency, enhance durability, and reduce costs. In this regard, carbon materials possess the advantages of good chemical stability, large specific surface area, good electrical conductivity, and fast charge transfer speed, which have broad application prospects in many fields [4]. The facile interfacial modification and functionalization of collagen-derived carbon materials with different chemical functional groups and rich heteroatoms are beneficial to modulate their charge storage capacity and catalytic activity. In addition, these collagen-derived carbon materials have the ability to further adjust the charge transport and even the electrochemical behavior of the multifunctional structural adjustment of hierarchical porosity [5].

Collagen-Derived Materials: Synthesis and Applications in Electrochemical Energy Storage and Conversion, First Edition. Feng Wang, Yaqin Huang, and Jin Niu.
© 2022 WILEY-VCH GmbH. Published 2022 by WILEY-VCH GmbH.

9.1 As Electrode Materials for Supercapacitors

Supercapacitors (SCs) have ultrahigh power density and excellent cycling stability. Owing to their intended performance, SCs have become an important application for electrochemical energy-storage devices [6]. Unfortunately, their relatively low energy densities seriously hinder their large-scale applications [7]. The energy density (E) of SC is determined by the equation $E = \frac{1}{2}CV^2$. Increasing the specific capacitance of the electrode (C) and expanding the working voltage (V) are effective ways to increase the energy densities of SCs [8]. The specific capacitance can be improved by optimizing the electrode materials, and the working voltage can be enlarged by choosing proper electrolytes with wide potential windows [9]. According to the different energy-storage mechanisms, the capacitances of SCs can be divided into electric double-layer capacitor (EDLC) and pseudocapacitance [10]. The former is caused by electrostatic charge adsorption at the electrolyte–electrode interface. The latter is caused by the reversible faradaic reaction on or near the electrode surface [11]. These two energy-storage mechanisms are closely related to the characteristics of the electrode materials, including their conductivity, pore structures, and specific surface areas (SSAs). In addition to these factors, the high wettability of electrode materials also plays a key role in delivering efficient SSA for electrochemical reactions, which is closely related to the surface functional groups of the electrode materials [12–14]. Collagen-derived porous carbons from different sources of collagen-rich biomass precursors (e.g. pig bone [15], cattle bone [16–19], fish scale [15, 20, 21], and fish skin [22]) have good pore structures and plentiful surface heteroatoms, and are used to assemble high-performance SCs.

Huang and coworkers firstly used fish scale–derived porous carbon network (FHLC) as the electrode material for an SC in a 6 M KOH aqueous electrolyte [15]. Due to its hierarchical porosity and high specific surface area, the FHLC electrode exhibited a specific capacitance of $158\,F\,g^{-1}$ at $0.05\,A\,g^{-1}$, which remained as high as $109\,F\,g^{-1}$ at $10\,A\,g^{-1}$. Nevertheless, FHLC fabricated via high-temperature pyrolysis is generally hydrophobic, which leads to low efficiency of charge storage in aqueous electrolytes. In addition, the presence of a large number of meso-pores/macropores in the layered porous structure leads to low density, resulting in poor bulk performance.

Heteroatom doping can simultaneously improve the wettability of the electrode while introducing additional pseudocapacitance. Furthermore, two-dimensional (2D) carbon materials with dominated micropores have a higher density than hierarchically porous three-dimensional (3D) carbon materials with similar SSA and can exhibit good rate performance without introducing mesoporosity/macroporosity [23–25]. Given these considerations, using a fish skin–derived porous carbon nanosheet (MPCNS), Wang et al. prepared a new electrode material for an SC in a 6 M KOH aqueous electrolyte (Figure 9.1a) [22]. The abundant surface heteroatoms (S, O, and N) endowed the MPCNS electrode with a higher pseudocapacitance and greatly enhanced its wettability to the aqueous electrolyte. Figure 9.1b shows that the electrolyte contact angle of MPCNS is smaller than that of the bulk porous carbon prepared by conventional KOH activation (MPCB) and the porous carbon

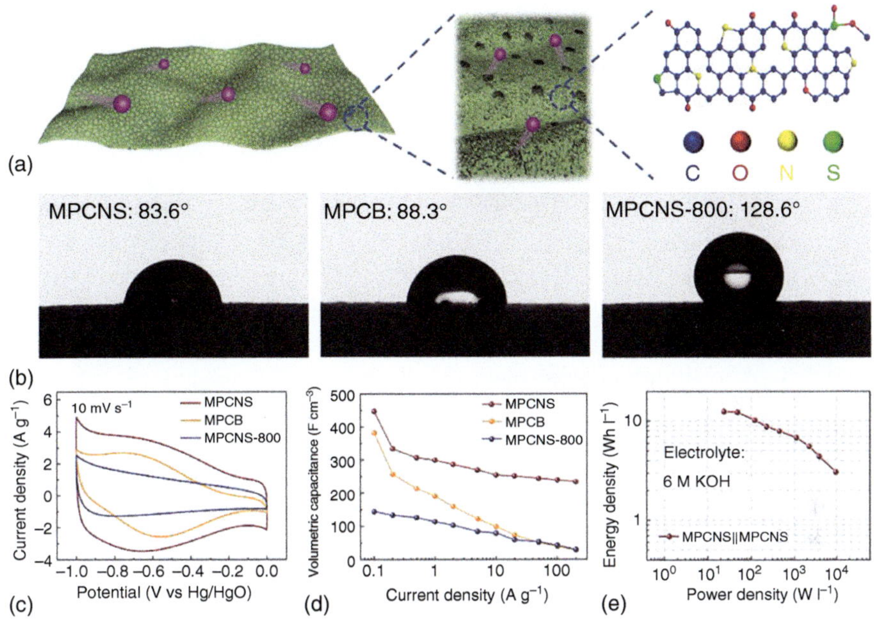

Figure 9.1 (a) The mechanism of energy storage and mass transfer in heteroatom-doped porous carbon nanosheet electrodes. (b) Electrolyte contact angle measurements (6 M KOH aqueous electrolyte) of the MPCNS electrode and referential electrodes. Electrochemical performance of SCs in 6 M KOH aqueous electrolyte: (c) CV curves at $10\,mV\,s^{-1}$. (d) Volumetric capacitances at increasing current densities. (e) Ragone plot of MPCNS||MPCNS SC. Source: Reproduced with permission from Niu et al. [22]. 2018, Elsevier.

nanosheet heat-treated at 800 °C (MPCNS-800). The good wettability of the MPCNS electrode also provides efficient SSA for charge storage, thereby enabling a large EDLC. Therefore, compared with the MPCB and MPCNS-800 electrodes, the cyclic voltammetry (CV) curve of the MPCNS electrode shows a clear redox peak and a larger loop (Figure 9.1c). Owing to the rational compact density, the MPCNS electrode shows a high gravimetric capacitance of $438\,F\,g^{-1}$ and a volumetric capacitance of $447\,F\,cm^{-3}$ at $0.1\,A\,g^{-1}$, even at a current density of $200\,A\,g^{-1}$. The ultrahigh current density also maintains up to $230\,F\,g^{-1}$ and $235\,F\,cm^{-3}$ (Figure 9.1d). The SC based on MPCNS electrodes provides high energy and power densities of $12.5\,Wh\,l^{-1}$ and $9.67\,kW\,l^{-1}$, respectively (Figure 9.1e). Moreover, the cycling performance of MPCNS||MPCNS SC is superior. Even after 20 000 cycles, there is no significant decay.

As reported in Zhang's work [26], gelatin was selected as the natural biomass, and oxygen-rich carbon with a layered sedimentary rock structure was successfully prepared (named as GA-temperature). Using a simple physical mixing method of gelatin and citric acid without a template, a layered sedimentary rock structure with rich microporous carbon was prepared, as shown in Figure 9.2. As the supercapacitor electrode, it has ultrahigh rate performance and excellent cycling stability.

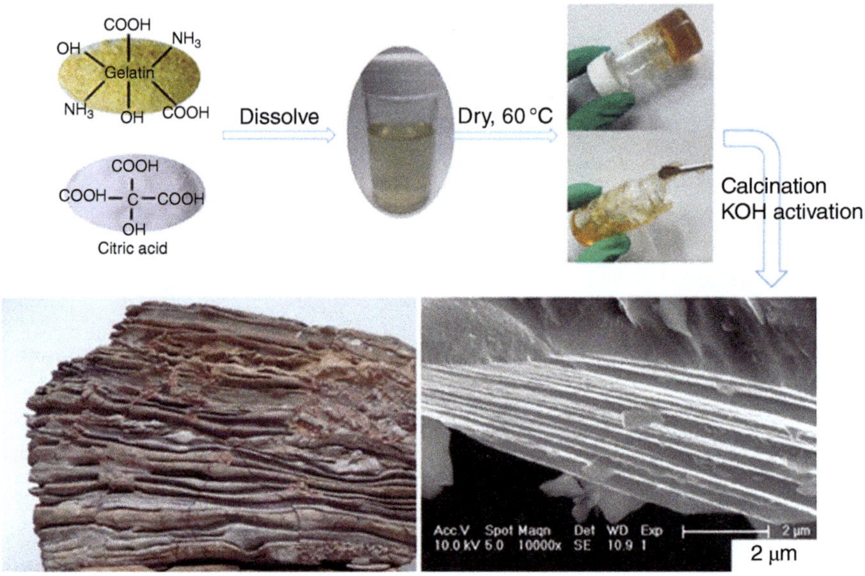

Figure 9.2 Schematic diagram of the preparation of the oxygen-enriched activated carbon with layered sedimentary rock structure. Source: Zhang et al. [26]/with permission of © American Chemical Society.

Figure 9.3a shows the CV curves of the electrode materials obtained in the symmetrical supercapacitor at $10\,mV\,s^{-1}$ in 1 M H_2SO_4 aqueous solution. As expected, the rectangular and almost symmetrical CV curve shows its ideal capacitive behavior. GA650//GA650 symmetrical supercapacitors show a wide reversible peak at 0.2 V. The CV curve is a slightly distorted rectangle with the largest integral area, indicating that the double-doped O and N heteroatom-activated carbon with a layered sedimentary rock structure has good pseudocapacitance performance. As the scan rate increases from $1\,mV\,s^{-1}$ to an ultrahigh $500\,mV\,s^{-1}$, the CV curve of GA650//GA650 remains rectangular, indicating that it has fast ion transfer and ion diffusion, good rate performance, and close to ideal capacitance behavior. The capacitance of GA650 at $1\,mV\,s^{-1}$ is $292.4\,F\,g^{-1}$, while the capacitances of GA750 and GA850 are 196.2 and $98.9\,F\,g^{-1}$, respectively. Nevertheless, when the scan rate increases to $500\,mV\,s^{-1}$, the capacitance of GA650 is still $128.4\,F\,g^{-1}$, while the capacitances of GA750 and GA850 are only 90.1 and $47.1\,F\,g^{-1}$, respectively. Figure 9.3b shows the specific capacitances calculated according to the discharge curves. The specific capacitance of the GA650//GA650 symmetrical supercapacitor at $0.5\,A\,g^{-1}$ is $181.7\,F\,g^{-1}$, which is much higher than the KOH-activated banana electrode ($74\,F\,g^{-1}$). Even at $30\,A\,g^{-1}$ (the current density increased 60 times from 0.5 to $30\,A\,g^{-1}$), the capacitance remained at $140.4\,F\,g^{-1}$, and the capacitance retention rate was 77.3%, suggesting its good rate performance. Figure 9.3c shows the cycling stability of GA650//GA650 electrode at $2\,A\,g^{-1}$. Even if the charge–discharge process exceeds 10 000 cycles, the capacitance retention rate is still as high as 100% with almost no capacitance decay, indicating superior electrochemical cycling stability.

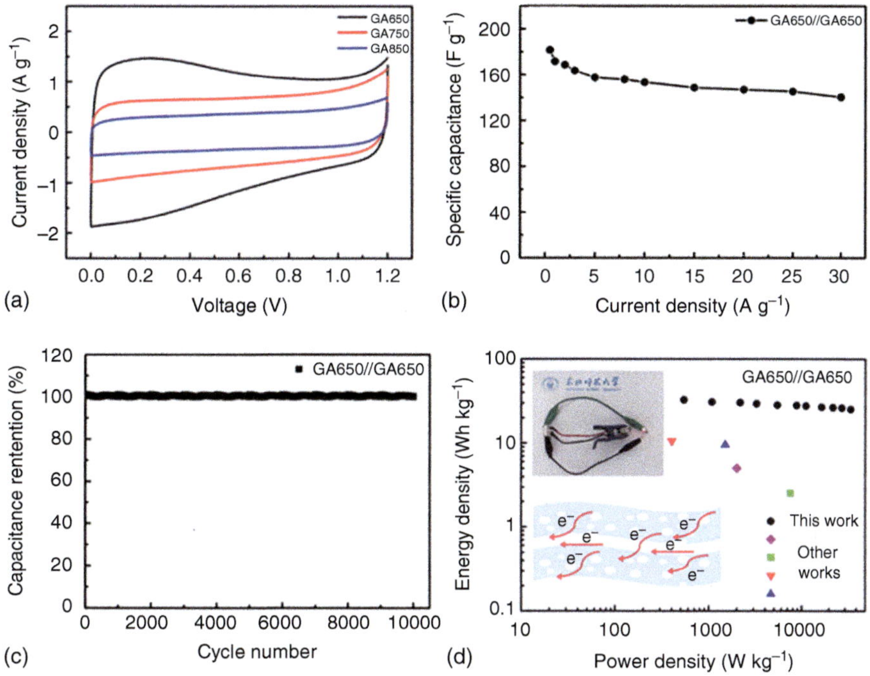

Figure 9.3 Electrochemical two-electrode performance of the symmetric supercapacitor. (a) CV curves of GA650, GA750, and GA850 at a scan rate of $10\,mV\,s^{-1}$. (b) The specific capacitance of the symmetric supercapacitance at various current densities. (c) Cycling stability in the symmetric supercapacitor at a current density of $2\,A\,g^{-1}$ over 10 000 cycles. (d) Ragone plot related to energy and power densities of GA650//GA650 electrode in symmetric supercapacitor compared with others [27–30]. The inset is an optical image showing a red LED powered by two supercapacitors in series. Source: Zhang et al. [26]/with permission of © American Chemical Society.

The GA650//GA650 symmetrical supercapacitor is assembled in an aqueous solution to determine its energy and power density. The maximum energy density of a symmetrical supercapacitor at a power density of $540.0\,W\,kg^{-1}$ is $32.7\,Wh\,kg^{-1}$, and the maximum energy density at a higher power density of $34.7\,kW\,kg^{-1}$ remains at $25.3\,Wh\,kg^{-1}$. As shown in Figure 9.3d, this is comparable to previously reported carbon-based symmetric supercapacitors. The above results show that the oxygen-enriched activated carbon with layered sedimentary rock structure prepared from gelatin and citric acid is a promising high-performance, high-rate electrochemical supercapacitor electrode material.

Studies have shown that organic electrolytes can be used to increase the energy density of SCs because they have a wider operating voltage window (up to 2.8 V) than aqueous electrolytes [8]. In nonaqueous electrolytes, abundant heteroatoms are unnecessary as they cannot improve the electrolyte wettability and generally lead to minimal pseudocapacitance [31]. However, good electrical conductivity and high SSA are relatively important prerequisites for ion storage and charge transfer. Moreover, layered porosity is advantageous because nonaqueous electrolytes have

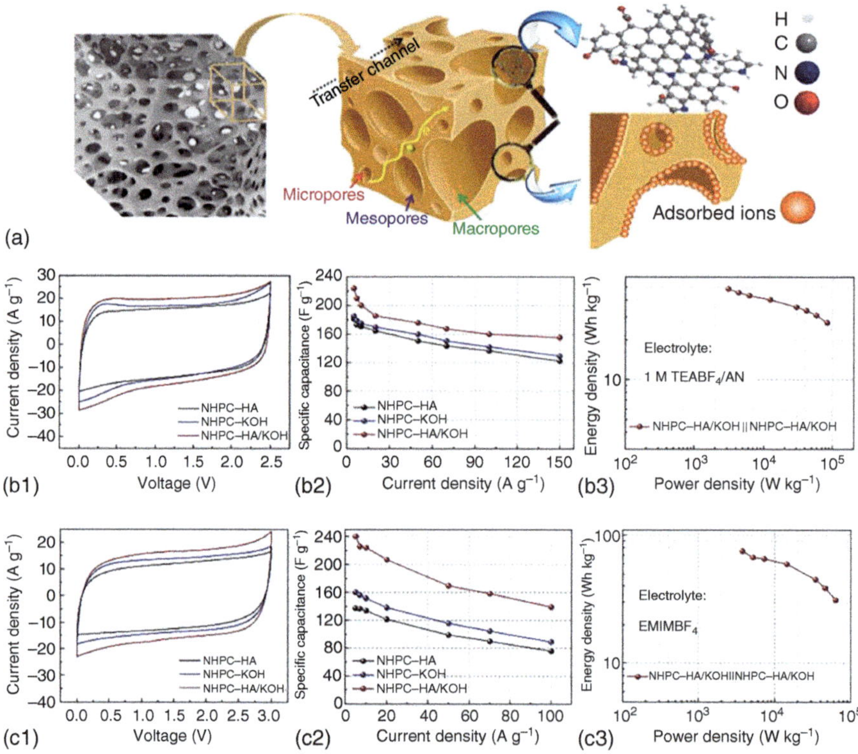

Figure 9.4 (a) The mechanism of energy storage and mass transfer in a heteroatom-doped porous carbon network electrode. Electrochemical performance of SCs in 1 m TEABF4/AN electrolyte: (b1) CV curves at a scan rate of 200 mV s^{-1}, (b2) specific capacitances at different current densities, and (b3) Ragone plot of NHPC–HA/KOH||NHPC–HA/KOH SC. Electrochemical performance of SCs in neat EMIMBF4 electrolyte: (c1) CV curves at a scan rate of 200 mV s^{-1}, (c2) specific capacitances at different current densities, and (c3) Ragone plot of NHPC–HA/KOH||NHPC–HA/KOH SC. Source: Reproduced with permission from He et al. [18]; © 2018, American Chemical Society.

a larger ion size than aqueous electrolytes. Therefore, mesopores/macropores with larger pore sizes than micropores can enhance ion diffusion and storage in nonaqueous electrolytes (Figure 9.4a) [17]. Based on this, Wang et al. used a porous carbon network derived from cattle bone (NHPC–HA/KOH) as the electrode for an SC in a 1 M $TEABF_4$/AN electrolyte [19]. As the NHPC–HA/KOH electrode had good electrical conductivity, mesopore/macropore-predominant porosity, and a larger specific surface area, it exhibited the largest CV loop area of all NHPC electrodes (Figure 9.4b1). When the current density increased to 5 A g^{-1}, NHPC–HA/KOH electrodes exhibited a high specific capacitance of 224 F g^{-1} (Figure 9.4b2). The SCs prepared with NHPC–HA/KOH electrodes had a high energy density of 48.6 Wh kg^{-1} and a power density of 84 kW kg^{-1} (Figure 9.4b3). In addition, the NHPC–HA/KOH||NHPC–HA/KOH SC still exhibited high capacitance retention of 96.2% even after 10 000 cycles.

Ionic liquid electrolytes have wider operating voltage windows (up to 4.0 V) and higher safety than organic electrolytes [31, 32]. Therefore, the NHPC–HA/KOH electrode was used to assemble an SC with a neat $EMIMBF_4$ electrolyte [19]. The NHPC–HA/KOH electrode exhibited a larger CV loop area at 200 mV s^{-1} than the comparative electrodes (Figure 9.4c1). It provided a high specific capacitance of 240 F g^{-1} at 5 A g^{-1} and maintained a high specific capacitance of 140 F g^{-1} even at 100 A g^{-1} (Figure 9.4c2). The as-assembled SC exhibited a high energy density of 75 Wh kg^{-1} at 3.75 kW kg^{-1}. Even at an ultrahigh power density of 63.5 kW kg^{-1}, the SC could still deliver an energy density of 31.3 Wh kg^{-1} (Figure 9.4c3). Apart from its high power and energy properties, the SC also showed good cycling performance with capacitance retention of 89.2% after 10 000 cycles.

9.2 As Electrode Materials for Hybrid Capacitors

Manufacturing energy-storage devices with high power and energy densities is not only related to the electrode material but also depends on the device configuration [33, 34]. As hybrid capacitors, in addition to high energy and high power, lithium-ion capacitors (LICs) can provide a long lifetime as well [35]. Typically, the high energy density of LICs is due to the Faraday process on the cell-type anode (e.g. surface charge transfer process and Li-ion insertion process) and the wide operating voltage window provided by the organic electrolyte. The long lifetime and high power density of LICs originate from non-Faraday processes (electrical double-layer effect) on capacitor-type cathodes [36]. Despite considerable progress being made in the development of electrode materials for LICs, there are still a series of problems with anode materials, such as short cycling life and limited rate performance, while the capacity of cathode electrode materials needs to be further improved [37]. By improving the conductivity of the LIC anode materials, adjusting their morphology, and introducing hierarchical porosity, cycling stability and rate capacity of anode materials for LICs can be improved [38]. For capacitor-type cathodes, the improvement of EDLC can be achieved by adjusting the porosity and increasing the specific surface area. At the same time, the pseudocapacitance can be increased by introducing surface heteroatoms [36].

Due to their extensive heteroatom doping, large SSA, controlled nanostructure, and good electrical conductivity, collagen-derived electrode materials have promising applications for high-performance LICs [39]. Wang et al. used cattle bone–derived porous carbon bulk (HDMPC) as both cathode and anode materials for LICs. The abundant defects, high SSA, good electrical conductivity, and mesopore-dominant porosity give the HDMPC electrode its battery-capacitive storage properties. As a cathode material, HDMPC exhibited a high capacity of 109 mAh g^{-1} at 0.1 A g^{-1} that remained as high as 47 mAh g^{-1} at 50 A g^{-1}. As an anode material, HDMPC provided 999 and 329 mAh g^{-1} at current densities of 1 and 20 A g^{-1}, respectively. In addition to the excellent rate performance, the HDMPC anode didn't show any capacity degradation even after 1000 cycles at 10 A g^{-1}.

On this basis, the researchers used the same HDMPC as the anode and cathode materials for the LIC for the first time (Figure 9.5a). An optimized mass ratio of 4 between the anode and cathode materials was used to ensure the good rate performance of the HDMPC||HDMPC LIC. In addition, to avoid oxidative decomposition of the electrolyte, operating voltage windows of 1.0 and 4.0 V were chosen. Figure 9.5b shows the CV curve of the HDMPC||HDMPC LIC. Because of the combination of the two different storage mechanisms of the HDMPC||HDMPC LIC, it is slightly different from the ideal rectangular shape of a double-layer capacitor. Similarly, the shape of galvanostatic charge–discharge curves of the HDMPC||HDMPC LIC also differed from the highly linear slope of an ideal capacitor (Figure 9.5c). The shape of the CV curves remains constant even at high scan rates of 200 mV s^{-1} and shows a small IR drop in the galvanostatic charge–discharge curves at high current densities, which suggest the good rate performance of the HDMPC||HDMPC LIC. Figure 9.5d shows the energy and power densities calculated from the galvanostatic profiles. The HDMPC||HDMPC LIC could obtain a high energy density of 106.4 Wh kg^{-1} and a power density of 500 W kg^{-1}. In addition, it could also exhibit an ultrahigh power density of 88.8 kW kg^{-1}. The energy density of 10.2 Wh kg^{-1} could still be achieved in a very short time of 0.4 seconds. As shown in Figure 9.5d, the combined high energy and power performance of the HDMPC||HDMPC LIC is superior to the previously reported work. Additionally, the HDMPC||HDMPC LIC has excellent long cycling stability (Figure 9.5e), with capacitance retention of 88.3% after 8000 cycles at 2 A g^{-1}. Inspired by this work, Liu et al. assembled LICs using a tuna bone–derived porous carbon network (TBC-900) as the anode and cathode materials [40]. The assembled TBC-900||TBC-900 LIC has a wide voltage window of 0–4 V and delivers a high energy density of 131 Wh kg^{-1} and an impressive power density of 62 kW kg^{-1}. What's more, the LIC maintains a high-energy-density retention of 82% even after 20 000 cycles at 5 A g^{-1}.

Sodium is a widely distributed and abundant resource on Earth. Na-ion capacitors (NICs) began to receive attention in 2012, although they are still in their infancy [37]. The energy-storage behaviors of NICs and LICs are similar, but NICs require different anode materials than LICs. Owing to the larger size of Na^+ ions (102 pm), hard carbon with low graphitization and large layer spacing is more conducive to reversible storage of Na^+ ions in NICs [41]. As far as hard carbons are concerned, collagen-derived materials are all hard carbons with low graphitization degrees and high defect density.

According to design principles for controlling NIC anode and cathode materials, Figure 9.6a shows that Wang et al. employed two fish scale–derived porous carbon networks (HPC-550 and HPC-800) as anode and cathode materials for the NIC [42]. The HPC-550 anode possessed abundant surface heteroatoms and large carbon-layer spacing, which resulted in a high reversible capacity of 305 mAh g^{-1} at 35 mA g^{-1} and good cycling performance. There were no capacity drops even after 8000 cycles at 7 A g^{-1}. Moreover, HPC-800 possessed hierarchical porosity and ultrahigh SSA, making it an appropriate cathode material for NICs. The HPC-800 cathode exhibits a high capacity of 128 mAh g^{-1} at 0.7 A g^{-1} and a long lifetime.

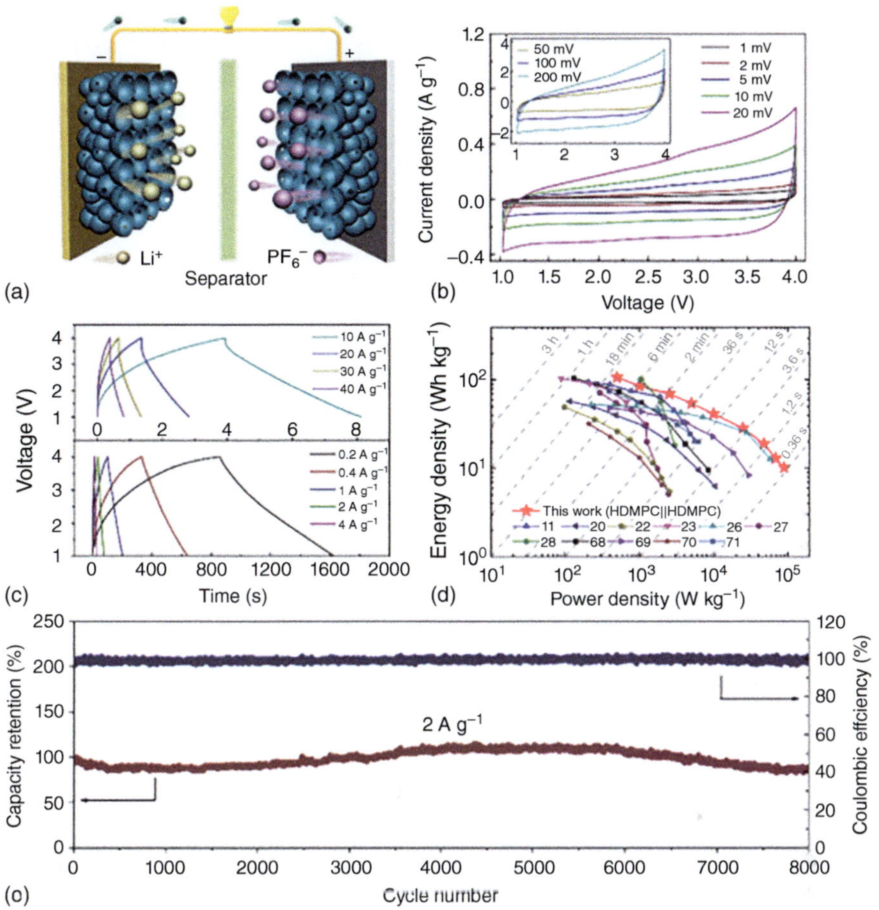

Figure 9.5 Electrochemical performance of HDMPC||HDMPC LIC. (a) Schematic illustration of the discharge process. (b) CV curves at various scan rates from 1 to 200 mV s^{-1}. (c) Charge–discharge profiles at different current densities from 0.2 to 40 A g^{-1}. (d) Ragone plots in comparison with other literature work. (e) Cycling performance at a current density of 2 A g^{-1}. Source: Reproduced with permission from Niu et al. [39]; © 2018, Elsevier.

Given the good performance of the half-cells, the NICs based on the HPC-550 anode and the HPC-800 cathode were assembled and evaluated over a voltage range of 0–4 V (Figure 9.6b). The CV curve shape of the HPC-550//HPC-800 NIC deviates slightly from the ideal rectangular shape due to the combination of two kinds of storage behaviors (Faraday and non-Faraday processes). The initial coulombic efficiency of the HPC-550//HPC-800 NIC is 91.3% at 0.7 A g^{-1}. To compare with other works, the calculation of the energy and power density of the HPC-550//HPC-800 NIC is done on the basis of the total active weight of the two electrode materials measured according to galvanostatic charge–discharge (GCD). It should be pointed out that in a practical cell, the energy and power densities are reduced when the weight of the electrolyte is considered. When the power

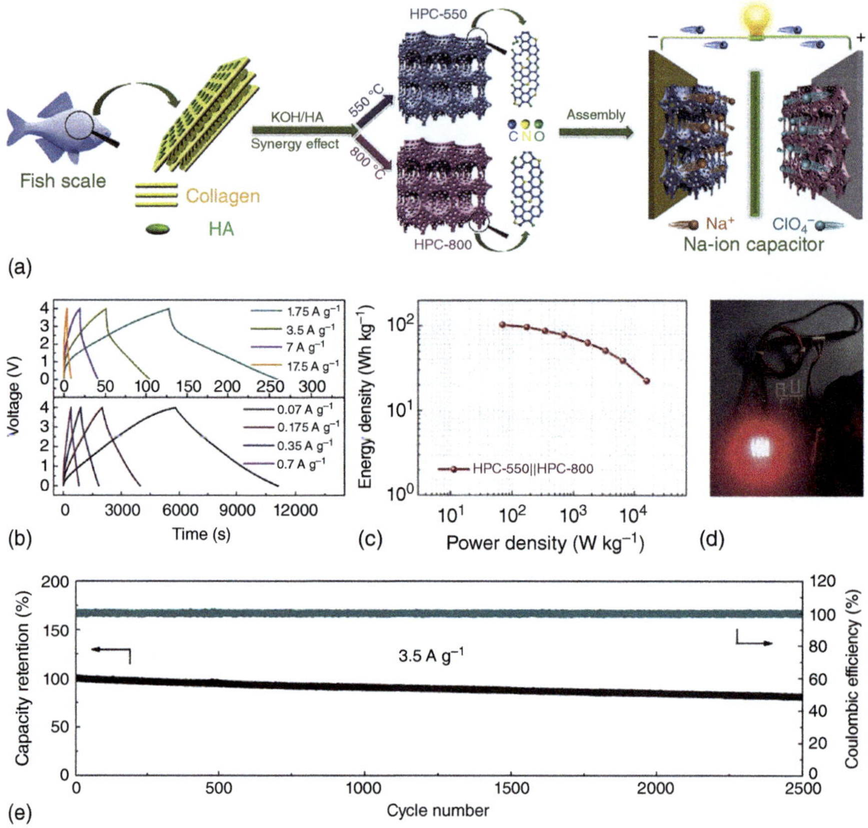

Figure 9.6 (a) Schematic illustration for the preparation of hierarchically porous carbons (HPCs) with N/O heteroatom-doping electrochemical performance of assembled NIC comprising HPC-550 anode and HPC-800 cathode. (b) GCD profiles at various current densities. (c) Ragone plots. (d) Photograph of 16 LEDs lighted by one HPC-550//HPC-800 NIC device. (e) Cycling performance at 3.5 A g^{-1}. Source: Liu et al. [42]/with permission of © Elsevier Ltd.

density was 70 W kg^{-1}, the energy density could be as high as 103.2 Wh kg^{-1}. Even at an ultrahigh power density of 15.9 kW kg^{-1}, the HPC-550//HPC-800 NIC could still deliver an energy density of 22.3 Wh kg^{-1}. Ragone plots show that the energy and power density of the HPC-550//HPC-800 NIC are better than most previously reported NICs (Figure 9.6c). With its high energy and power density, the assembled HPC-550//HPC-800 NIC can easily light up to 16 light-emitting diodes (LEDs) after being charged to 4.0 V (Figure 9.6d). Additionally, the HPC-550//HPC-800 NIC has a long life of more than 2500 cycles and retains 81.1% of its initial capacity at 3.5 A g^{-1} (Figure 9.6e).

To obtain higher power densities, Wang and coworkers also used a 2D hierarchical porous carbon nanosheet (HPCNS) as the anode material for a NIC (Figure 9.7a) [41]. In a half-cell, the tremella-like HPCNS anode exhibited a high reversible capacity of 362 mAh g^{-1} at 0.035 A g^{-1}. Even at an ultrahigh current density of 17.5 A g^{-1},

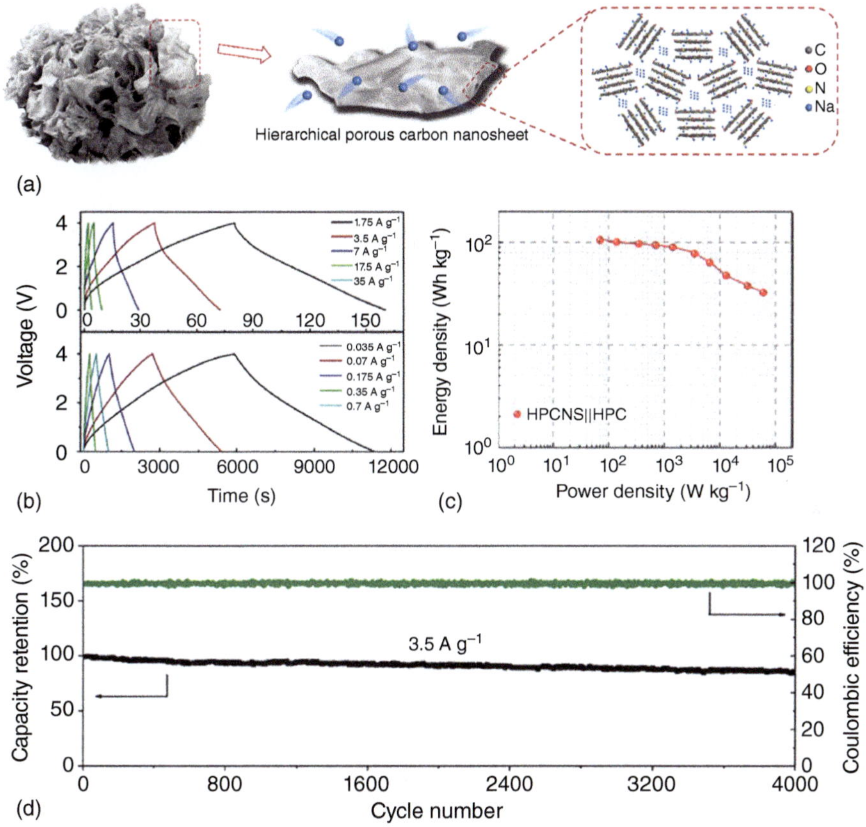

Figure 9.7 (a) The schematic illustration of the discharge process. Electrochemical performance of assembling NICs using HPCNS and HPC as the anode material and cathode material, respectively. (b) Galvanostatic charge–discharge curves at different current densities. (c) Ragone plots. (d) Cycling performance at the current density of 3.5 A g^{-1}. Source: Niu et al. [41]/with permission of © Elsevier Ltd.

the HPCNS anode also showed a capacity of 104 mAh g^{-1}. Particularly, a capacity of 151 mAh g^{-1} was still maintained after 10 000 cycles at 7 A g^{-1}.

A NIC was assembled using HPCNS as the anode material and a cattle bone-derived HPC as the cathode material. As shown in Figure 9.7b–d, the electrochemical performance of the HPCNS||HPC NIC was introduced. According to Figure 9.7b, the shape of the galvanostatic charge–discharge curve of the HPCNS||HPC NIC was not totally the same as the highly linear slope of an ideal supercapacitor, which is a result of the combination of two different types of storage behaviors (faradaic and non-faradaic reactions). In addition, the HPCNS||HPC NIC only showed a small IR drop even at high current density, which indicated the rapid transport of electrons and ions within the high-performance electrode materials. According to the galvanostatic profiles, the power and energy density were calculated as shown in Figure 9.7c. The HPCNS||HPC NIC could exhibit a high energy density of 105.2 Wh kg^{-1} at a power density of 70 W kg^{-1} and maintains an energy density of 32.5 kW kg^{-1} even at an ultrahigh power density of 60.4 Wh kg^{-1}. Compared with the state-of-the-art works, the integrated high energy and power performance of the HPCNS||HPC NIC was not much different or even better (Figure. 9.7d). Moreover, the HPCNS||HPC NIC indicated good long-term cycling stability. At the current density of 3.5 A g^{-1}, the capacitance retention remained at 85.8% after 4000 cycles.

The cuttlebone-derived activated carbon (CBCS) has been synthesized as an anode material for Na-HEC by Guo et al. [43]. Cuttlebone was carbonized under an inert atmosphere and then activated under air at 300 °C for three hours to give porous carbon (CBCS-Ax, x denotes the activation time). For the cathode material, the carbonized product was further activated with NaOH and KOH in a 1 : 1 : 2 ratio (NaOH : KOH : carbonized product) at 700 °C under N_2 for one hour and had a very high specific surface area of 3229 m^2 g^{-1}. In addition, the assembly of Na-HEC with CBCS-A as the anode material and CBC as the cathode provided an extremely high energy density and significant cycling stability (Figure 9.8).

To assess the Na^+ storage performance, electrochemical measurements of the half-cells with sodium metal were carried out. As shown in Figure 9.8a, the CV curves for the CBCS-A3 and CBCS half-cells are introduced. In the first CV scan for CBCS-A3, due to the formation of the solid electrolyte interface layer and the decomposition of the electrolyte, a broad reduction peak of 0.8 V typically appears. In subsequent cycles, this broad peak disappears, and the resulting CV curve shows a rectangular-like shape, indicating the existence of pseudocapacitance Na^+ storage behavior. The CV curve of the CBCS-A3 sample clearly shows a larger rectangular region compared to the CV curve of the CBCS, suggesting a larger pseudocapacitance for energy storage. Based on this capacitance contribution analysis, 59.7% of the total capacity of CBCS-A3 can be attributed to the surface capacitance mechanism (Figure 9.8b), which is the highest among all samples (Figure 9.8c). This is mainly due to the increased N and O fraction after air oxidation. As can be seen in Figure 9.8d, the initial charge and discharge capacities of CBCS-A3 reach 1350 and 695 mAh g^{-1}, respectively, which are even comparable to Li^+ charge storage. The initial Coulombic efficiency of 51.5% is higher than that of CBCS at 35.4%, further proving the positive effect of the air oxidation activation (AOA) treatment.

As shown in Figure 9.8e, the Ragone plots for the CBCS-A3//CBC-C NIC and CBCS//CBC-C NIC with different mass ratios are introduced, and the tests are carried out between 0 and 4.0 V. On this basis, the power density and gravimetric energy are calculated. It is obvious that the CBCS-A3//CBC-C NICs with a mass ratio of 1 : 3 show the best combination of energy and power. These devices offer a high energy density of 95 Wh kg^{-1} at a power density of 1000 W kg^{-1}. Even at the excellent power density of 53000 W kg^{-1}, an energy density of 36 Wh kg^{-1} can be maintained. This means that a high energy density of 70 Wh kg^{-1} can be achieved when the NIC device is charged for only 25 seconds. These results represent the most advanced NICs and are even comparable to most LICs with carbon-based anodes and cathodes (Figure 9.8e). Moreover, the CBCS-A3//CBC-C NICs showed a good cycling performance. After 8000 cycles at 5 A g^{-1}, the capacitance retention remained 90.5% (Figure 9.8f), indicating that they have broad practical application prospects.

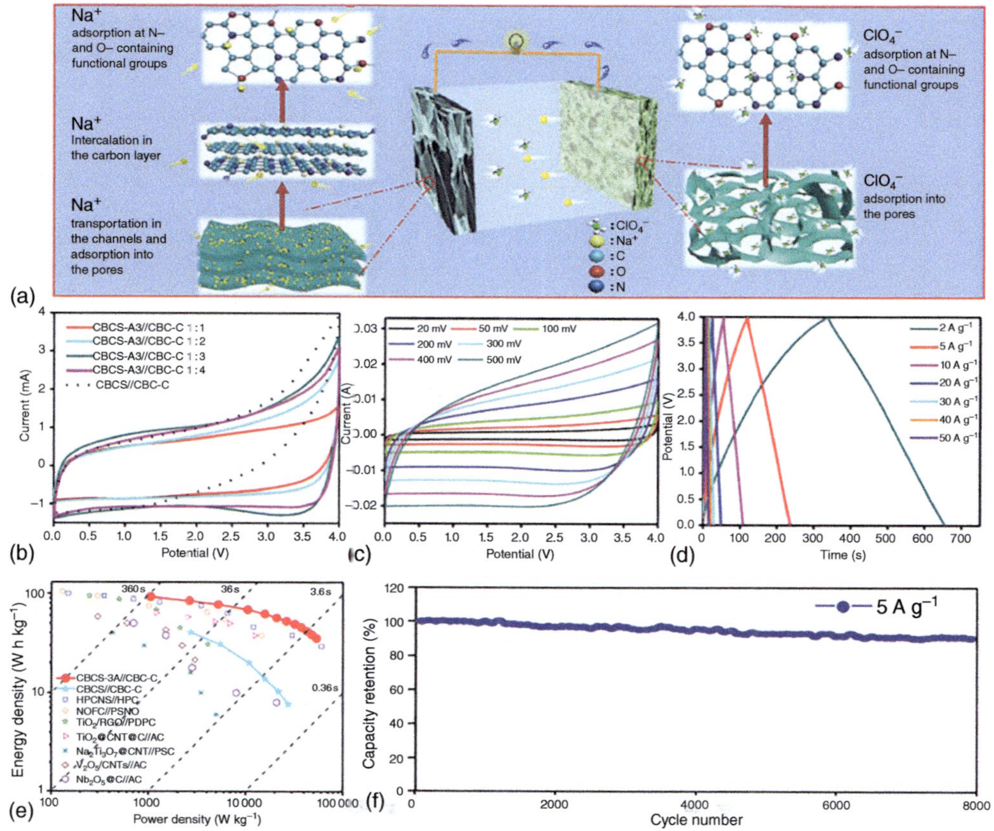

Figure 9.8 (a) Schematic illustration of the charge-storage mechanisms for the CBCS//CBC-C NIC; (b) CV curves of CBCS-A3//CBC-C NICs with various mass ratios and CBCS//CBC-C NIC devices; (c) CV curves of CBCS-A3//CBC-C NICs at various scan ratios; (d) galvanostatic profiles of CBCS-A3//CBC-C NIC device; (e) Ragone plots of CBCS-A3//CBC-C NIC and CBCS//CBC-C NIC, and energy–power density performance comparison with state-of-the-art reported energy-storage systems in literature; and (f) cycling performance of CBCS-A3//CBC-C NIC. Source: Reproduced with permission from Guo et al. [43]; © 2018, American Chemical Society.

9.3 As Anode Materials for Alkali-Ion Batteries

Alkali metal–ion batteries are used in various applications, such as electric vehicles, portable electronics, and grid energy storage. Benefiting from the advantages of good electrical conductivities, high safeties, and low costs, carbon-based materials are considered to be prospective anode materials for alkali metal–ion batteries. Collagen can be used as a precursor for the preparation of high-defect, low-cost porous carbon materials. In general, defects such as vacancies, edges, and heteroatoms are created during pyrolysis. What's more, activation or template methods are also feasible ways to attain porous structures. The unique structural compositions endow collagen-derived carbon anodes with good performance in alkali metal–ion batteries [44].

Advanced LIBs are seen as the most prospective electrochemical energy-storage device compared to conventional lead–acid, nickel–iron, and nickel–metal hydride. Rechargeable LIBs can be used for energy storage and conversion, so they are regarded as an alternative energy source to fossil fuels. Due to the low reduction potential, easy accommodation, and fast ion mobility of Li in most solids, LIBs generally exhibit higher specific energy, volumetric energy density, and higher power density than other metal–ion batteries. With over 25 years of technological development and innovation, Li-ion batteries have been successfully put into commercialization and participated in all aspects of our daily lives with a wide range of applications, from digital cameras to pure electric vehicles [45]. Nevertheless, LIBs still suffer from a number of drawbacks (i.e. high power density but low energy density for the former, and the anode/cathode electrode materials for the latter still need to be improved in terms of rate capability and long cycling stability).

Niu et al. directly carbonized cattle bones without additional activators and templates to prepare porous carbons with high defect density and specific surface area (Figure 9.9a). Due to the self-activation of hydroxyapatites within the cattle bones, the high-defect porous carbons prepared at 1100 °C (PC-1100) possess good electrical conductivity (5141 $S\,m^{-1}$), a narrow mesopore size distribution centered at approximately 4.0 nm, a maximum mesopore volume (1.829 $cm^3\,g^{-1}$), and a high specific surface area (2096 $m^2\,g^{-1}$).

PC-1100 possesses a unique pore structure and numerous defect sites, which make it show good performance at both low and high current densities. The reversible capacities of PC-1100 at 1, 2, 5, 10, and 20 $A\,g^{-1}$ are 1230, 951, 682, 491, and 343 $mAh\,g^{-1}$, respectively (Figure 9.9b,c). It still maintains a reversible capacity of 281 $mAh\,g^{-1}$ even at 30 $A\,g^{-1}$. In addition, the PC-1100 shows excellent long cycling stability. PC-1100 exhibited a high reversible capacity of 661 $mAh\,g^{-1}$ after 1500 cycles at 10 $A\,g^{-1}$, as shown in Figure 9.9d. The dominant mesoporous structure and good electrical conductivity of PC-1100 can provide a rich pathway for rapid ion and electron transport, leading to good cycling performance and high coulombic efficiency. In addition, the long cycling life is the result of the fiber bundle structure, which results in good mechanical stability of PC-1100. Even after long cycling, the PC-1100 retains its integrated structure.

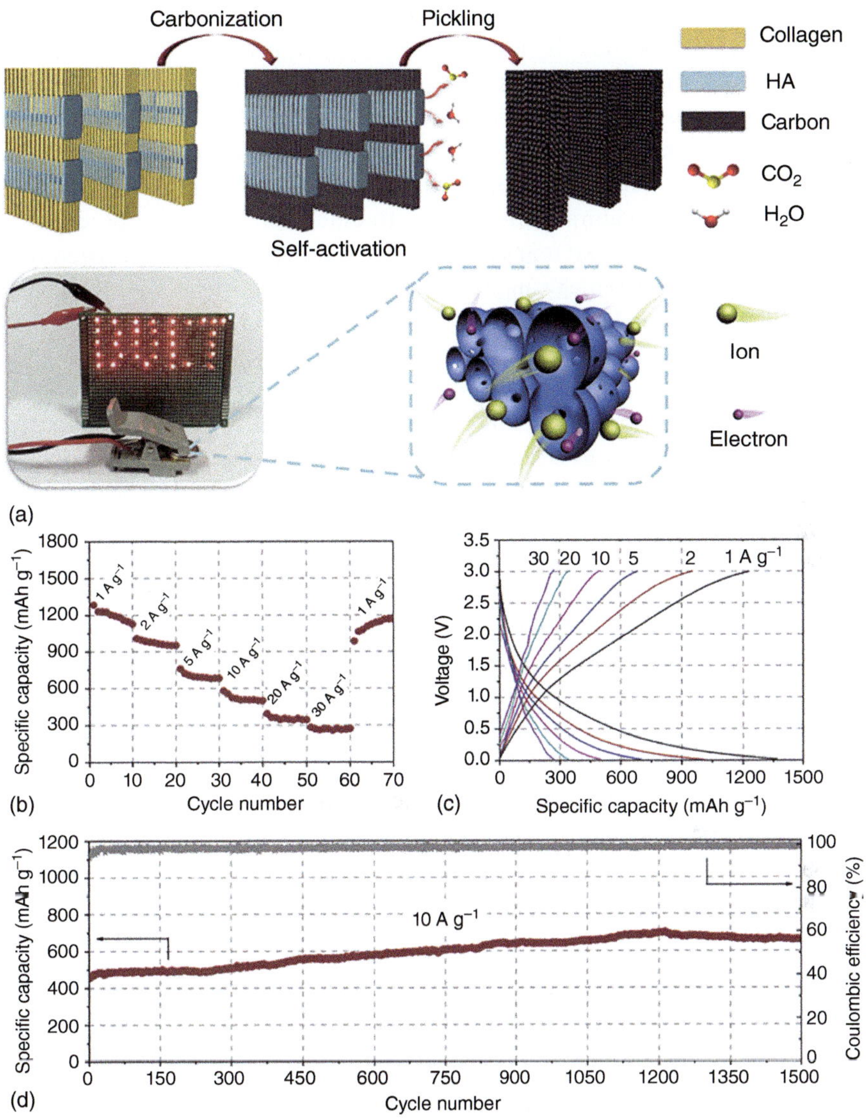

Figure 9.9 (a) Illustration of synthesis procedure and photograph of derived Li-ion cell. Electrochemical performance of half-cell using PC-1100 as the electrode and Li metal as the counter electrode in 1 M $LiPF_6$ electrolyte. (b) Rate capacity and (c) charge–discharge processes of PC-1100 electrode at various current densities from 1 to 30 A g^{-1}. (d) Cycling performance of PC-1100 electrode at 10 A g^{-1}. Source: Wang et al. [35]/with permission of © Elsevier Ltd.

As we all know, K^+ is readily intercalated in graphite through the $KC_{24} \rightarrow KC_{16} \rightarrow KC_8$ transition [46]. In the process of K^+ intercalation/deintercalation, excessive K^+ ionic radius might cause the notorious volume expansion on the anode side, leading to destroyed electrode structure, serious safety hazard, and rapid capacity attenuation [47]. Compared to conventional graphite anodes, biomass carbon materials, as one of the hard carbon members, offer intrinsic structural advantages, such as a naturally disordered structure rich in layered pores, an in situ self-doped elemental system consisting of sulfur/phosphorus/nitrogen, and increased layer spacing, which can provide quite a lot of active sites for K^+ storage and enhance the pseudocapacitive behavior in potassium ion capacitor (PIB) systems [48, 49].

Yuan et al. synthesized animal bone–derived hierarchical porous carbons (HPCs) with colony-like microstructures, with considerable forest-like pores and wide layer spacing without the need to add any active agent (Figure 9.10a) [47]. Thanks to their unique surface chemistry, the assembled PIBs had excellent rate behavior and cycling stability. Therefore, the HPCs prepared in this study provided a gravimetric specific capacity of 231.5 mAh g^{-1} and good cycling stability (95% after 2000 charge–discharge cycles) at a current density of 0.2 C.

Figure 9.10b,c further illustrates the CV curves for a range of scan rates from 0.1 to 1.0 mV s^{-1}. The increase in current response as the scan rate increases demonstrates the capacitance-controlled mechanism of the 3D-HPCs. Figure 9.10d shows the selected GCD profiles of the 3D-HPCs electrode obtained from different cycles. Because of the surface-driven K^+ storage behavior, the initial discharge capacity is close to 470 mAh g^{-1}, which shows a near-sloped line. In addition, the GCD results show good reversible stability at 0.2 C (1 C = 290 mA g^{-1}), and almost no capacity decay occurs until the 100th cycle. Electrochemical impedance spectroscopy (EIS) characterization was also performed to assess the ion/electron transfer resistance of the 3D-HPCs electrodes (Figure 9.10e). According to the two defined regions at a particular Nyquist plot, the semicircle represents the charge-transfer resistance (R_{ct}), and the Warburg sloping line represents the ion diffusion. It can be found that the R_{ct} of the 3D-HPCs is approximately 1300 Ω at the initial cycle.

As shown in Figure 9.10f, the cycling performance of the 3D-HPCs anode was performed. Notably, the resulting depot capacity was maintained at 205 mAh g^{-1} after 450 galvanostatic cycles at a current density of 0.2 C, indicating that it had excellent cycling stability. The bone-derived HPCs porous carbon exhibited enhanced cycling capacity after repeated charging and discharging processes compared to the chitinous/bamboo charcoal/potato/sugarcane/cyanobacteria/silica-derived biocarbon-based anode [48, 49]. Moreover, as shown in Figure 9.10g, the 3D-HPCs anode exhibited significant rate performance, providing specific capacities of 314, 192, 155, and 113 mAh g^{-1} at 0.2, 0.5, 1.0, and 2.0 C, respectively. It's worth noting that when the current density was again reset to 0.2 C, the corresponding capacities recovered to 273 mAh g^{-1}. This phenomenon demonstrates that the 3D-HPCs electrodes possess excellent reversibility, and their rate performance has been improved compared with the reported carbonaceous anodes for PIBs.

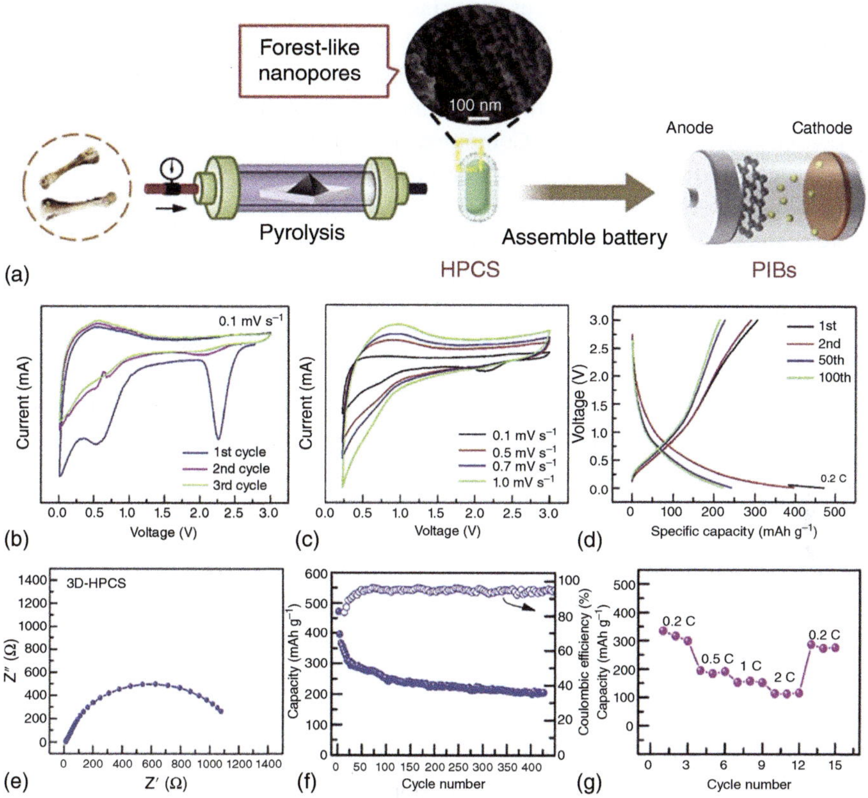

Figure 9.10 (a) Preparation process of the 3D-HPCs hard-carbon materials, (b) cyclic voltammograms of 3D-HPCs electrode for the first three cycles, (c) CV curves at different scan rates, (d) 1st, 2nd, and 50th galvanostatic charge discharge profiles of 3D HPCs electrode at 0.2 C, (e) EIS analysis, (f) cycling performance of 3D-HPCs electrode at the rate of 0.2 C, (g) specific capacities of 3D-HPCs electrode under different rates. Source: Yuan et al. [47]/with permission of © Elsevier Ltd.

9.4 As Anode Modifiers for Alkali-Ion Batteries

Currently, high-capacity anode materials for sodium-ion batteries include metal Sb/Sn alloys, phosphorus, and metal carbides/oxides/sulfides. These cathode materials are diffusion-controlled and cannot be loaded with high current densities, making it difficult to improve their rate performance. Additionally, due to the large radius of sodium ions, the insertion of sodium ions in large quantities would lead to structural disintegration or pulverization of anode materials, resulting in poor contact between the anode electrode and the current collectors as well as a short cycling life. The large difference in rate and cycling performance between the anode and cathode, therefore, limits further improvements in NIC power and lifetime [50, 51].

In view of this, many researchers have used nanostructures to improve the rate performance and cycling stability of NIC anode materials. Also, disordered or amorphous structures can be effective in improving the electrochemical performance of electrode materials. Amorphous materials are arranged in a long-range disorder but maintain a short-range order over a few atoms. An intrinsically isotropic amorphous structure can provide plentiful percolation channels for the rapid diffusion of Na ions and accommodate the volume change of anode materials that occurs during charge–discharge processes. Compared with crystalline materials, amorphous materials can provide better cycling and rate performance for anode electrodes in electrical energy-storage (EES) devices. However, amorphous and subnanomaterials are prone to agglomeration and recrystallization due to their intrinsic instability. Moreover, harsh reaction conditions and complicated synthesis processes are usually required to synthesize amorphous nanostructures, which seriously raise the production cost and limit the practical applications of subnano or amorphous anode materials in EES devices.

Wang and coworkers [52] attached amorphous SnO_2 subnanoclusters to N,O-doped porous carbon nanosheets (A-SnO_2@PCNS). The carbon carrier is a HPCNS prepared from bovine bone, which has abundant surface N and O heteroatoms. The schematic illustration of the synthesis for the amorphous SnO_2 is shown in Figure 9.11a. Due to the nano-restriction effect of the micropores of HPCNS, the amorphous SnO_2 (A-SnO_2) clusters are controlled in subnanometer size. In addition, due to the abundant surface heteroatoms of HPCNS, A-SnO_2 subnanoclusters are firmly fixed on the porous carbon nanosheets through numerous chemical bonds. Owing to the combination of nanopore confinement and strong chemical bonding, the structural stability of A-SnO_2 subnanoclusters can be enhanced. Thanks to the amorphous subnanostructure and excellent carbon support, the as-prepared A-SnO_2@PCNS has excellent rate capability and cycling performance and can be used as an anode for Na-ion storage. The NIC based on the A-SnO_2@PCNS anode delivered high energy and power densities with good cycling stability, which provides a new idea for the practical applications of NICs.

The A-SnO_2@PCNS anode was pre-sodiated to improve the coulombic efficiency at first. The A-SnO_2@PCNS anode and cattle bone–derived layered porous carbon cathode are assembled into NIC (A-SnO_2@PCNS||HPC NIC, Figure 9.11b). As shown in Figure 9.11c, the GCD curves of the NIC are not highly linear. Similarly, the CV curves of the NIC are not rectangular, which suggests hybrid EES features. According to the GCD profiles, the energy density of the NIC is calculated to be as high as 196.4 $Wh\,kg^{-1}$ with a power density of 100.0 $W\,kg^{-1}$. Even at an ultrahigh power density of 28.1 $kW\,kg^{-1}$, it can still provide a high energy density of 59.3 $Wh\,kg^{-1}$. The energy density and power density of the A-SnO_2@PCNS||HPC NIC are much higher than those of the HPCNS||HPC NIC reported in our previous work, [41] demonstrating that the A-SnO_2@PCNS anode is better than the heteroatom-doped carbon anode. In addition, the performance of the A-SnO_2@PCNS||HPC NIC is better than that of the state-of-the-art NICs reported previously (Figure 9.11d). The assembled battery can light up 68 LEDs because of its high-power properties (Figure 9.11e). Additionally, the NIC also showed a good cycling performance. The initial capacity remains at about 72%, even after 5000 cycles at a current density of 5 $A\,g^{-1}$ (Figure 9.11f), which indicates that the NIC had good potential in practical applications in the future.

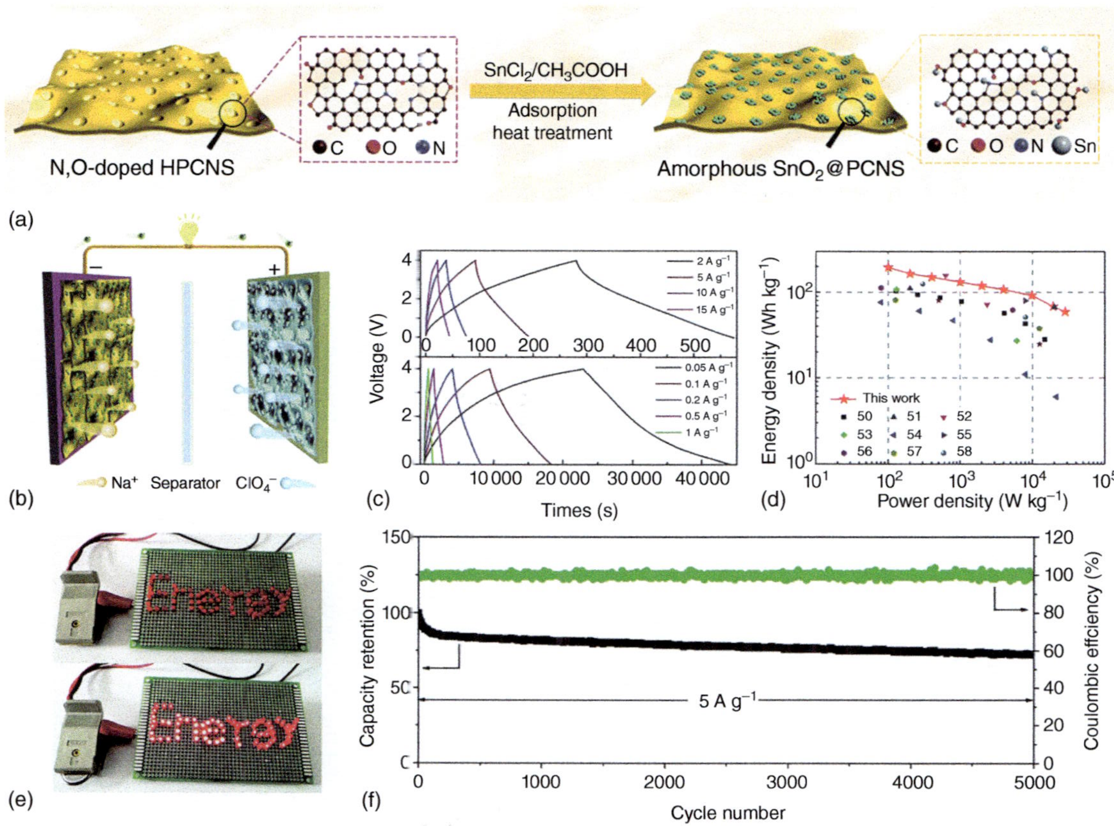

Figure 9.11 (a) Schematic illustration of the synthesis of amorphous SnO_2 subnanoclusters confined in micropores of N,O-doped porous carbon nanosheets (A-SnO_2@PCNS). Electrochemical performance of the A-SnO_2@PCNS||HPC NIC. (b) Schematic illustration of the discharge process. (c) GCD curves at different current densities. (d) Ragone plots and comparison with other reports in the literature. (e) Photograph of 68 LEDs powered by one NIC. (f) Cycling performance at a current density of $5\,A\,g^{-1}$. Source: Niu et al. [52]/with permission of © The Royal Society of Chemistry.

Shao et al. used biomass and silicon nanoparticles as precursors to synthesize silicon-based anode materials by coating and low-temperature pyrolysis [53]. The gelatin-derived carbon has many heteroatoms. In addition to acting as a "conductive backbone" to enhance electron/ion transfer, these heteroatoms can also simultaneously accommodate the Si expansion and immobilize the Si on the current collector via "dual-interfacial bonding." The obtained bender-free Si-based anode has the advantages of high initial columbic efficiency (85.3%), high gravimetric capacity (3160 $mAh\,g^{-1}$ at 0.2 $A\,g^{-1}$), large areal capacity (2.81 $mAh\,cm^{-2}$ at 0.18 $mA\,cm^{-2}$, approaching the commercial LIB requirement), good rate capability (1613 $mAh\,g^{-1}$ at 5 $A\,g^{-1}$), and cycling performance. Therefore, an assembled Li-ion capacitor utilizing the Si-based anode exhibits high energy density (213 $Wh\,kg^{-1}$), high power density (22.3 $kW\,kg^{-1}$), low discharge rate, and long lifetime. The scalable synthetic method combined with high performance makes this Si-based anode promising for practical applications. The structure of the precursor and synthesis of the Si@GC anode are shown in Figure 9.12a.

The traditional Si-based anode is shown in Figure 9.12b. The carbon black is severely agglomerated and separated from Si NPs by the insulating binder, which hinders the transfer of electrons and Li-ion. For Si@GC anode (Figure 9.12c), the interconnected GC coating on Si surface can improve the electronic conductivity and facilitate the Li-ion transportation. Furthermore, the strong interfacial bonding between Si NPs and GC allows the volume expansion of Si to be adjusted, and the solid state electrolyte interface (SEI) on the silicon surface is stabilized during the charge–discharge process. Additionally, the strong interfacial bonding between GC and Cu foil can enhance the adhesion of Si@GC on Cu foil, improving the transfer of electrons from Cu foil to Si@GC and preventing Si@GC from being removed from Cu foil during the cycling. Because of the concomitant conductive network construction and dual-interfacial engineering, Si@GC anode exhibited excellent performance for Li-ion storage. Significantly, the synthesis strategy of Si@GC anode was scalable and practicable. More importantly, the subsequent experiments showed that the strategy could be extended to use commercial biomass precursors to make other Si-based anodes (e.g. carboxymethylcellulose sodium, sodium alginate, and carboxylation chitosan). In terms of Li-ion storage, these binder-free Si-based anodes had better performance than traditional Si anodes, which also demonstrated the practicality and scalability of this simple strategy.

9.5 As Modifying Agents for Li–S Batteries

Since their theoretical energy density is as high as 2600 $Wh\,kg^{-1}$, Li–S batteries are considered to be one of the promising candidates for the next generation of energy-storage devices [54]. Furthermore, sulfur cathodes have other attractive merits, such as high natural abundance, low cost, and environment friendliness. Yet, the poor electrical conductivity of sulfur ($5 \times 10^{-30}\,S\,cm^{-1}$ at 25 °C) coupled with the high solubility of polysulfides leads to low reversible capacity and poor cycling performance of Li–S batteries, which hinder their widespread application [55]. At

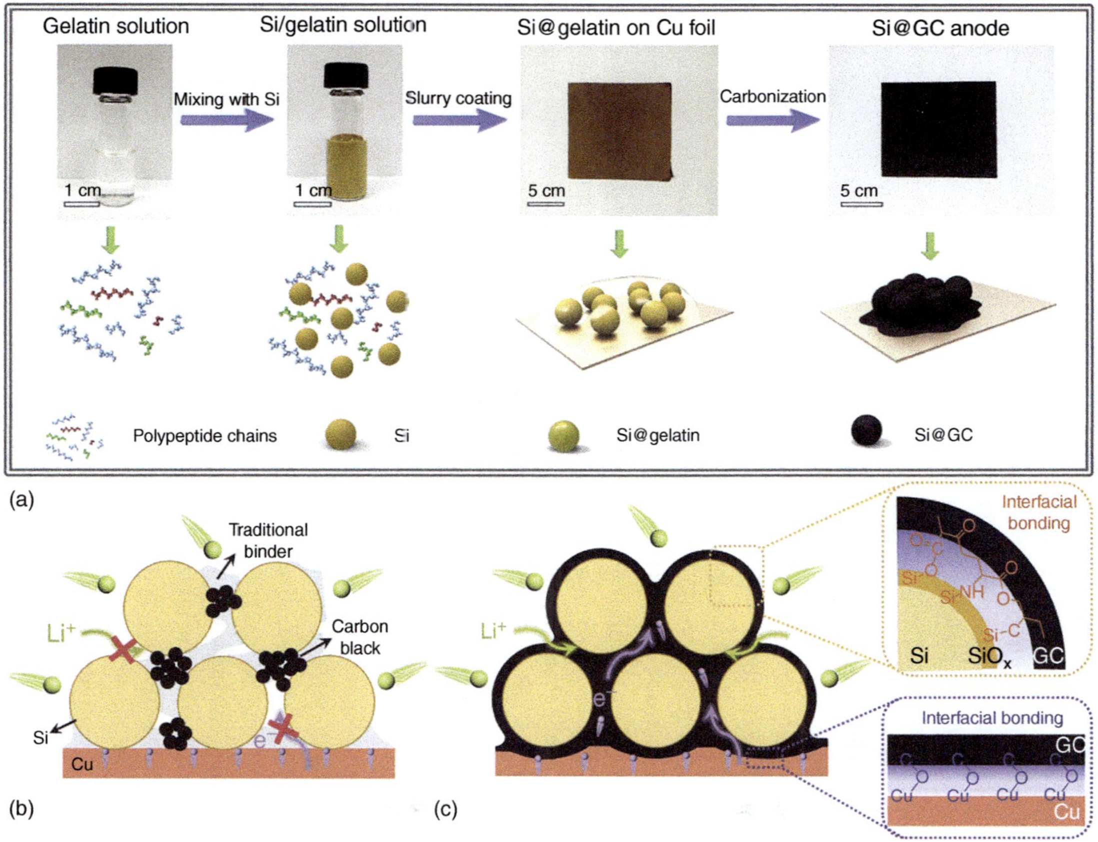

Figure 9.12 (a) Schematic diagram of the procedure to prepare the Si@GC anode. (b) Schematic illustration of conventional Si-based anode with traditional binder and conductive agent. (c) Schematic illustration of Si@GC anode with conductive network and enhanced dual interface. Source: Shao et al. [53]/with permission of © Elsevier Ltd.

this stage, a variety of strategies have been proposed to improve the performance of Li–S batteries, such as the introduction of porous materials into the cathode, the design of advanced composite cathodes, and modifications to the separators [56–59]. Since porous carbons have the advantages of high SSA, excellent electrical conductivities, and well-defined porosities, they have been widely used to enhance the performance of Li–S batteries.

Particularly, collagen-derived materials have been effectively used for the modification of Li–S batteries in recent years. In a typical case, Huang et al. enhanced the Li–S batteries performance using a pig bone–derived porous carbon network (BHPC-850) as an additive [60]. BHPC-850 has a high pore volume and SSA, resulting in a significant increase in the adsorption of polysulfide ions on the cathode during the discharge process. Furthermore, owing to the good electrical conductivity of BHPC-850, the conductivity of the modified sulfur electrode has also been enhanced. Therefore, the obtained sulfur cathode exhibited better performance than the conventional sulfur electrode, having a high discharge capacity of 1265 mAh g^{-1}, which remained at 643 mAh g^{-1} after 50 cycles. To further inhibit the dissolution of polysulfides in the electrolyte, a fish-scale-derived porous carbon network was introduced as an inhibitor to the sulfur cathode [61]. In addition to being used as an additive to trap polysulfides, more importantly, the collagen-derived materials coating also acts as a physical barrier on the electrode (Figure 9.13a). UV–vis absorption spectra show that the collagen-derived materials layer has a strong absorption capacity (Figure 9.13b). After 50 cycles at 0.1 C, the capacity retention of the cathode was 15% higher than that of its counterpart. It is hypothesized that the high surface area of the fish-scale-based porous carbon allows polysulfides to be adsorbed mainly in the cathode region during the discharge process, thus inhibiting the dissolution of polysulfides in the electrolyte (Figure 9.13c). The cathode coated with the collagen-derived materials layer has almost no change in the upper and lower discharge voltage platforms during the discharge process, exhibiting a stable voltage platform (Figure 9.13d), demonstrating that the collagen-derived materials layer can excellently prevent the migration of polysulfide ions. Furthermore, after 100 cycles at 1 C, it showed capacity retention of approximately 80% and a coulombic efficiency of approximately 97%. Thus, the collagen-derived materials–coated cathode exhibited enhanced electrochemical performance with a high discharge capacity of 1426 mAh g^{-1} that remained at 990 mAh g^{-1} after 100 cycles at 1 C (Figure 9.13e).

Combining sulfur and collagen-derived materials to form a composite material is another effective way to improve the performance of Li–S batteries. Huang and coworkers prepared porous sulfur/carbon nanocomposites by heat-treating sulfur with a fish-scale-derived porous carbon network (Figure 9.14a) [55]. Collagen-derived materials have a high specific surface area, which improved the reactivity and reduced the loss of active material. In addition, the large pore volume of collagen-derived materials can accommodate the volume expansion of the sulfur and alleviate the shuttling of polysulfide ions. The addition of sulfur results in a significant decrease in SSA from 2441 to 229 $m^2 g^{-1}$ and in pore volume from 1.69 to 0.3 $cm^3 g^{-1}$, demonstrating that sulfur has been successfully loaded into the collagen-derived material pores. The voltage profile shows that the nanocomposite

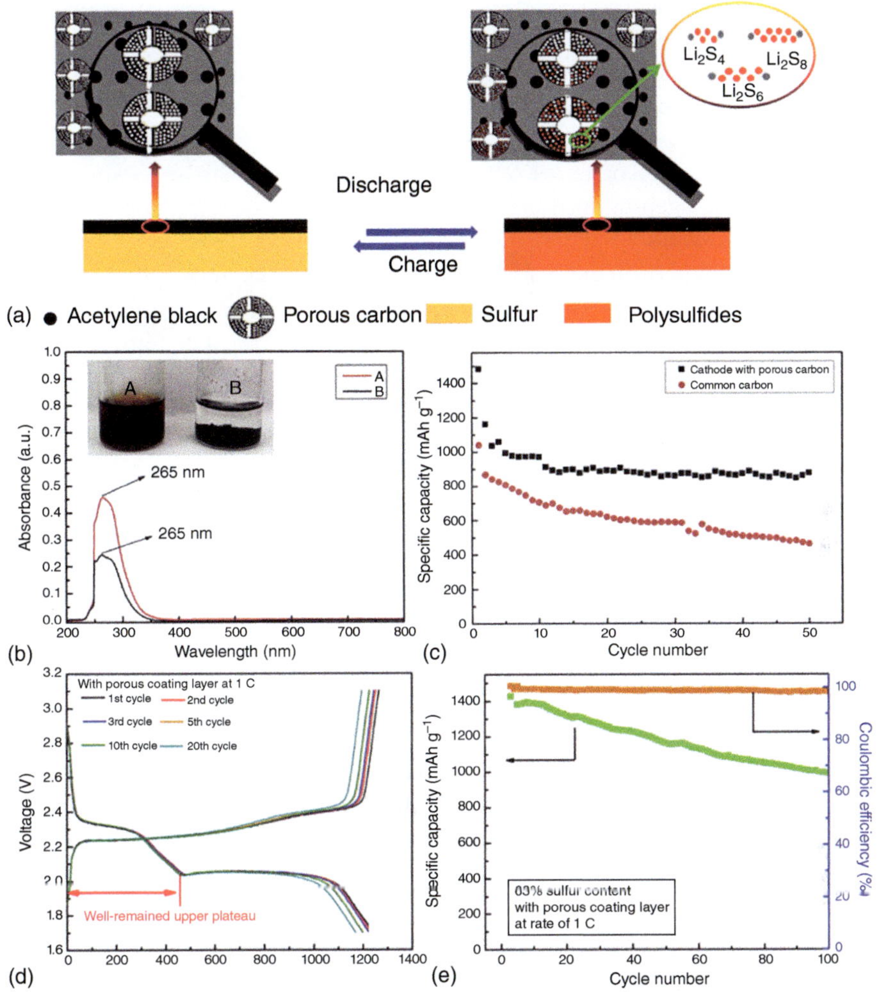

Figure 9.13 Electrochemical performance of the sulfur cathode with porous carbon coating layer: Schematic diagram illustrating the role of the porous coating layer during the electrochemical reaction. (a) Schematic diagram illustrating the role of porous coating layer during the electrochemical reaction. (b) The UV-visible absorption spectra of Li_2S_6-containing electrolyte without (A) and with (B) porous carbon, and the inset shows the comparison between A and B; (c) Cycle performance of the cathode with porous carbon and common cathode at rate of 0.1 C. (d) The discharge/charge curve of the cathode with the porous coating layer at the 1st, 2nd, 3rd, 5th, 10th and 20th cycle at 1 C. (e) 63% sulfur content cathode coated with porous layer at 1 C. Source: Gao et al. [61]/with permission of © Elsevier Ltd.

electrodes displayed high discharge capacities of 1370 and 686 $mAh\,g^{-1}$ at 0.1 and 3 C (Figure 9.14b), respectively. The lithium–sulfur battery (LSB) based on the nanocomposite cathode exhibits a high initial discharge capacity of 1039 $mAh\,g^{-1}$, which remains at 1023 $mAh\,g^{-1}$ after 70 cycles at 1 C (Figure 9.14c), much higher than 66.7% (605 $mAh\,g^{-1}$) of the nanocomposite. Assuming that as the sulfur

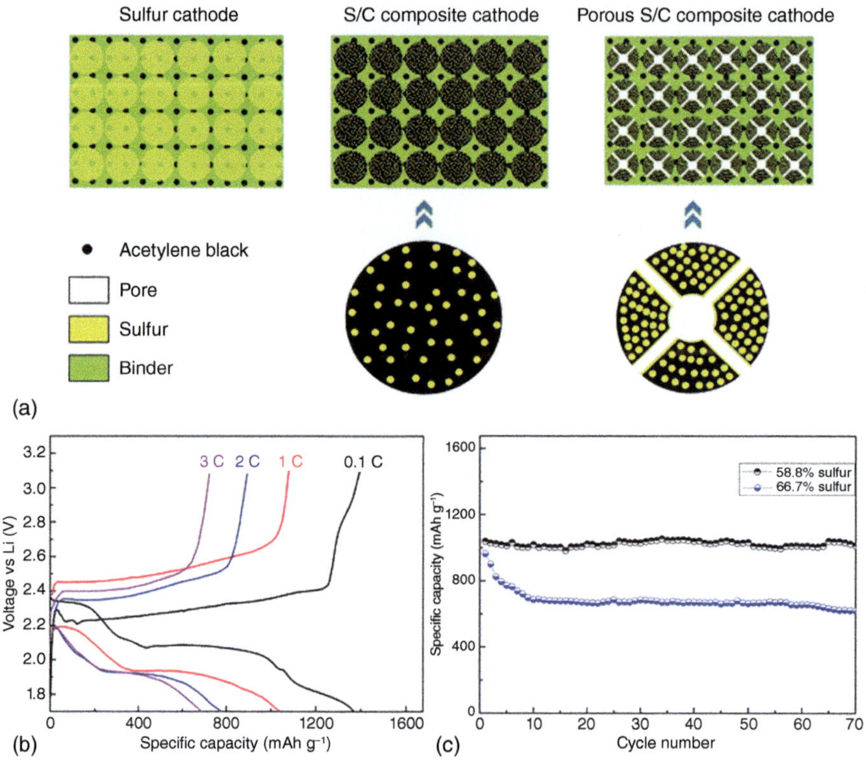

Figure 9.14 Electrochemical performance of the sulfur/carbon nanocomposite cathode: (a) Figure of the sulfur/carbon nanocomposite for improving cathode performance. (b, c) Electrical performance. Source: Reproduced with permission from Zhao et al. [55]; © 2013 The Royal Society of Chemistry.

content of the nanocomposite increases to 66.7%, there will not be enough room to accommodate the volume expansion of sulfur generated during the electrochemical reaction.

In addition to optimizing the positive electrode, diaphragm modification is also an effective method to suppress the polysulfide ion shuttle effect [62]. Fish bone–derived porous carbon mesh (FBPC) had a large specific surface area and a hierarchical pore structure, which can effectively enhance the physical blocking of polysulfides. Huang et al. prepared a flexible separator by coating a layer of FBPC on the surface of a conventional polypropylene diaphragm (Figure 9.15a) [63]. Furthermore, FBPC has abundant surface heteroatoms and provides abundant adsorption sites, which significantly improve the reactivation of adsorbed polysulfides. The synergistic effect of the physical blocking and chemisorption of polysulfides resulted in a high discharge capacity of Li–S batteries using FBPC-modified separator, reaching 1237.5 mAh g^{-1} at 0.2 C and 954.2 mAh g^{-1} at 1 C (Figure 9.15b). Furthermore, the Li–S batteries manufactured using the FBPC-modified spacer showed good cycling performance, and the low-capacity decay level was 0.064% per cycle (Figure 9.15c). Also, when using FBPC-modified diaphragms, the sulfur content can be further

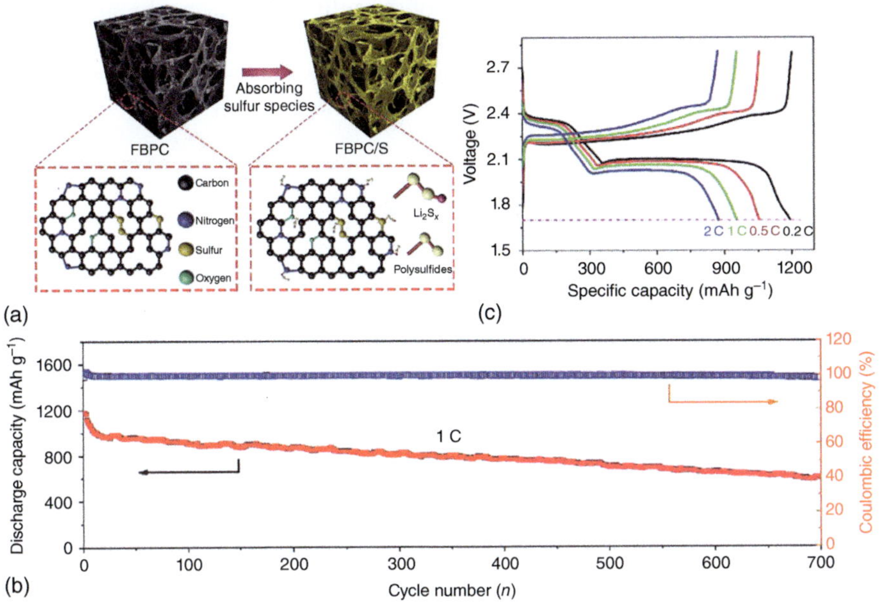

Figure 9.15 Electrochemical performance of Li–S batteries using FBPC-modified separator: (a) Figure of the FBPC-modified separator for improving cathode performance. (b, c) Electrical performance. Source: Reproduced with permission from Ai et al. [63]; © 2017, Elsevier Ltd.

increased to 84 wt%, and a high reversible capacity of 1011.3 mAh g^{-1} can be utilized with high sulfur content.

In addition, according to recent studies, surface functional groups (such as nitrile group, oxygenated framework, and oxygen-containing groups) or doped heteroatoms (such as S, B, and N) in the porous carbon play a key role in enhancing the electrochemical performances of the sulfur–carbon composite cathodes [64]. Among them, nitrogen is one of the most attractive doping heteroatoms. It enhances the electrical conductivity of carbon, improves the wettability of carbon, inhibits the diffusion of soluble polysulfides, and hinders the shuttle effect of lithium–sulfur batteries. At the same time, the most commonly used method to obtain nitrogen-rich porous carbon materials is to pyrolyze nitrogen-containing chemicals, such as polyaniline, polypyrrole, and polyacrylonitrile.

Qu and coworkers [27] prepared highly ordered nitrogen-rich mesoporous carbon (HNMC) by pyrolysis of gelatin through a simple template process, and sulfur-HNMC composite based on HNMC was synthesized for lithium–sulfur batteries by a melt-diffusion method. According to the analysis, elemental sulfur was found to be uniformly dispersed within the mesopores of HNMC. According to electrochemical tests, the sulfur-HNMC (53.3 wt% sulfur) cathode exhibited high rate performance and long cycling stability. The initial discharge capacity of the sulfur-HNMC (53.3 wt% sulfur) cathode was up to 1209 mAh g^{-1} and remained up to 600 mAh g^{-1} after 200 cycles at 1 C. Furthermore, the sulfur-HNMC also exhibits long-term cycling stability of 300 cycles at 3 C. The results show that the high rate capability and long-term cycling stability of the sulfur cathode can be significantly improved by encapsulating sulfur into highly ordered nitrogen-rich mesoporous carbon nanostructures, and that HNMC nanostructures are expected to become the carbon matrices for high-performance lithium–sulfur batteries.

9.6 As Metal-Free Electrocatalysts

Porous carbon materials are widely used in biochemical processes, adsorption, catalysis, oxygen reduction reactions, and energy storage due to their excellent chemical/thermal/mechanical stability. In electrochemical catalytic systems, aqueous-based electrolytes are often used, so electrodes are required to be hydrophilic in nature. However, porous carbon materials are usually highly hydrophobic and have low electronic conductivity, which limits the accessibility of specific active sites for catalytic processes. By introducing heteroatoms (e.g. B, O, and N) into the carbon backbone, relevant properties in sensing or catalysis can be improved. The electron-rich atoms contained in heteroatom-doped carbon materials lead to a complete change in the electronic properties of the material, which is essential for applications in electronic devices, optics, and fuel cell catalysts [65].

Since collagen-derived materials have a porous structure and abundant surface heteroatoms, even if no transition metals are introduced, they can also show catalytic activity. Song et al. reported a simple and low-cost method of synthesis for the first time [66]. N-doped porous carbon with various pore sizes was prepared using only chicken bone as a precursor for carbon, nitrogen, and a template. The morphology and electrocatalytic properties of the resulting porous carbon materials were explored. The results showed that the obtained N-doped porous carbon had high oxygen reduction reaction (ORR) activity, which was close to that of commercial Pt/C catalysts. The Brunauer-Emmett-Teller (BET) surface area of chicken bone–derived carbon materials (CCB) is $769\,m^2\,g^{-1}$, and the thermally stable compound $Ca_{10}(PO_4)_6(OH)_2$, present in the skeleton, plays an important role in the formation of a high surface area material with microporous pores. CV curves of CCB show the pseudocapacitive behavior of nitrogen-doped carbon, which indicates high electronic conductivity and effective charge propagation through the pores. The resulting CCB exhibits high ORR electrocatalytic activity in KOH solutions, as evidenced by its high current density, low onset potential, low half-wave potential, and strong methanol tolerance. Thus, CCB has high ORR activity and stability and is a prospective non-precious metal cathode that can be used in fuel cells.

Guo et al. directly converted protein-rich fish-scale biowaste into hierarchically porous 3D network nanocarbon (FS-Temperature) [67]. Interestingly, compared with the commercial 20 wt% Pt/C catalysts, this material had better electrocatalytic activity and stability for oxygen reduction in alkaline and acidic solutions, which were closely related to the internal pore structure, specific surface area, chemical state of its nitrogen atoms, and BET surface area. The ORR activity of the FS350, 20 wt% Pt/C, and NC catalysts was determined in a $0.1\,mol\,dm^{-3}$ KOH solution saturated with N_2 and O_2. The excellent ORR performance of FS350Z-900 compared to commercial Pt/C catalysts can be attributed to five aspects: (i) FS350Z-900 has strong interactions between the electrocatalytic active site and the C—N covalent bond in the graphite layer; (ii) the surface properties of FS350Z-900 can be maintained well during cycling; (iii) the nature of the nitrogen-containing active sites is different from that of the Pt/C catalysts; (iv) silicates from glass corrosion can block the active sites of the Pt/C catalysts, thus reducing their stability; and (v) Pt nanoparticles in Pt/C catalyst are easy to suffer aggregation and migration, resulting in the decrease of ORR activity and stability. The above findings suggest that this N-doped porous carbon nanocatalyst derived from FS biowaste is a very prospective Pt/C catalyst due to its excellent ORR activity and electrochemical CV stability in both alkaline and acidic solutions.

Wang et al. reported a porous carbon network derived from cattle bone as a metal-free catalyst for ORR [68]. Nevertheless, there was a significant gap between the catalytic activities of collagen-derived materials and commercial Pt/C (0.827 mV). To improve the performance of collagen-derived materials, they introduced additional N and P atoms as dopants for the cattle bone–derived porous carbon network using phytate and dicyandiamide, respectively (Figure 9.16a). N (3.2 at%) and P (4.0 at%) were uniformly doped into the collagen-derived materials by controlled pyrolysis (Figure 9.16b,c). Since the as-prepared metal-free catalyst (N,P-HPC) contains a large number of active sites and a hierarchical porous structure with high SSA, it shows significant ORR activity with a half-wave potential of 0.853 V, a Tafel plot slope of 63 $mV\,dec^{-1}$, and a kinetic current density of 38.2 $mA\,cm^{-2}$ at 0.8 V in an alkaline electrolyte (Figure 9.16d,e). In addition, the N,P-HPC electrocatalyst exhibited better durability and methanol tolerance than commercial Pt/C catalysts as well (Figure 9.16f).

Cazetta et al. prepared N- and S-co-doped nanoporous carbon (NSC), which had high surface area, large density of heteroatom dopants, and a very good electrocatalytic activity for hydrazine oxidation reaction (HzOR) [69]. Furthermore, through chemical analysis of both materials, the effectiveness of bone char (BC) to serve as a precursor of N and S doping atoms has been determined. The results have shown that the presence of both N and S heteroatom dopants in the amorphous alloy lattice is mainly related to BC. Their high surface area combined with their pyridinic N, pyrrolic N, graphitic S, and thiophenic S groups allows NSC to exhibit high electrocatalytic activity toward HzOR with low overpotential while giving high current density. Therefore, the material has great prospects to serve as a sustainable metal-free electrocatalyst for HzOR in direct hydrazine fuel cells.

Yang and coworkers prepared N,P-doped hierarchical porous carbon catalysts by direct pyrolysis of gelatin/phytic acid supramolecular aggregates without an additional activation process [70]. Compared with the samples prepared by chitosan/phytic acid supramolecular aggregates, the optimized catalyst material NPC1000 had the highest N and P content with the higher specific surface area, mesopore volume, and total pore volume. Therefore, the sample NPC1000 had the best electrocatalytic activity, which was almost a direct four-electron reaction. The sample also had higher stability and methanol tolerance compared with the

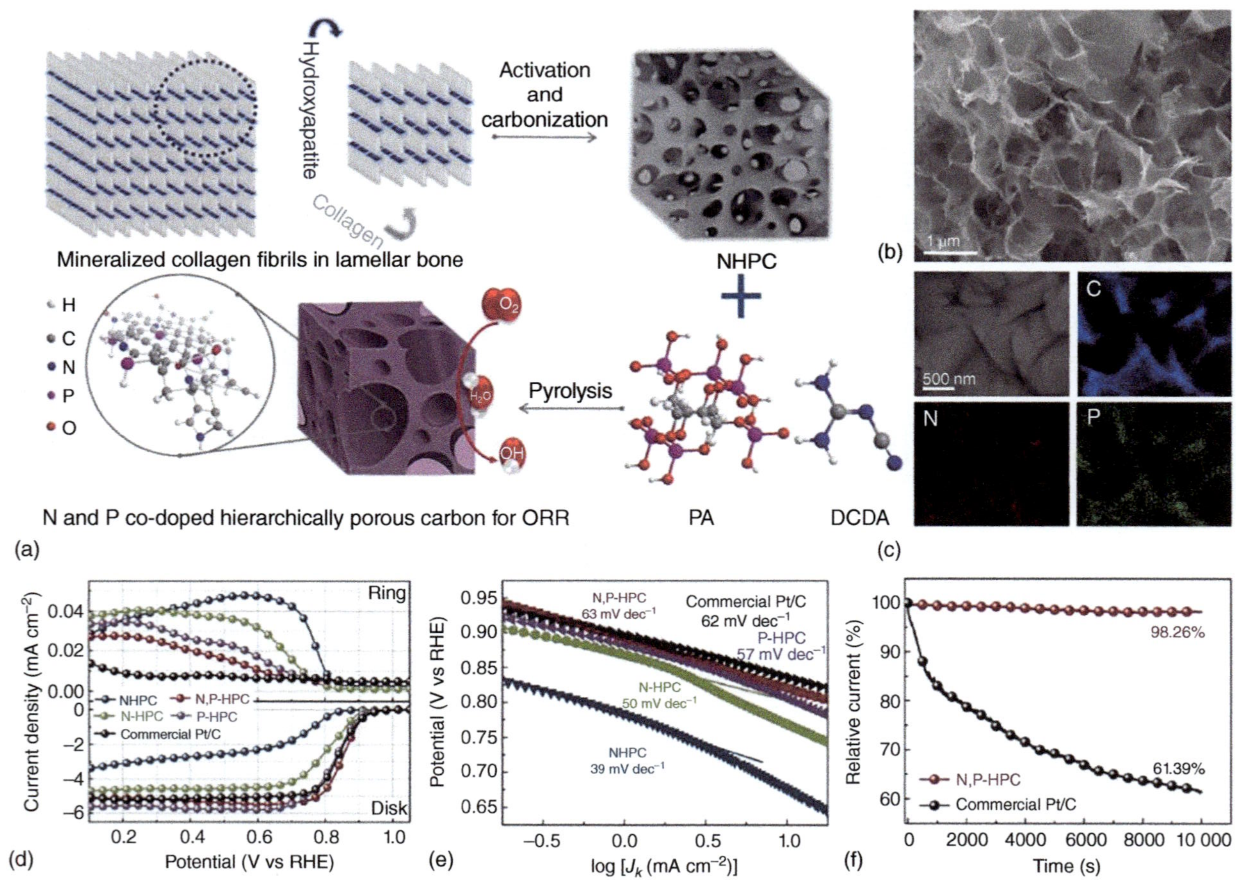

Figure 9.16 (a) Schematic illustration of the synthetic process for the N,P-HPC electrocatalyst. (b) Scanning electron microscopy (SEM) image of the N,P-HPC. (c) Transmission electron microscopy (TEM) and corresponding elemental mapping images of the N,P-HPC. (d) RRDE voltammograms of the NHPC, N-HPC, N,P-HPC, P-HPC, and Pt/C electrocatalysts in O_2-saturated 0.1 M KOH (scan rate: 5 mV s^{-1}, rotation rate: 1600 rpm). (e) Tafel plots and the corresponding slopes in the high potential region. (f) Chronoamperometric responses of the N,P-HPC and Pt/C in O_2-saturated 0.1 M KOH with a rotation rate of 900 rpm at 0.85 V. Source: Zan et al. [68]/with permission of © The Royal Society of Chemistry.

Pt/C catalyst. This work provides a simple, economical, and environment-friendly method to prepare nonprecious porous carbon catalysts with the desired properties.

Benefiting from the synergetic effect of N,P co-doping and unique hierarchically porous structure, the graphic NPC1000 shows high electrocatalytic activity and better durability. Many reports believe that pyridine N and graphite N (responsible for the good conductivity of the material) located at the edges or defect sites of carbon are the active sites in the ORR process [71, 72]. In addition to N doping, two structures (P single bond O and P single bond C) doped by P are also beneficial to ORR [10]. In addition to affecting the environment of carbon atoms, the introduction of P (NPC1000) can also change the porous structure of NC1000. Compared with NPC-1000 (chitosan), the catalytic performance of layered porous NPC1000 (gelatin) is further enhanced, which may be due to the increase in total pore volume and mesopore ratio and the total content of N and P. There is no unique relationship between the ORR activity of NPC samples and the content of N/P, and it is the same as the previous report [73]. Although the total content of N/P is the lowest, NPC1000 shows the best catalytic performance among the NPC samples.

Under the conditions of 0.7 V (vs RHE (reversible hydrogen electrode)) and 1600 rpm, the chronoamperometry test was performed (Figure 9.17). After 10 000 seconds of continuous operations, the current density of NPC1000 is still at 92.4%, while the Pt/C catalyst drops to 88.4% under the same conditions. The results show that NPC1000 has better stability than Pt/C. The anti-methanol poisoning performance of the catalyst was tested. It can be seen from Figure 9.17g that after adding 4 ml of methanol to the electrolyte solution for 500 seconds, there is no significant change in NPC1000 in the presence of methanol. In contrast, the rapid decrease in the current density was observed on the Pt/C electrode, demonstrating that NPC1000 also has better resistance to methanol poisoning than Pt/C.

9.7 As Support Materials for Hybrid Electrocatalysts

Wang and coworkers exploited the hierarchical structure of the collagen-derived materials surface and the abundant N atoms to achieve Pt nanoparticle loading in FHLC [74]. Pt nanoparticles were uniformly dispersed on the porous surface of FHLC with a homogeneous size distribution (average 2.0 nm). Compared to Pt loaded on carbon black, it showed higher current density, lower overpotential, and better catalytic activity when used for ORR in acidic media. Later, they used a porous carbon network derived from pig bone as the carbon support of Pt nanoparticles and obtained a material with better ORR catalytic activity than carbon black–loaded Pt nanoparticles, which confirmed that the high SSA, hierarchical porous structure, and N-doped surface of collagen-derived materials significantly improve the catalytic performance of Pt.

In the existing researches, chlorinated paraffin had been effectively used as carbon supports for Pt nanoparticles, while the weight content of Pt accounted for 20% of the total hybrid catalyst, which was compared to that of commercial Pt/C. For future large-scale applications, the weight content of Pt needs to be

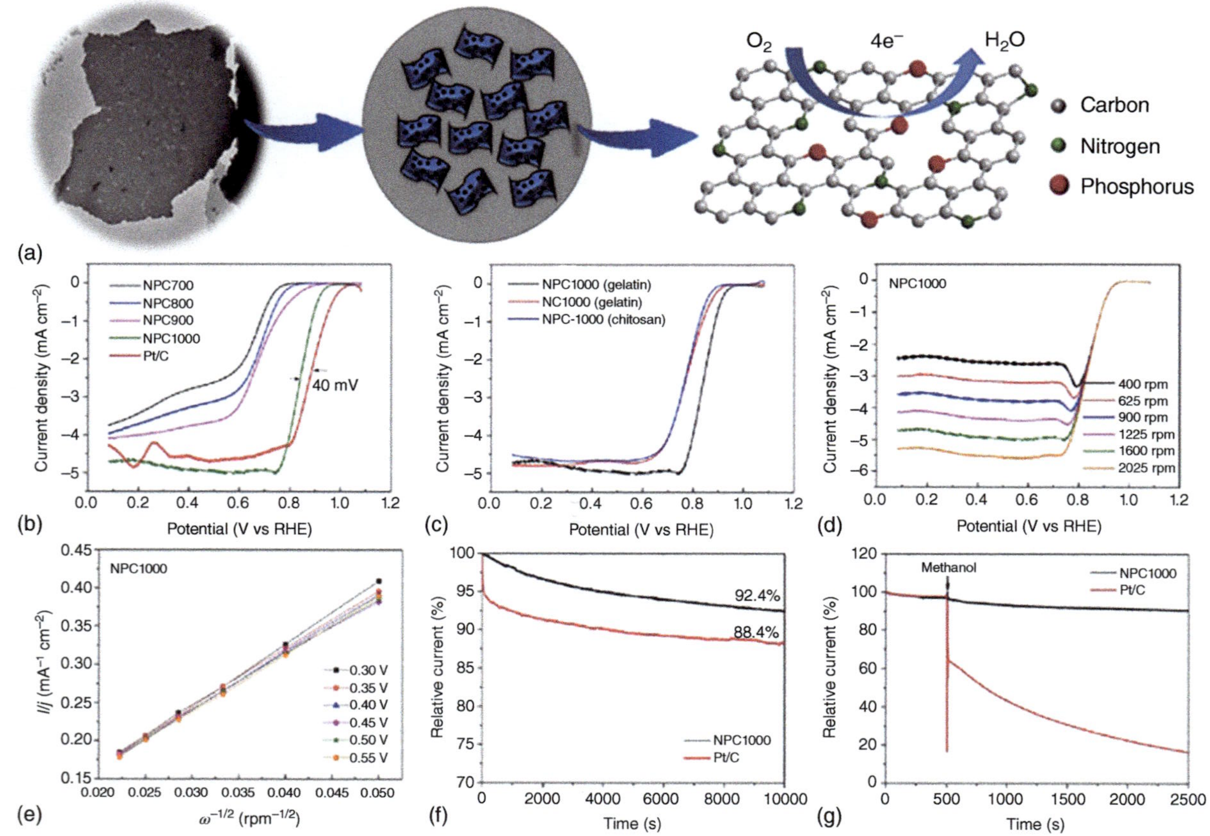

Figure 9.17 (a) Schematic illustration of the synthesis process of porous carbon frameworks from gelatin and phytic acid. The electrochemical characterization of catalysts. (b) LSV curves of the NPCs and Pt/C. (c) LSV curves of the NPC1000, NC1000, and NPC-1000. (d) LSV curves of NPC1000 at different rotation rates. (e) K–L plots at different potentials for NPC1000. (f) Chronoamperometric responses for NPC1000 and Pt/C at 0.7 V and 1600 rpm. (g) Comparison of the I–t curves over 2500 seconds with the addition of methanol at 500 seconds. Source: Reproduced with permission from Mao et al. [70]; © 2019, Elsevier.

further reduced while maintaining high performance. The design and synthesis of single-atom catalysts are considered to be a feasible approach to realize both maximum utilization and minimum consumption of metal catalysts [75]. Compared to traditional particle catalysts, single-atom catalysts typically exhibit better performance than their native structural counterparts [76]. As shown in Figure 9.18a, via the photochemical solid-phase reduction method, Liu and colleagues used an N-doped porous carbon network derived from cattle bone (Pt_1/NPC) to prepare well-defined single Pt atoms [74]. The formation of Pt—N—C bonds promoted a high-degree dispersion of Pt atoms on the collagen-derived materials carriers with a loading of 3.8% (Figure 9.18b–d). The Pt1/NPC catalysts showed ultrahigh hydrogen evolution reaction (HER) electrocatalytic activity with an overpotential of 25 mV at 10 mA cm^{-2} (Figure 9.18e), a mass activity of 2.86 A mg^{-1} (24 times higher than commercial Pt/C), and good durability (Figure 9.18f). Furthermore, the Pt_1/NPC catalyst also exhibited excellent catalytic activity for ORR (Figure 9.18g–i). The mass activity of single Pt atoms supported by collagen-derived materials was 4.3 times that of commercial Pt/C. Benefiting from the good chemical and electronic structure of Pt–N_4 coordination, Pt_1/NPC catalyst has excellent electrocatalytic performance. In this structure, a single Pt atom underwent significant electron transfer to an adjacent coordinated N atom.

In addition to improving the catalytic activity of noble metals, collagen-derived materials can also enhance the performance of transition metal oxides. As shown in Figure 9.19a, Wang and coworkers used a porous carbon network derived from cattle bone as a carrier for Co_3O_4 nanoparticles [77]. Due to the well-defined hierarchical porous structure, large SSA, and high N-dopant content of collagen-derived materials, the Co_3O_4 nanoparticles are uniformly dispersed on the N-doped collagen-derived materials (Figure 9.19b,c). Based on the synergistic interaction between Co_3O_4 and collagen-derived materials, the prepared catalyst (Co_3O_4-NHPC) exhibits significant catalytic activity in alkaline electrolytes for both ORR (better than commercial Pt/C) and oxygen evolution reaction (OER) (comparable to commercial RuO_2). Furthermore, a primary Zn-air battery with the Co_3O_4-NHPC cathode shows outstanding performance with an open-circuit potential of 1.39 V, a specific capacity of 795 mAh g^{-1} (at 2 mA cm^{-2}), and a peak power density of 80 mW cm^{-2}.

9.8 As Support Materials for Metal–Nitrogen–Carbon Electrocatalysts

TM–N–C electrocatalysts are considered to be one of the most prospective non-noble metal electrocatalysts that can replace Pt electrocatalysts. It has been proved that collagen-derived materials are an excellent carrier for the synthesis of TM–N–C catalysts [78]. The precursors of transition metal compounds can be adsorbed by uniformly dispersed micropores, and the plentiful N atoms on the surface constrain them to form TM–N active sites instead of metal aggregates [79]. In addition, the

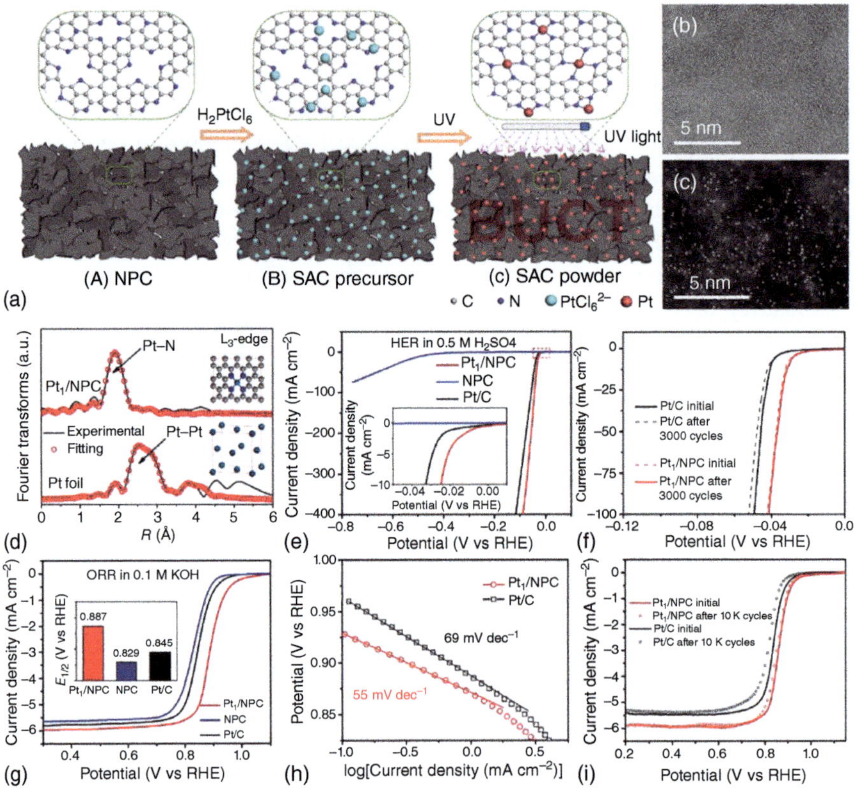

Figure 9.18 (a) Schematic illustration of the synthetic process for the Pt_1/NPC electrocatalyst. (b) TEM and (c) high-resolution, high-angle annular dark-field scanning transmission electron microscope (HAADF-STEM) images of Pt_1/NPC. (d) Fourier transform of extended X-ray absorption fine structure (EXAFS) spectra at the Pt L_3-edge and corresponding R-space fitting curves for the Pt_1/NPC catalyst and Pt foil. The insets show the atomic structure models for the Pt_1/NPC and Pt foil. (e) The HER polarization curves of the Pt_1/NPC, NPC, and 20% Pt/C catalysts recorded at a scan rate of 5 mV s^{-1} in N_2-saturated 0.5 M H_2SO_4 solution. The inset shows the enlarged curves at the onset potential region of the HER. (f) Polarization curves recorded in N_2-saturated 0.5 M H_2SO_4 before and after 3000 CV cycles. (g) ORR polarization curves of Pt_1/NPC, NPC, and 20% Pt/C catalysts recorded at a rotation rate of 1600 rpm and a scan rate of 5 mV s^{-1} in O_2-saturated 0.1 M KOH solution. The inset shows the half-wave potentials for the different catalysts. (h) Tafel plots for the above catalysts. (i) Polarization curves recorded in O_2-saturated 0.1 M KOH before and after 10 000 CV cycles. Source: Li et al. [74]/with permission of © American Chemical Society.

hierarchical porosity of collagen-derived materials can enhance the transport of reactants/electrolytes to or away from the active sites [80, 81].

According to reports, Wang and coworkers prepared Co–N–C electrocatalysts for the first time using porous carbon networks derived from cattle bones as carrier materials [82]. They used vitamin B_{12} to crack collagen-derived materials to prepare Co–N–C with well-defined pores and a specific surface area as high as 859 m^2 g^{-1}. When used as an ORR catalyst, Co–N–C is superior to commercial Pt/C in terms of catalytic activity (half-wave potential 0.835 V), operational stability, and

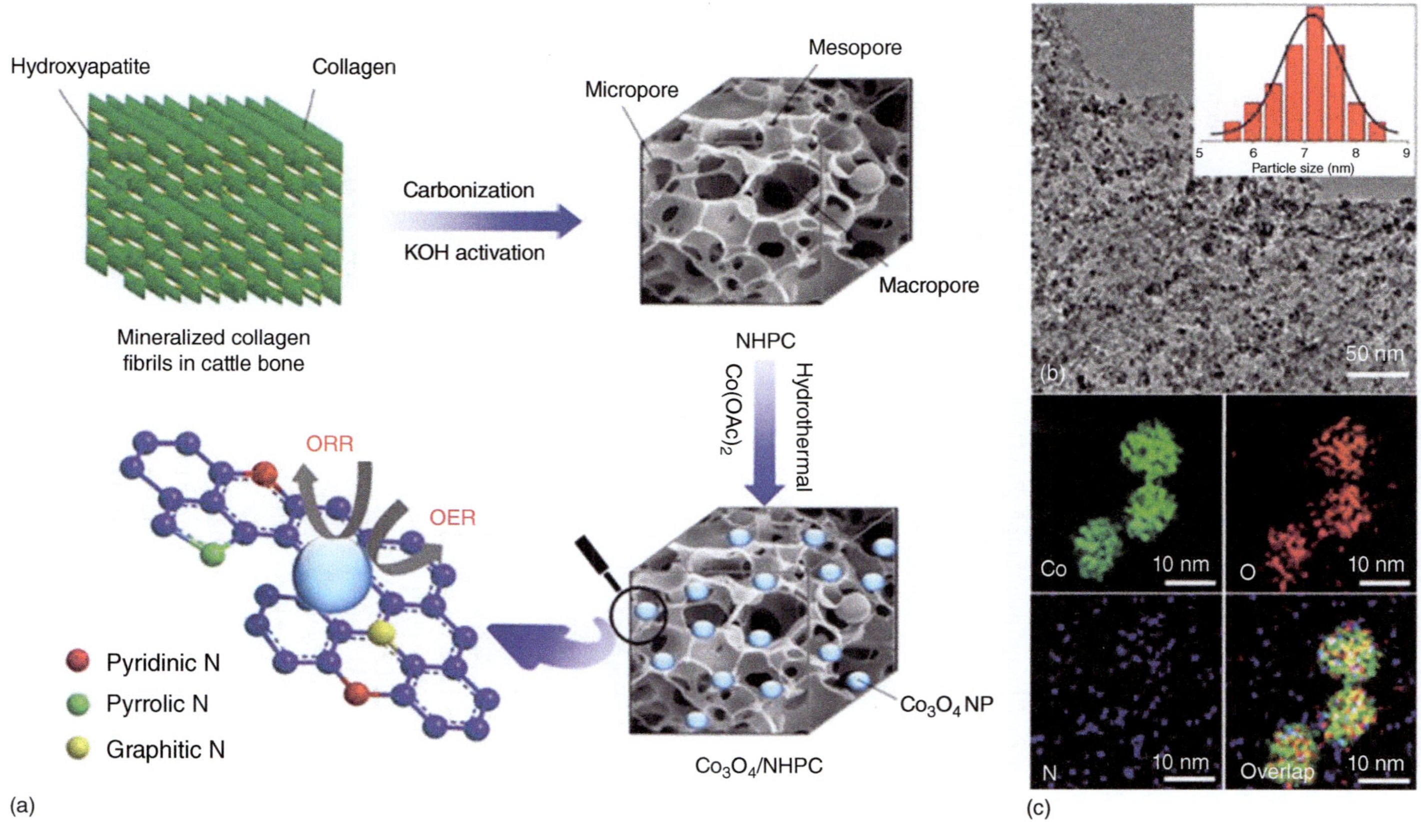

Figure 9.19 (a) Schematic illustration of the synthetic process for the Co_3O_4-NHPC electrocatalyst. (b) TEM image of the Co_3O_4-NHPC. (c) Dark-field high-resolution STEM elemental mapping images of Co, O, N, and overlap. Source: Guan et al. [77]/with permission of © American Chemical Society.

fuel selectivity. To obtain a catalyst with higher catalytic activity, they also used 5,10,15,20-tetrakis(4-phenyl)-porphyrin iron(III) chloride to pyrolyze the porous carbon network of pig bone to prepare an Fe–N–C catalyst. Unlike the synthesis of the Co–N–C catalyst, the pyrolysis product is further processed by acid washing and second heat treatment to adjust the Fe–N molecules at the atomic level. Benefiting from the existence of active D1 (Fe^{II}–N_4) and D3 (N–Fe^{II}–N_{2+2}) sites, as well as the large SSA and its 3D porous structure, the active sites were fully utilized, and mass transport for ORR was enhanced. Consequently, Fe–NC electrocatalyst exhibits excellent ORR activity; in alkaline medium, the half-wave potential is 0.87 V (20 mV higher than Pt/C) and the kinetic current density is 10.9 mA cm^{-2} at 0.85 V (2.3 times of Pt/C).

Currently, most of the ORR catalysts with excellent activity are achieved in alkaline media. However, catalysts with good activity under acidic conditions are essential for proton exchange membrane (PEM) fuel cells. But such catalysts are still rare. To address this issue, Wang et al. prepared an Fe–N–C catalyst (SA-Fe-HPC) that exhibited good ORR catalytic performance in both acidic and alkaline media. The porous carbon network derived from cattle bone was used as the carbon support, and the iron phthalocyanine/unsubstituted phthalocyanine (FePc/Pc) complex was used as the Fe precursor (Figure 9.20a) [83]. It is worth noting that Pc molecules were deliberately coassembled with FePc molecules to provide a steric hindrance for the demetallation of Fe. Additionally, the N-rich coordination structure in the FePc/Pc composite is also used to modulate the coordination environment of Fe atoms during the pyrolysis process. At the same time, the Fe atoms were efficiently restrained and stabilized by in situ-doped N atoms and a large number of micropores. As a result, single Fe atoms were uniformly dispersed on the collagen-derived materials (Figure 9.20b,c). Fourier transform extended X-ray absorption fine-structure spectroscopy (Figure 9.20d), and Mossbauer spectroscopy (Figure 9.20e) showed that single-Fe atoms were coordinated by N atoms in the form of Fe–N_4. In alkaline media, the ORR activity of the SA-Fe-HPC catalyst ($E_{1/2}$ = 0.89 V) was superior to that of commercial Pt/C electrodes (Figure 9.20f,g). Even in acidic media, the ORR activity of SA-Fe-HPC catalyst is comparable to that of commercial Pt/C electrode ($E_{1/2}$ = 0.81 V), and by comparison, it has better long-term electrochemical stability and fuel selectivity (Figure 9.20h,i).

An approach to inducing Fe–Nx site remodeling on pig bone–derived nitrogen-doped layered porous carbon was proposed by Li and coworkers [84]. Thus, they synthesized a highly efficient non-noble metal electrocatalyst for oxygen reduction reactions. The synthetic process of the Fe-N-HPC-AH electrocatalyst is shown in Figure 9.21. Firstly, the pig bones are pre-carbonized and then KOH activation and acid washing are performed to remove inorganic hydroxyapatite crystals and synthesize HPC. Later, the as-prepared HPC adsorbed FeTPPCl to form the FeTPPCl-adsorbed–HPC composite through electrostatic attraction and π–π interactions.

The electrochemical stability of Fe-N-HPC-AH is also investigated between 0.6 and 1.1 V, with a scan rate of 100 mV s^{-1}, in N_2-saturated alkaline electrolyte. The Fe-N-HPC-AH exhibits remarkable electrochemical stability with only 3 mV

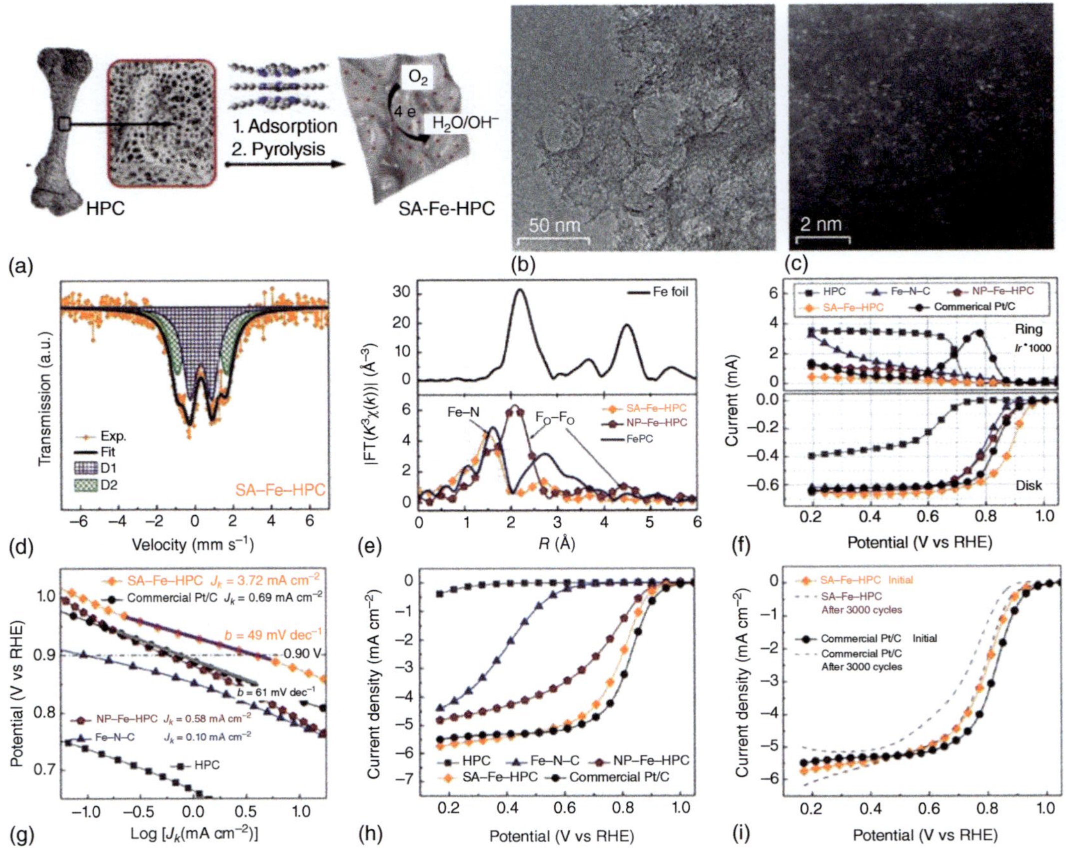

Figure 9.20 (a) Schematic illustration of the synthetic process for the SA-Fe-HPC electrocatalyst. (b) TEM and (c) high-resolution HAADF-STEM images of the SA-Fe-HPC. (d) The ^{57}Fe Mossbauer transmission spectra measured at 293 K for the SA-Fe-HPC. (e) Fourier transforms of k^3-weighted $\chi(k)$-function of the EXAFS spectra for the SA-Fe-HPC and NP-Fe-HPC samples with iron foil and FePc as references. (f) The corresponding rotating ring-disk electrode (RRDE) voltammograms recorded in O_2-saturated 0.1 M KOH at 1600 rpm. The disk potential was scanned at 5 mV s^{-1}, and the ring potential was constant at 1.5 V. (g) Tafel plots derived by the mass-transport correction from the corresponding LSV data. (h) LSV voltammograms recorded in O_2-saturated 0.1 M H_2SO_4 at 1600 rpm with a sweep rate of 5 mV s^{-1}. (i) ORR polarization curves of the SA-Fe-HPC and Pt/C before and after 3000 potential cycles over 0.6–1.1 V with a sweep rate of 100 mV s^{-1} in N_2-saturated 0.1 M H_2SO_4. Source: Zhang et al. [83]/with permission of © John Wiley & Sons.

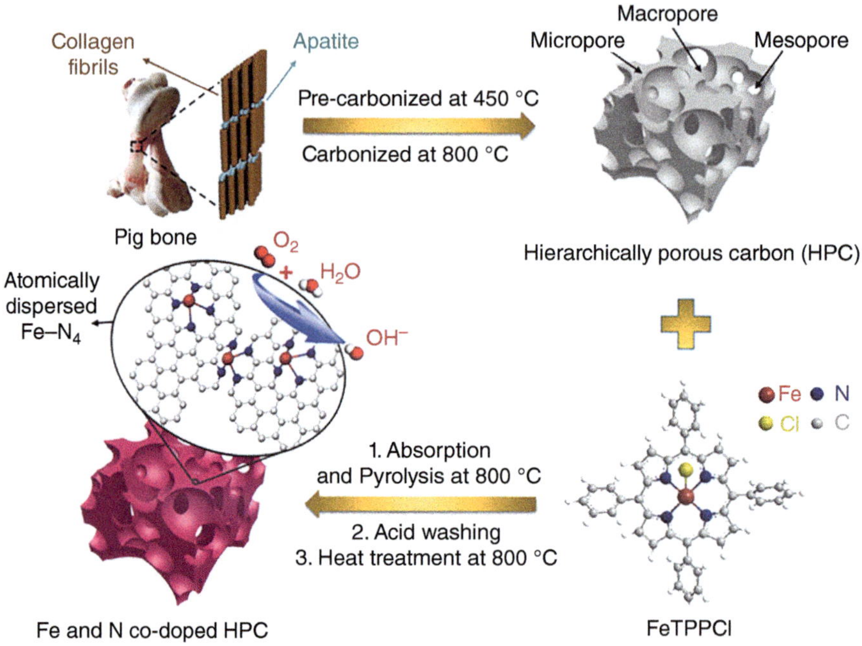

Figure 9.21 Schematic representation of the synthetic process for Fe-N-HPC-AH. Source: Zhang et al. [83]/with permission of © John Wiley & Sons.

negative shift of $E_{1/2}$ after 3000 cycles. By comparison, the commercial Pt/C shows a 27 mV negative shift of $E_{1/2}$ after the identical test, demonstrating a severe decrease in ORR activity. Furthermore, Fe-N-HPC-AH had excellent methanol tolerance. When 1 M methanol is added to 0.1 M KOH, the change of the linear sweep voltammetry (LSV) curves is negligible, while under the same conditions, the Pt/C anode current increases significantly.

Wang and coworkers also achieved an environment-friendly and efficient synthesis of ultrathin 2D microporous/mesoporous carbon nanosheets (CNSs) with Fe–N_4 (Fe-N-CNS) as the primary active site for high-performance electrocatalysis of ORR using an ice/salt-assisted method [85]. An ice/salt-assistant approach was used to synthesize the ultrathin 2D Fe and N co-doped carbon nanosheets (Fe-N-CNS), which included the freeze-drying of FeTCPPCl mixed with KCl followed by a KCl-confined pyrolysis process (Figure 9.22a). Firstly, the CV measurement was performed in 0.1 M KOH electrolyte at a scan rate of 20 mV s^{-1} to evaluate the electrochemical performance of the ORR electrocatalyst (Figure 9.22b). No obvious reduction peak was detected in the potential range of 0–1.3 V for all the samples in N_2-saturated KOH electrolyte. When Fe-N-CNS is saturated with O_2 in KOH, an obvious reduction peak appears at 0.85 V, indicating that oxygen reduction has occurred. The peak potential of Fe-N-CNS is much larger than the peak potentials of Fe-N-C-(I) (0.59 V) and Fe-N-C-(K) (0.74 V), suggesting that ORR activity is improved. The onset potential (E_{onset}) and half-wave potential ($E_{1/2}$) in the LSV curves are 1.0 and 0.87 V, respectively, which exceed the onset potential (E_{onset}) and

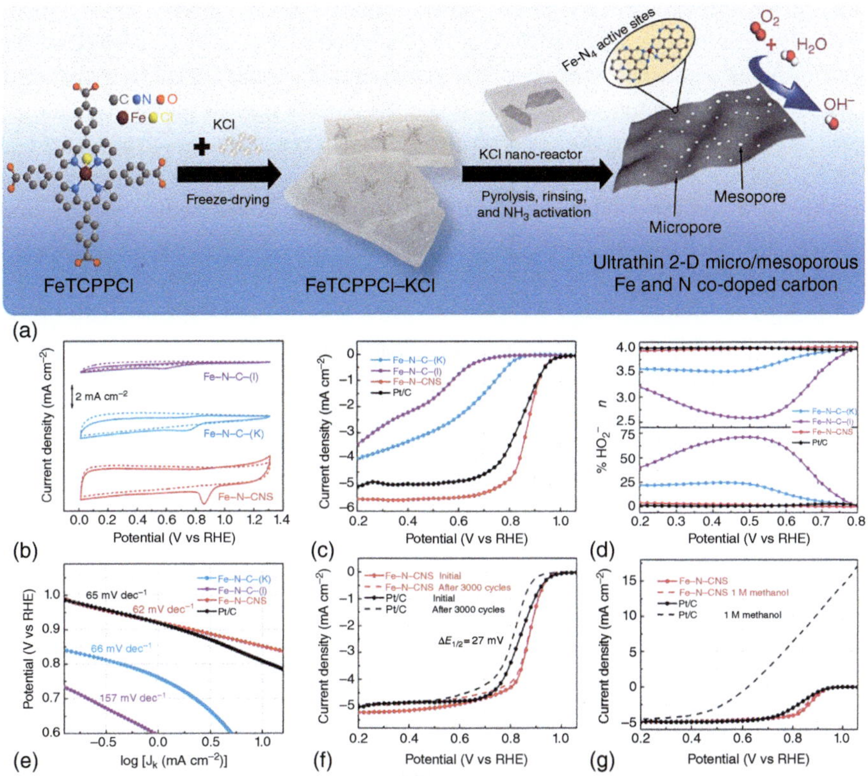

Figure 9.22 (a) Schematic representation of the synthetic process. (b) Cyclic voltammograms of Fe–N–C-(I), Fe–N–C-(K), and Fe-N-CNS. (c) LSV curves of Fe–N–C-(I), Fe–N–C-(K), Fe-N-CNS, and commercial Pt/C in O_2-saturated 0.1 M KOH at a scan rate of 5 mV s^{-1} and a rotation rate of 1600 rpm. (d) Electron transfer number and peroxide percentage calculated from the corresponding RRDE curves (scan rate: 5 mV s^{-1}, ring potential: 1.5 V). (e) Tafel plots derived from the corresponding LSV data of (c). (f) LSV curves of Fe-N-CNS and Pt/C in O_2-saturated 0.1 M KOH before and after 3000 cycles between 0.6 and 1.1 V. (g) LSV curves of Fe-N-CNS and Pt/C in O_2-saturated 0.1 M KOH with or without 1 M methanol. Source: Reproduced with permission from Li et al. [85]; © 2019, John Wiley & Sons.

half-wave potential ($E_{1/2}$) of commercial Pt/C. The starting potential of the former is almost the same, and the $E_{1/2}$ value of the latter is 20 mV higher. These test results further confirm the greater ORR activity of Fe-N-CNS (Figure 9.22c). The kinetic current density (J_k) was calculated to be 4 mA cm^{-2} at 0.88 V for Fe-N-CNS, which was significantly higher than that of Pt/C (2.6 mA cm^{-2}). According to the calculation results of the rotating ring-disk electrode (RRDE), a transferred electron number higher than 3.9 with a very low percentage (<4 %) of peroxide yield (% HO_2^-) was found for Fe-N-CNS, similar to Pt/C, thus indicating a four-electron pathway for ORR catalyzed by Fe-N-CNS (Figure 9.22d). The small Tafel slope (62 mV dec^{-1}) obtained in the Tafel plot of Fe-N-CNS reflected a rapid kinetic process for ORR (Figure 9.22e). The ORR activity of Fe-N-CNS was also evaluated in

0.1 M $HClO_4$ electrolyte. LSV curves showed that Fe-N-CNS possessed a relatively good ORR activity with $E_{1/2}$ of 0.74 V but not as good as that of Pt/C. Fe-N-CNS had an ultrathin 2D nanosheet structure. The active sites of D1 and D3 $Fe–N_4$ were highly exposed and had obvious mesoporous structures, which improved the utilization of active sites and facilitated the mass transport of ORR reactants and products.

Fe-N-CNS exhibited excellent electrochemical durability with a negligible negative shift of $E_{1/2}$ after 3000 cycles (Figure 9.22f), which is better than Pt/C, whose negative shift of $E_{1/2}$ was 27 mV after the same test. After adding 1 M methanol, when O_2 was preferentially adsorbed on the $Fe–N_4$ sites, the LSV curves of Fe-N-CNS only slightly oscillated, which indicated that Fe-N-CNS has better methanol resistance. On the contrary, the Pt/C electrocatalyst had a significant positive current density under the same conditions (Figure 9.22g), demonstrating a poor tolerance to methanol crossover. These results highlighted the excellent performance of the ultrathin 2D Fe-N-CNS in terms of durability and methanol tolerance.

References

1 Yang, Z., Ren, J., Zhang, Z. et al. (2015). Recent advancement of nanostructured carbon for energy applications. *Chemical Reviews* 115: 5159–5223.

2 Wang, J., Nie, P., Ding, B. et al. (2017). Biomass derived carbon for energy storage devices. *Journal of Materials Chemistry A* 5: 2411–2428.

3 Sevilla, M. and Mokaya, R. (2014). Energy storage applications of activated carbons: supercapacitors and hydrogen storage. *Energy & Environmental Science* 7: 1250–1280.

4 Yang, M. and Zhou, Z. (2017). Recent breakthroughs in supercapacitors boosted by nitrogen-rich porous carbon materials. *Advanced Science* 4, 1600408.

5 Kong, D., Gao, Y., Xiao, Z. et al. (2019). Rational design of carbon-rich materials for energy storage and conversion. *Advanced Materials* 31, 1804973.

6 Raza, W., Ali, F., Raza, N. et al. (2018). Recent advancements in supercapacitor technology. *Nano Energy* 52: 441–473.

7 Lu, H. and Zhao, X.S. (2017). Biomass-derived carbon electrode materials for supercapacitors. *Sustainable Energy & Fuels* 1: 1265–1281.

8 Zhong, C., Deng, Y., Hu, W. et al. (2015). A review of electrolyte materials and compositions for electrochemical supercapacitors. *Chemical Society Reviews* 44: 7484–7539.

9 Wang, Y., Song, Y., and Xia, Y. (2016). Electrochemical capacitors: mechanism, materials, systems, characterization and applications. *Chemical Society Reviews* 45: 5925–5950.

10 Yu, Z., Tetard, L., Zhai, L., and Thomas, J. (2015). Supercapacitor electrode materials: nanostructures from 0 to 3 dimensions. *Energy & Environmental Science* 8: 702–730.

11 Liu, J., Wang, J., Xu, C. et al. (2018). Advanced energy storage devices: basic principles, analytical methods, and rational materials design. *Advanced Science* 5, 1700322.

12 Xie, L., Li, X., Deng, J. et al. (2018). Sustainable and scalable synthesis of monodisperse carbon nanospheres and their derived superstructures. *Green Chemistry* 20: 4596–4601.

13 Lei, J., Guo, Q., Yin, D. et al. (2019). Bioconcentration and bioassembly of N/S co-doped carbon with excellent stability for supercapacitors. *Applied Surface Science* 488: 316–325.

14 Lei, J., Guo, Q., Yao, W. et al. (2018). Bioconcentration of organic dyes via fungal hyphae and their derived carbon fibers for supercapacitors. *Journal of Materials Chemistry A* 6: 10710–10717.

15 Chen, W., Zhang, H., Huang, Y., and Wang, W. (2010). A fish scale based hierarchical lamellar porous carbon material obtained using a natural template for high performance electrochemical capacitors. *Journal of Materials Chemistry* 20: 4773–4775.

16 Goodman, P.A., Li, H., Gao, Y. et al. (2013). Preparation and characterization of high surface area, high porosity carbon monoliths from pyrolyzed bovine bone and their performance as supercapacitor electrodes. *Carbon* 55: 291–298.

17 Niu, J., Shao, R., Liang, J. et al. (2017). Biomass-derived mesopore-dominant porous carbons with large specific surface area and high defect density as high performance electrode materials for Li-ion batteries and supercapacitors. *Nano Energy* 36: 322–330.

18 He, D., Niu, J., Dou, M. et al. (2017). Nitrogen and oxygen co-doped carbon networks with a mesopore-dominant hierarchical porosity for high energy and power density supercapacitors. *Electrochimica Acta* 238: 310–318.

19 Shao, R., Niu, J., Liang, J. et al. (2017). Mesopore- and macropore-dominant nitrogen-doped hierarchically porous carbons for high-energy and ultrafast supercapacitors in non-aqueous electrolytes. *ACS Applied Materials & Interfaces* 9: 42797–42805.

20 Wang, J., Shen, L., Xu, Y. et al. (2015). Lamellar-structured biomass-derived phosphorus- and nitrogen-co-doped porous carbon for high-performance supercapacitors. *New Journal of Chemistry* 39: 9497–9503.

21 Liu, M., Niu, J., Zhang, Z. et al. (2018). Potassium compound-assistant synthesis of multi-heteroatom doped ultrathin porous carbon nanosheets for high performance supercapacitors. *Nano Energy* 51: 366–372.

22 Niu, J., Liu, M., Xu, F. et al. (2018). Synchronously boosting gravimetric and volumetric performance: biomass-derived ternary-doped microporous carbon nanosheet electrodes for supercapacitors. *Carbon* 140: 664–672.

23 Kumar, K.S., Choudhary, N., Jung, Y., and Thomas, J. (2018). Recent advances in two-dimensional nanomaterials for supercapacitor electrode applications. *ACS Energy Letters* 3: 482–495.

24 Xu, B., Hou, S., Cao, G. et al. (2012). Sustainable nitrogen-doped porous carbon with high surface areas prepared from gelatin for supercapacitors. *Journal of Materials Chemistry* 22: 19088–19093.

25 Subramanian, V., Luo, C., Stephan, A.M. et al. (2007). Supercapacitors from activated carbon derived from banana fibers. *The Journal of Physical Chemistry C* 111: 7527–7531.

26 Zhang, L.-L., Li, H.-H., Shi, Y.-H. et al. (2016). A novel layered sedimentary rocks structure of the oxygen-enriched carbon for ultrahigh-rate-performance supercapacitors. *ACS Applied Materials & Interfaces* 8: 4233–4241.

27 Qu, Y., Zhang, Z., Zhang, X. et al. (2015). Highly ordered nitrogen-rich mesoporous carbon derived from biomass waste for high-performance lithium-sulfur batteries. *Carbon* 84: 399–408.

28 Zhu, L., Gao, Q., Tan, Y. et al. (2015). Nitrogen and oxygen co-doped microporous carbons derived from the leaves of *Euonymus japonicas* as high performance supercapacitor electrode material. *Microporous and Mesoporous Materials* 210: 1–9.

29 Zhao, J., Lai, H., Lyu, Z. et al. (2015). Hydrophilic hierarchical nitrogen-doped carbon nanocages for ultrahigh supercapacitive performance. *Advanced Materials* 27: 3541–3545.

30 Zuo, Z., Kim, T.Y., Kholmanov, I. et al. (2015). Ultra-light hierarchical graphene electrode for binder-free supercapacitors and lithium-ion battery anodes. *Small* 11: 4922–4930.

31 Sun, J., Niu, J., Liu, M. et al. (2018). Biomass-derived nitrogen-doped porous carbons with tailored hierarchical porosity and high specific surface area for high energy and power density supercapacitors. *Applied Surface Science* 427: 807–813.

32 Guo, N., Li, M., Wang, Y. et al. (2016). Soybean root-derived hierarchical porous carbon as electrode material for high-performance supercapacitors in ionic liquids. *ACS Applied Materials & Interfaces* 8: 33626–33634.

33 Huang, W., Zhang, H., Huang, Y. et al. (2011). Hierarchical porous carbon obtained from animal bone and evaluation in electric double-layer capacitors. *Carbon* 49: 838–843.

34 Zuo, W., Li, R., Zhou, C. et al. (2017). Battery-supercapacitor hybrid devices: recent progress and future prospects. *Advanced Science* 4, 1600539.

35 Wang, G., Lu, C., Zhang, X. et al. (2017). Toward ultrafast lithium ion capacitors: a novel atomic layer deposition seeded preparation of $Li_4Ti_5O_{12}$/graphene anode. *Nano Energy* 36: 46–57.

36 Li, B., Zheng, J., Zhang, H. et al. (2018). Electrode materials, electrolytes, and challenges in nonaqueous lithium-ion capacitors. *Advanced Materials* 30, 1705670.

37 Wang, H., Zhu, C., Chao, D. et al. (2017). Nonaqueous hybrid lithium-ion and sodium-ion capacitors. *Advanced Materials* 29, 1702093.

38 Han, C., Li, H., Shi, R. et al. (2018). Nanostructured anode materials for non-aqueous lithium ion hybrid capacitors. *Energy & Environmental Materials* 1: 75–87.

39 Niu, J., Shao, R., Liu, M. et al. (2018). Porous carbon electrodes with battery-capacitive storage features for high performance Li-ion capacitors. *Energy Storage Materials* 12: 145–152.

40 Li, X., Liu, J., Zhang, Y. et al. (2012). High concentration nitrogen doped carbon nanotube anodes with superior Li+ storage performance for lithium rechargeable battery application. *Journal of Power Sources* 197: 238–245.

41 Niu, J., Liang, J., Shao, R. et al. (2017). Tremella-like N,O-codoped hierarchically porous carbon nanosheets as high-performance anode materials for high energy and ultrafast Na-ion capacitors. *Nano Energy* 41: 285–292.

42 Liu, M., Niu, J., Zhang, Z. et al. (2019). Porous carbons with tailored heteroatom doping and well-defined porosity as high-performance electrodes for robust Na-ion capacitors. *Journal of Power Sources* 414: 68–75.

43 Guo, Y., Liu, W., Wu, R. et al. (2018). Marine-biomass-derived porous carbon sheets with a tunable n-doping content for superior sodium-ion storage. *ACS Applied Materials & Interfaces* 10: 38376–38386.

44 Selvamani, V., Ravikumar, R., Suryanarayanan, V. et al. (2015). Fish scale derived nitrogen doped hierarchical porous carbon – a high rate performing anode for lithium ion cell. *Electrochimica Acta* 182: 1–10.

45 Wu, F., Maier, J., and Yu, Y. (2020). Guidelines and trends for next-generation rechargeable lithium and lithium-ion batteries. *Chemical Society Reviews* 49: 1569–1614.

46 Hwang, J.-Y., Myung, S.-T., and Sun, Y.-K. (2018). Recent progress in rechargeable potassium batteries. *Advanced Functional Materials* 28, 1802938.

47 Yuan, X., Zhu, B., Feng, J. et al. (2021). Biomass bone-derived, N/P-doped hierarchical hard carbon for high-energy potassium-ion batteries. *Materials Research Bulletin* 139: 111282.

48 Wu, X., Chen, Y., Xing, Z. et al. (2019). Advanced carbon-based anodes for potassium-ion batteries. *Advanced Energy Materials* 9, 1900343.

49 Rajagopalan, R., Tang, Y., Ji, X. et al. (2020). Advancements and challenges in potassium ion batteries: a comprehensive review. *Advanced Functional Materials* 30, 1909486.

50 Odoom-Wubah, T., Rubio, S., Tirado, J.L. et al. (2020). Waste Pd/Fish-collagen as anode for energy storage. *Renewable and Sustainable Energy Reviews* 131, 109968

51 Chen, H., Liu, H., Guo, Y. et al. (2018). Hierarchically ordered mesoporous TiO_2 nanofiber bundles derived from natural collagen fibers for lithium and sodium storage. *Journal of Alloys and Compounds* 731: 844–852.

52 Niu, J., Liang, J., Gao, A. et al. (2019). Micropore-confined amorphous SnO_2 subnanoclusters as robust anode materials for Na-ion capacitors. *Journal of Materials Chemistry A* 7: 21711–21721.

53 Shao, R., Niu, J., Zhu, F. et al. (2019). A facile and versatile strategy towards high-performance Si anodes for Li-ion capacitors: concomitant conductive network construction and dual-interfacial engineering. *Nano Energy* 63: 103824.

54 Gao, M., Shih, C.-C., Pan, S.-Y. et al. (2018). Advances and challenges of green materials for electronics and energy storage applications: from design to end-of-life recovery. *Journal of Materials Chemistry A* 6: 20546–20563.

55 Zhao, S., Li, C., Wang, W. et al. (2013). A novel porous nanocomposite of sulfur/carbon obtained from fish scales for lithium-sulfur batteries. *Journal of Materials Chemistry A* 1: 3334–3339.

56 Li, J., Li, S., Liu, Q. et al. (2019). Synthesis of hydrogen-substituted graphyne film for lithium-sulfur battery applications. *Small* 15, 1805344.

57 Li, S., Cen, Y., Xiang, Q. et al. (2019). Vanadium dioxide-reduced graphene oxide binary host as an efficient polysulfide plague for high-performance lithium–sulfur batteries. *Journal of Materials Chemistry A* 7: 1658–1668.

58 Shao, H., Huang, B., Liu, N. et al. (2016). Modified separators coated with a $Ca(OH)_2$-carbon framework derived from crab shells for lithium–sulfur batteries. *Journal of Materials Chemistry A* 4: 16627–16634.

59 Shao, H., Ai, F., Wang, W. et al. (2017). Crab shell-derived nitrogen-doped micro-/mesoporous carbon as an effective separator coating for high energy lithium–sulfur batteries. *Journal of Materials Chemistry A* 5: 19892–19900.

60 Wei, S., Zhang, H., Huang, Y. et al. (2011). Pig bone derived hierarchical porous carbon and its enhanced cycling performance of lithium–sulfur batteries. *Energy & Environmental Science* 4: 736–740.

61 Gao, M., Li, C., Liu, N. et al. (2014). Inhibition on polysulfides dissolve during the discharge–charge by using fish-scale-based porous carbon for lithium–sulfur battery. *Electrochimica Acta* 149: 258–263.

62 Peng, Q., Yu, F., Wang, W. et al. (2019). Ultralight polyethylenimine/porous carbon modified separator as an effective polysulfide-blocking barrier for lithium–sulfur battery. *Electrochimica Acta* 299: 749–755.

63 Ai, F., Liu, N., Wang, W. et al. (2017). Heteroatoms-doped porous carbon derived from tuna bone for high performance Li–S batteries. *Electrochimica Acta* 258: 80–89.

64 Zhou, G., Yin, L.-C., Wang, D.-W. et al. (2013). Fibrous hybrid of graphene and sulfur nanocrystals for high-performance lithium–sulfur batteries. *ACS Nano* 7: 5367–5375.

65 Wang, G., Peng, H., Qiao, X. et al. (2016). Biomass-derived porous heteroatom-doped carbon spheres as a high-performance catalyst for the oxygen reduction reaction. *International Journal of Hydrogen Energy* 41: 14101–14110.

66 Song, H., Li, H., Wang, H. et al. (2014). Chicken bone-derived N-doped porous carbon materials as an oxygen reduction electrocatalyst. *Electrochimica Acta* 147: 520–526.

67 Guo, C., Hu, R., Liao, W. et al. (2017). Protein-enriched fish "biowaste" converted to three-dimensional porous carbon nano-network for advanced oxygen reduction electrocatalysis. *Electrochimica Acta* 236: 228–238.

68 Zan, Y., Zhang, Z., Liu, H. et al. (2017). Nitrogen and phosphorus co-doped hierarchically porous carbons derived from cattle bones as efficient metal-free electrocatalysts for the oxygen reduction reaction. *Journal of Materials Chemistry A* 5: 24329–24334.

69 Cazetta, A.L., Zhang, T., Silva, T.L. et al. (2018). Bone char-derived metal-free N- and S-co-doped nanoporous carbon and its efficient electrocatalytic activity for hydrazine oxidation. *Applied Catalysis B: Environmental* 225: 30–39.

70 Mao, X., Cao, Z., Chen, S. et al. (2019). Facile synthesis of N,P-doped hierarchical porous carbon framework catalysts based on gelatin/phytic acid

supermolecules for electrocatalytic oxygen reduction. *International Journal of Hydrogen Energy* 44: 5890–5898.

71 You, C., Wang, L., Huang, Y. et al. (2018). High porosity nitrogen and phosphorous Co-doped carbon nanosheets as an efficient catalyst for oxygen reduction. *International Journal of Hydrogen Energy* 43: 9749–9756.

72 Deng, L., Yuan, H., Cai, X. et al. (2016). Honeycomb-like hierarchical carbon derived from livestock sewage sludge as oxygen reduction reaction catalysts in microbial fuel cells. *International Journal of Hydrogen Energy* 41: 22328–22336.

73 Luo, H., Jiang, W.-J., Zhang, Y. et al. (2018). Self-terminated activation for high-yield production of N,P-codoped nanoporous carbon as an efficient metal-free electrocatalyst for Zn-air battery. *Carbon* 128: 97–105.

74 Li, T., Liu, J., Song, Y., and Wang, F. (2018). Photochemical solid-phase synthesis of platinum single atoms on nitrogen-doped carbon with high loading as bifunctional catalysts for hydrogen evolution and oxygen reduction reactions. *ACS Catalysis* 8: 8450–8458.

75 Yang, X.-F., Wang, A., Qiao, B. et al. (2013). Single-atom catalysts: a new frontier in heterogeneous catalysis. *Accounts of Chemical Research* 46: 1740–1748.

76 Liu, J. (2017). Catalysis by supported single metal atoms. *ACS Catalysis* 7: 34–59.

77 Guan, J., Zhang, Z., Ji, J. et al. (2017). Hydrothermal synthesis of highly dispersed Co_3O_4 nanoparticles on biomass-derived nitrogen-doped hierarchically porous carbon networks as an efficient bifunctional electrocatalyst for oxygen reduction and evolution reactions. *ACS Applied Materials & Interfaces* 9: 30662–30669.

78 Liu, H., Cao, Y., Wang, F. et al. (2014). Pig bone derived hierarchical porous carbon-supported platinum nanoparticles with superior electrocatalytic activity towards oxygen reduction reaction. *Electroanalysis* 26: 1831–1839.

79 Lefèvre, M., Proietti, E., Jaouen, F., and Dodelet, J.-P. (2009). Iron-based catalysts with improved oxygen reduction activity in polymer electrolyte fuel cells. *Science* 324: 71.

80 Liang, H.-W., Zhuang, X., Brüller, S. et al. (2014). Hierarchically porous carbons with optimized nitrogen doping as highly active electrocatalysts for oxygen reduction. *Nature Communications* 5: 4973.

81 Liang, H.-W., Brüller, S., Dong, R. et al. (2015). Molecular metal-Nx centres in porous carbon for electrocatalytic hydrogen evolution. *Nature Communications* 6: 7992.

82 Dou, M., He, D., Shao, W. et al. (2016). Pyrolysis of animal bones with vitamin B12: a facile route to efficient transition metal–nitrogen–carbon (TM–N–C) electrocatalysts for oxygen reduction. *Chemistry - A European Journal* 22: 2896–2901.

83 Zhang, Z., Sun, J., Wang, F., and Dai, L. (2018). Efficient oxygen reduction reaction (ORR) catalysts based on single iron atoms dispersed on a hierarchically structured porous carbon framework. *Angewandte Chemie International Edition* 57: 9038–9043.

84 Li, H., Zhang, Z., Dou, M., and Wang, F. (2018). Towards high-performance electrocatalysts for oxygen reduction: inducing atomic-level reconstruction of Fe–Nx site for atomically dispersed Fe/N-doped hierarchically porous carbon. *Chemistry - A European Journal* 24: 8848–8856.

85 Li, H., Zhang, Z., Iyoda, T. et al. (2019). Ice/salt-assisted synthesis of ultrathin two-dimensional micro/mesoporous iron and nitrogen co-doped carbon as an efficient electrocatalyst for oxygen reduction. *Chemistry - A European Journal* 25: 5768–5776.

10

Challenges and Opportunities

10.1 Principles for Material Design

The general principles governing the design of collagen-derived materials play important roles in their future synthesis and application. In situ experiments, theoretical calculations, and machine learning can be further used to fully understand the design principles of collagen-derived materials.

10.1.1 In situ Characterization and Analysis

The development of advanced electrochemical energy storage and conversion systems requires optimizing the current electrode materials and finding new electrode materials. Therefore, it is important to get in-depth understanding of the relationship between structural and composition changes in materials during electrochemical processes using a variety of qualitative techniques under specific test conditions. In the same way, in situ characterization and analysis are also beneficial to the exploration of new collagen-derived materials for electrochemical energy storage and conversion. Unfortunately, in situ studies on collagen-derived materials and their applications are very scarce. Herein, we use graphite as a substitute for collagen-derived materials (collagen-derived carbons) and choose Li-ion batteries as typical applications to illustrate the roles of in situ characterization and analysis played in material development.

Due to the sensitivity of the working electrode to environment, direct in situ measurements cannot accurately reflect electrochemical processes. Therefore, it is very important to obtain the related information under real working conditions. Recently, tremendous progress has been made toward the development of advanced characterization techniques for alkali metal–ion batteries under in situ conditions. Here, "in situ" means that the experiment is performed under the actual working conditions. Each in situ technique can be used to reveal specific behaviors and mechanisms of the working electrodes. For battery electrodes, the in situ techniques can provide information about solid electrolyte interphase (SEI) formation, structural change, redox mechanisms, alkali metal–ion transport and diffusion, and possible side reactions under real working conditions [1].

Collagen-Derived Materials: Synthesis and Applications in Electrochemical Energy Storage and Conversion, First Edition. Feng Wang, Yaqin Huang, and Jin Niu.
© 2022 WILEY-VCH GmbH. Published 2022 by WILEY-VCH GmbH.

Table 10.1 Principle and characteristics of different characterization technologies.

Characterization technologies	Characterization characteristics
In situ transmission electron microscope (TEM)	(a) Observe the morphological changes (b) Distinguish the lithiation and delithiation phase (c) Distinguish the distribution of elements and content Measure the mechanical properties
In situ scanning electron microscope (SEM)	(a) Observe the morphological changes (b) Distinguish the distribution of elements and content (c) Measure the mechanical property
In situ X-ray diffraction (XRD)	(a) Monitor the crystallographic structural changes and stability during charge/discharge or cycling process
In situ X-ray absorption spectroscopy (XAS)	(i) Distinguish the species and content of elements (ii) Measure the oxidation state of elements (iii) Detect the information about the molecular structure (especially for amorphous materials)
In situ Raman	Determine the structural changes of electrodes or nanodevice during cycling
In situ infrared spectroscopy	Determine the species of the ions in electrolyte during cycling
In situ nuclear magnetic resonance (NMR)	Observe the local electronic environment around the nucleus

Source: Based on Lorie et al. [2].

The in situ characterization technologies can be divided into five categories, including X-ray, electron, neutron, optical, and scanning probe microscopy. Table 10.1 shows the principles and characteristics of the typical in situ methods. So far, there have been numerous in situ measurements to study the Li-ion intercalation process of graphite anode (Figure 10.1). The intercalation dynamics and different intercalated stages of graphite anode in different batteries can be characterized by in situ X-ray diffraction (XRD) [4–6], nuclear magnetic resonance (NMR) [2], and neutron diffraction [7]. In addition, the space information of interfacial kinetics and morphology changes can be confirmed by optical, Raman, and electron microscopic techniques [8, 9].

Reynier et al. used in situ XRD to characterize lithiated graphite anode at different stages [10]. The residual LiC_{12} (the stage II compound) is detected in the supposedly fully lithiated graphite anode (Figure 10.2a,b), which is not in accordance with the theoretical result. It is because that the chemical potential between stages I and II is too small to average the lithium composition (LiC_{12} and LiC_6) under real electrochemical conditions. The same result was also obtained by Sharma et al. who used

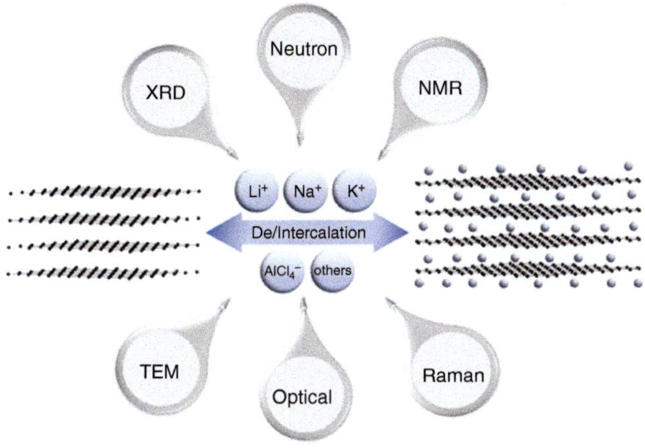

Figure 10.1 Schematic illustration of in situ structural characterization of the graphite intercalation process. Source: Reproduced with permission from Li and Su [3]; © 2019, John Wiley & Sons Australia, Ltd.

in situ neutron scattering to characterize the lithiation process of graphite anode (Figure 10.2c,d) [11].

In situ transmission electron microscope (TEM) has been used to study the graphite-based anode (bilayered graphene anode) [12]. As shown in Figure 10.3a, new device was designed to study the lithiation/delithiation processes (Figure 10.3b) by in situ TEM. Figure 10.3c,d shows the high-resolution TEM (HRTEM) images of the bilayered graphene anode during the lithiation and delithiation processes. It is revealed that multilayered close-packed Li phase is arranged between the two graphene layers. Notably, it is found that the lithium super-dense phase can form between the graphene bilayer, which could account for the reason why graphene anode and other low-graphitization anodes show higher capacity than graphite anode.

Harris et al. used in situ optical microscope to measure the Li-ions transport in graphite anode [13]. It is shown that the graphite anode in different staging phases exhibits different colors. The graphite anode in the phase of LiC_6 shows a gold color. Hu et al. combined the in situ optical microscopy and in situ Raman spectroscopy to investigate the graphite anode during electrochemical lithiation/delithiation (Figure 10.4a–c) [14]. It is found that the transmittance obviously increases in the lithiation process (Figure 10.4d–g). In the delithiation process, the transmittance can reversibly change (Figure 10.4h,i), which is corresponding to the in situ Raman result (Figure 10.4j). Figure 10.4k,l displays the in situ transmittance and voltage changes of graphite anode at different lithiation stages. With the decrease of the voltage and stage phase, Li-intercalated graphite optical transmission is increasing. The in situ optical characterization can also be used to study Li-ion intercalation process of other high-graphitization anodes.

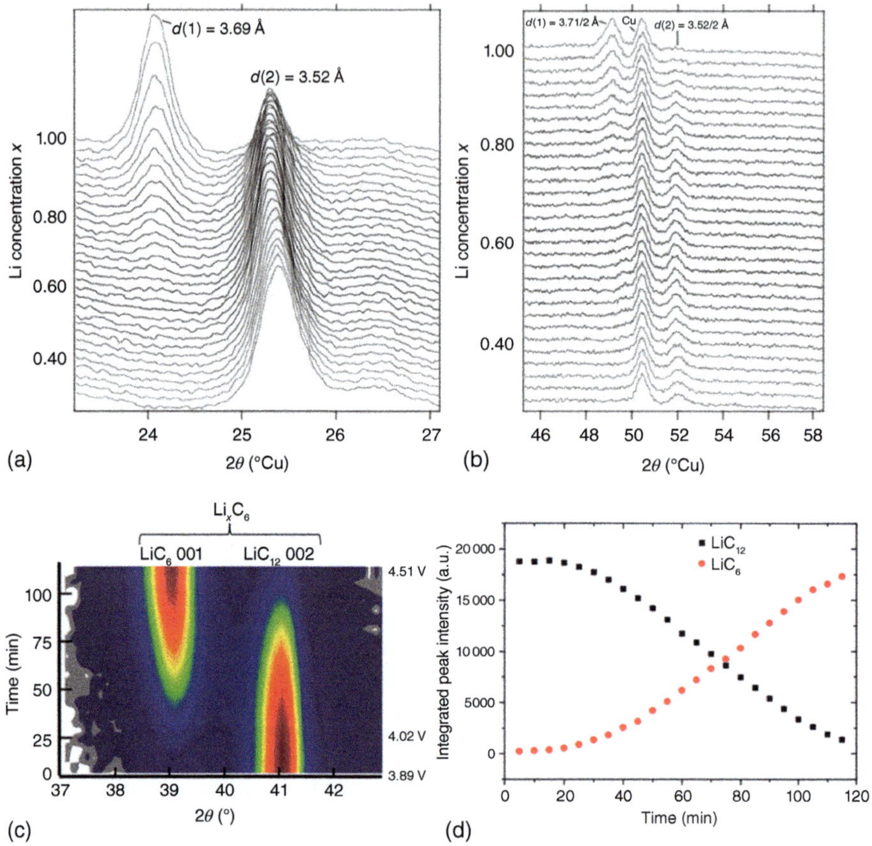

Figure 10.2 In situ XRD and neutron scattering profiles of graphite during Li-ion intercalation. In situ XRD profiles of (a) (002) and (b) (004) regions of graphite. (c) In situ neutron scattering patterns of the Li-ion battery that collated as a function time (the angular region $37° \leq 2\theta \leq 43°$). (d) Intensity of the LiC_{12} (002) (black squares) and LiC_6 (001) (red circles) reflections (from data shown in (a)) with Gaussian fitting. Source: (a, b) Reproduced with permission from Reynier et al. [10]; © 2007, Elsevier. (c, d) Reproduced with permission from Sharma and Peterson [11]; © 2013, Elsevier.

10.1.2 Theoretical Calculation and Simulation

The electrode materials greatly limit the development of high-performance energy-storage systems. It usually takes too much time to find the optimal electrode materials using conventional trial-and-error methods. Recently, computational techniques have shown great potential for use in rational design of novel materials. Computational materials science can provide a tool to predict the performance of electrode materials and reveal the relations between material structural composition and electrochemical performance. Figure 10.5 shows the different ranges of spatial and temporal scales covered by the simulation methods [15].

As one of the most powerful tools, first-principles (ab initio) modeling can be used to explore the material structures and properties using quantum mechanics.

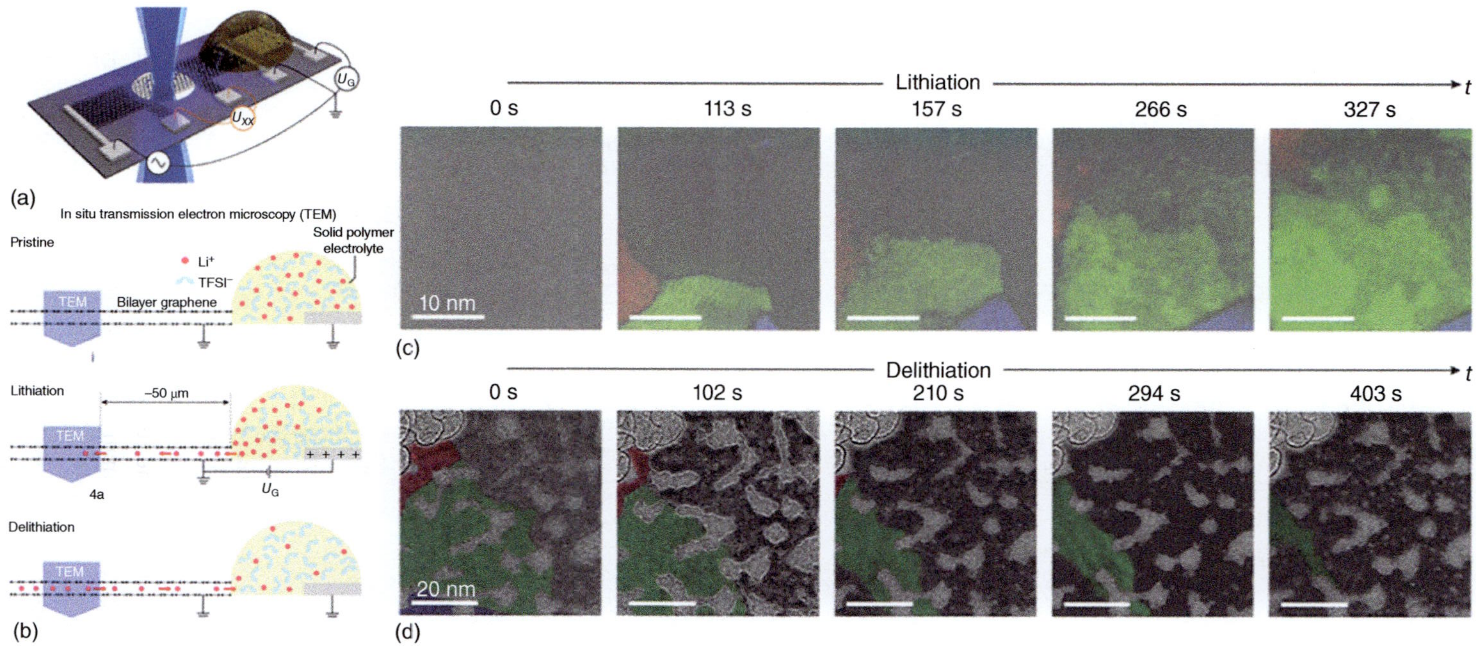

Figure 10.3 In situ TEM study on lithiation and delithiation of a bilayered graphene. (a) Schematic illustration of the device and (b) side view of the in situ device. (c, d) Time series of digital dark-field versions of the original images during lithiation (c) and (d) delithiation. Crystal grains of different in-plane orientations are colored in red, green, and blue. Source: Harris et al. [13]/with permission of © Nature Publishing Group.

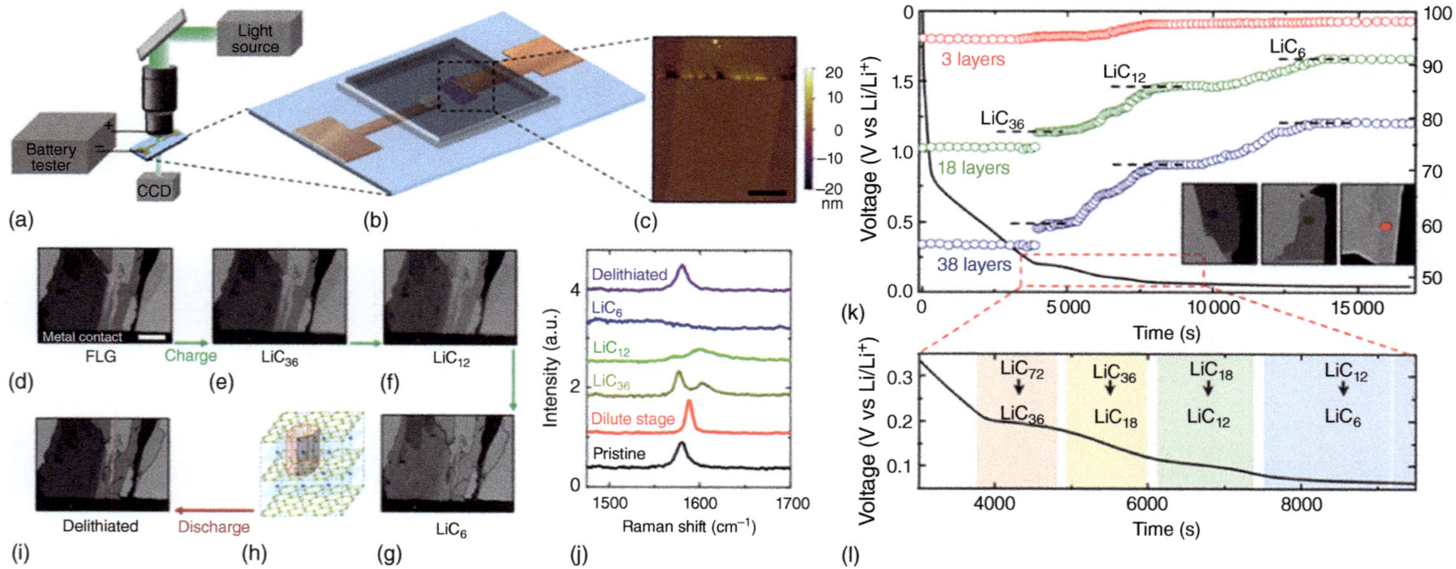

Figure 10.4 In situ optical and Raman microscopy study on lithiation. (a, b) Schematic illustration of in situ experimental setup. (c) An atomic force microscope image of ultrathin graphite sheet attached to the electrical contact. (d–i) Transmission optical microscope images of ultrathin graphite before intercalation and at different intercalation stages. (h) A schematic of the LiC_6 lattice structure. (j) Raman spectrum of Li-ions insertion into ultrathin graphite at different intercalation stages. (k) The profiles of optical transmittance (right) and electrochemical potential (left) during electrochemical Li intercalation process. (l) Detail of voltage profile vs time near the intercalation plateau. Source: Bao et al. [14]/with permission of © Nature Publishing Group.

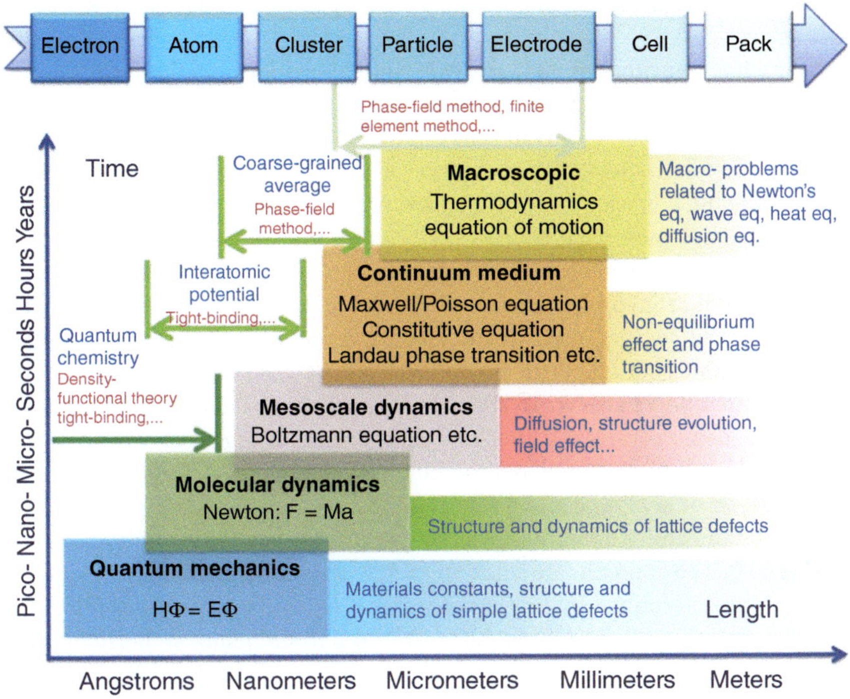

Figure 10.5 Multi-scale computation methods and their applications in Li-ion battery research and development. Source: Reproduced with permission from Shi et al. [15]; © 2016, Chinese Physical Society and IOP Publishing Ltd.

By first-principles modeling, many ground-state properties of the materials can be accurately predicted before the practical fabrication [16]. The applications of first-principles calculation in Li-ion batteries include investigating the electronic structures of the electrode materials, the energy barrier of the Li-ion migration, the structural stabilities during the electrochemical processes, electron conduction behaviors, and other dynamics performance and thermodynamics characteristics.

Molecular dynamics (MD) simulation can be used to study classical particle (e.g. atoms, ions, and molecules) movement in a certain macroscopic system [17]. The MD simulation is used based on Newton's equation. It can obtain the particle positions and momenta in each time step, thus monitoring the trajectory and energy change of the particles at a certain temperature in a certain time. For Li-ion batteries, MD simulation can provide the information about the Li-ion diffusion within the electrolyte/electrode, structural change, and phase transition of the electrode material during the charge–discharge process.

Monte Carlo (MC) simulation is similar to MD in terms of monitoring the particle trajectories and movements. The physical and chemical properties of the systems are determined by applying statistical procedures. However, the specific simulations are carried out in different ways [18]. MC simulation can be used to simulate the Li-ion diffusion process, structural stability, and phase change of the electrode material during the lithiation–delithiation process, and the electrode potential at different stages [19].

First-principles calculation, MC and MD simulations are suitable for use in the atom scale, while phase-field (PF) modeling has a wider span from atom scale to continuum scale. It is commonly used to simulate the kinetic behaviors of the electrodes [20]. The phase-field (PF) simulation is semi-phenomenological and based on thermodynamics, elastoplastic mechanics, and electrochemistry. Therefore, it is also considered as a mesoscale method [21].

Gao et al. built a phase-field model to predict the onset of Li plating of graphite anodes [22]. Figure 10.6a–c shows their model predictions. The results of the phase-field model for graphite anode predict the Li-ion concentration change within the graphite particle. The thick dashed lines represent the experimental result, while the colored areas correspond to the computational result. Additionally, Figure 10.6d,e demonstrates the fitted and experimental charge profiles and surface concentration in lithiation and delithiation process. Their model can correctly predict the onset of Li plating and the nucleation potential of Li deposition.

Subdividing the entire domain into small parts is the core idea of finite element method (FEM). The approximate solution for the complex equation over the whole domain can be determined by linking the simple equations over the subdomains [23]. For ion batteries, the FEM method can be used to simulate the thermal behavior during abuse, the stress/current/voltage distribution of the electrode during the charge–discharge process [24].

10.2 Strategies for Material Synthesis

Sustainability has attracted more and more attention for material synthesis and applications. In addition to "high performance," "green chemistry" and "green technology" also play important roles in material design [25–28]. As byproducts of the meat-processing industry, collagen-enriched biomass wastes (bones, skin, and scales) have a high output of nearly 100 million tons every year [29]. Due to the unique natural structures and compositions, gelatin and carbon materials derived from the collagen-enriched biomass have promising applications in electrochemical energy storage and conversion [30]. Although the collagen-derived materials are promising candidates of traditional nonsustainable materials for use in electrochemical energy storage and conversion, it is still a big challenge to realize the green and large-scale production of collagen-derived materials.

First of all, there should be enough precursors for the large-scale production. The collagen-enriched biomass should be effectively collected and classified to ensure large-scale synthesis. In addition, the high-efficiency equipment and technology for the synthesis of the collagen-derived materials should be developed, especially for the pilot-scale production. Moreover, the byproducts should be appropriately disposed or reused. For example, the calcium salts and phosphates produced by the synthetic processes of bone-derived materials can be used as fodders and fertilizers.

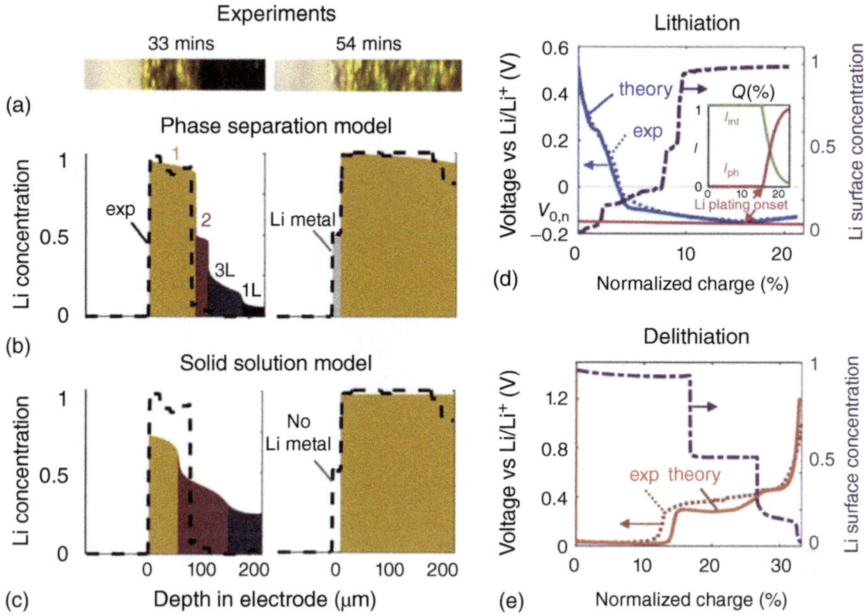

Figure 10.6 (a) Images of graphite particle used to extract the Li concentration and Li plating profiles. (b, c) Model predictions of Li concentration profiles compared with experimental data. (b) Predictions using the phase separation model. (c) Predictions using the solid-solution model. The model considering phase separating in graphite can capture the abrupt increase of c_s when it enters spinodal region, and therefore, gives accurate prediction of Li onset. In contrast, the solid-solution model underestimates c_s and fails to predict the onset of Li plating. (d, e) Voltage and predicted Li surface concentration during lithiation and delithiation of graphite. Inset in (d): current for Li plating and intercalation in the dimensionless form. The onset of Li plating is marked by the red arrow. The blue, purple, and orange arrows are used to indicate which y-axis the curve corresponds to. Source: Gao et al. [22]/with permission of © Elsevier Ltd.

In addition to high efficiency, sustainability and controllability are other two important targets for the synthesis of collagen-derived materials. Take the collagen-derived carbon as an example. Although the KOH-assistant method has been proposed, it is still not highly controllable to change the morphologies, porosities, and heteroatoms of collagen-derived carbons. KOH can not only change the material structures but also eliminate the heteroatoms during the pyrolysis process. Moreover, the utilization of numerous KOH not only requires high-level equipment but also harms the environment in the posttreatment process. As derivates of collagen, gelatin and amino acids can be obtained by treating the collagen-enriched biomass with hot water, acid solution, and alkaline solution. The good water solubility of gelatin and amino acids makes it possible to introduce sustainable water-soluble templates (e.g. NaCl) and dopants (e.g. phytic acid). By tailored treatments, collagen-derived carbons with controllable structures and compositions could be synthesized (Figure 10.7) [31]. More importantly, only water would be used in the carbon processing instead of organic solvents, acids, and alkalines [32].

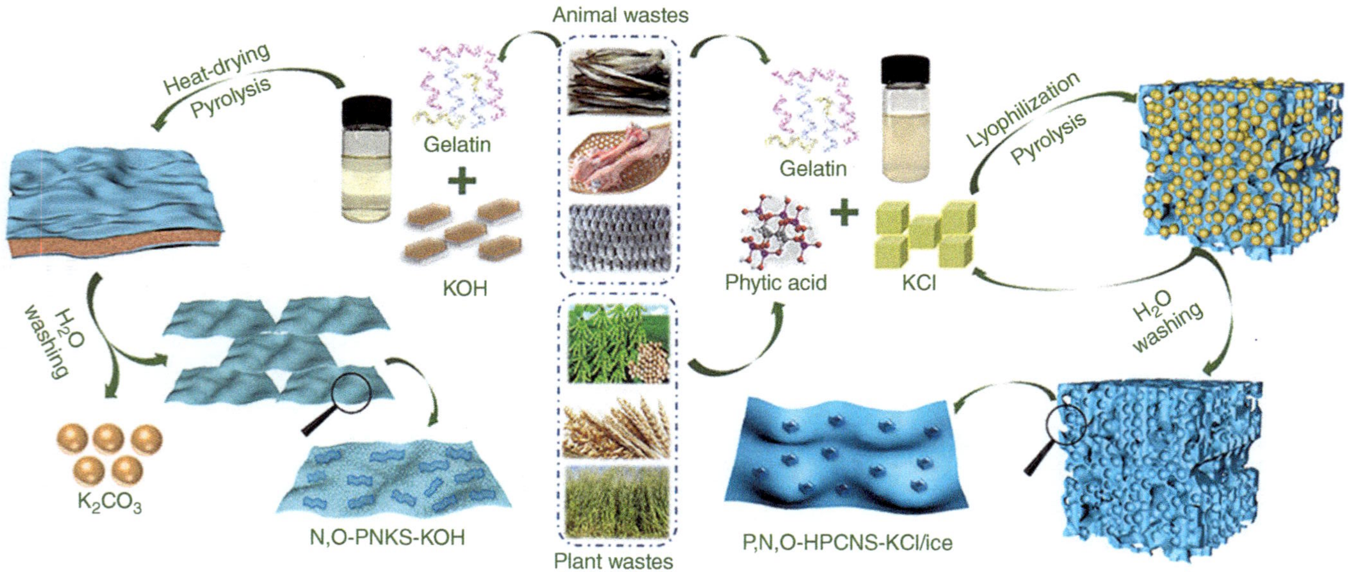

Figure 10.7 Schematic illustration of the synthesis of P,N-doped interconnected carbon nanosheets with hierarchical porosity using KCl/ice as dual-templates and N-doped carbon nanosheets with micropore-dominant porosity with the assistant of KOH. Source: Niu et al. [31]/with permission of © American Chemical Society.

10.3 Diversities for Material Applications

10.3.1 Capacitive Desalination

Although water is abundant on Earth, only 2.5% of the water is drinkable. The desalination of seawater and brackish water has been regarded as promising way to produce freshwater [33]. For capacitive desalination, saltwater passes through the electrodes with a high surface area. The ions in saltwater are absorbed on the electrode surface by applying voltage. Hydrogel and its composites show good performance in capacitive desalination [34, 35]. The hydrogel and its supporter (e.g. graphene or carbon nanotubes) can show high capacitive capability, good stability, and significant desalting capacity. As natural polymer, gelatin can be used as raw starting materials for hydrogel preparation due to the rich amino and carboxyl functional groups [36]. Therefore, the gelatin and its composites can also show good prospects for use in capacitive desalination.

10.3.2 Solar Steam Generation

In addition to the capacitive desalination, the polluted and salted water can be converted into purified steam by solar steam generation. Different from the conventional heating method, the solar steam generator only needs to heat the surface of the water, showing high generation efficiency and no energy consumption [37].

As a collagen-derived material, gelatin has a good adhesion property due to its rich amino and carboxyl functional groups, which form numerous hydrogen bonds. Hu et al. integrated black phosphorous (BP) nanosheets and porous polyurethane (PU) foam into a solar vapor generator using gelatin as a binder (Figure 10.8) [38]. The effects of crystal defect on the solar steam generation were revealed and the optimized solar vapor generator showed good performance. In addition to binder for solar vapor generator, collagen-derived carbons with high surface area and rich heteroatoms/defects are also promising materials for solar steam generation, presenting tremendous potential to address the freshwater shortage.

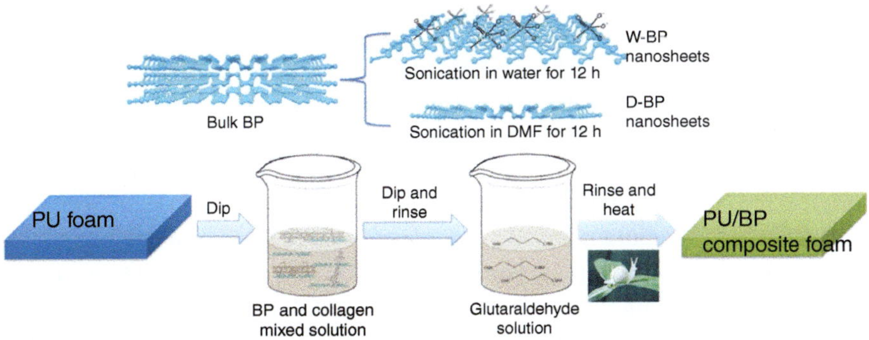

Figure 10.8 The schematic of the preparation of W-BP nanosheets and D-BP nanosheets as well as PU composite foam. BP, black phosphorous; PU, polyurethane. Source: Reproduced with permission from Cai et al. [38]; © 2020, Elsevier.

10.3.3 Gas Adsorption

Hydrogen has a high gravimetric energy density of 120 kJ g^{-1}. As a gas energy carrier, it is a promising clean energy [39]. Meanwhile, CO_2 capture is a promising way to solve global warming before clean energy is widely used. Therefore, high-efficiency gas-storage materials are much needed. Gao et al. prepared N-doped porous carbon with high surface area using pigskin as precursor. Different from the convention KOH activation method, they used $CaCO_3$ as active agent [40]. The $CaCO_3$ resulted in high surface area and numerous mesopores of the collagen-derived carbons. The effects of temperature on the structural compositions and adsorption capacities of CO_2 and H_2 were systematically studied. The optimized sample showed high adsorption capacity of CO_2 (1.27 mmol g^{-1} at 298 K under 0.15 bar, 4.4 mmol g^{-1} at 298 K under 1 bar) and adsorption capacity of H_2 (2.49 wt% at 77 K under 1.13 bar).

10.3.4 Microwave Adsorption

To increase the electrocommunication rates of the communication devices, the dominant frequency range has been shifted to a higher range. However, there are more and more applications of electromagnetic waves in the Ku band (12.5–18 GHz), including the data transmission by broadcast, radar, and satellite. Therefore, it is important to develop advanced materials for microwave absorbing on the Ku band. As one of the promising materials for microwave absorbing, collagen fibers have hierarchical structures that are beneficial to good microwave absorption. In addition, collagen has rich electric dipoles and molecular bounded charges, which enable it to be a biological electret with the polarization properties in electromagnetic field [41]. The wealthy merits of collagen fiber make it promising for use in the microwave absorption.

10.3.5 Photocatalysis

Heterogeneous photocatalysts play important roles in solar energy conversion and environmental remediation. As one of the most promising photocatalysts, nontoxic TiO_2 has high photocatalytic activity, good chemical stability, and low cost. However, there is still a big challenge for the practical applications of TiO_2, because of its broad bandgap and low quantum efficiency [42]. The nano-fibrous structures of collagen fibers have many advantages, including open porosity, good flexibility, and fast mass transfer. In addition, the collagen fibers have abundant functional groups (amino and carboxyl) that can spontaneously adsorb metal ions (e.g. Ti^{4+}), enabling the collagen fibers to be promising support materials for photocatalysts [43]. By using the collagen fiber as a supporter, Shi and coworkers successfully synthesized Ce-doped TiO_2 (Ce_x/TiO_2) nanofibers [44]. The photocatalytic activity of TiO_2 was obviously improved by incorporating Ce together with mesoporous nanofibers (Figure 10.9). The Ce_x/TiO_2 nanofibers were used as photocatalysts and further tested under visible light for the photocatalytic degradation of Rhodamine B. It is shown that the photocatalytic activities of the Ce_x/TiO_2 nanofibers were compared to the commercial photocatalyst.

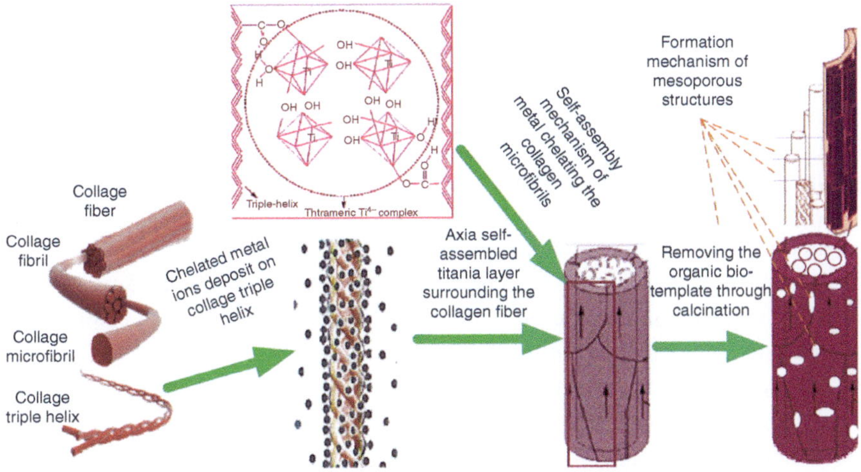

Figure 10.9 Schematic illustration for the preparation of the Ce_x/TiO_2 nanofibers. Source: Reproduced with permission from Xiao et al. [44]; © 2013, American Chemical Society.

10.4 Inspirations for Other Materials and Their Applications

Overall, the above contents about collagen and collagen-derived materials in this book present the unique synthesis and promising applications of these emerging materials. More importantly, the universal methodologies for the examination, characterization, and modeling of collagen and collagen-derived materials are also beneficial to study other biomass macromolecules (e.g. sodium alginate, sodium methylcellulose, and chitosan), functional carbons (e.g. heteroatom doped carbons, porous carbons, and two-dimensional carbons), and their hybrid materials. Moreover, the relations between the structural compositions and wide applications of collagen and collagen-derived materials also provide design principles and functional applications of other materials. In future, it is undoubted that more exciting related research works will be presented by the scientists about collagen, collagen-derived materials, and other materials, as more technologies enter these areas.

References

1 Liu, D., Shadike, Z., Lin, R. et al. (2019). Review of recent development of in situ/operando characterization techniques for lithium battery research. *Advanced Materials* 31, 1806620.

2 Lorie Lopez, J.L., Grandinetti, P.J., and Co, A.C. (2018). Enhancing the real-time detection of phase changes in lithium–graphite intercalated compounds through derivative operando (dOp) NMR cyclic voltammetry. *Journal of Materials Chemistry A* 6: 231–243.

3 Li, N. and Su, D. (2019). In-situ structural characterizations of electrochemical intercalation of graphite compounds. *Carbon Energy* 1: 200–218.
4 Kim, H., Hong, J., Yoon, G. et al. (2015). Sodium intercalation chemistry in graphite. *Energy & Environmental Science* 8: 2963–2969.
5 Seidl, L., Bucher, N., Chu, E. et al. (2017). Intercalation of solvated Na-ions into graphite. *Energy & Environmental Science* 10: 1631–1642.
6 Liu, J., Yin, T., Tian, B. et al. (2019). Unraveling the potassium storage mechanism in graphite foam. *Advanced Energy Materials* 9, 1900579.
7 Letellier, M., Chevallier, F., and Béguin, F. (2006). In situ ^{7}Li NMR during lithium electrochemical insertion into graphite and a carbon/carbon composite. *Journal of Physics and Chemistry of Solids* 67: 1228–1232.
8 Cohn, A.P., Share, K., Carter, R. et al. (2016). Ultrafast solvent-assisted sodium ion intercalation into highly crystalline few-layered graphene. *Nano Letters* 16: 543–548.
9 Wen, Y., He, K., Zhu, Y. et al. (2014). Expanded graphite as superior anode for sodium-ion batteries. *Nature Communications* 5: 4033.
10 Reynier, Y., Yazami, R., and Fultz, B. (2007). XRD evidence of macroscopic composition inhomogeneities in the graphite-lithium electrode. *Journal of Power Sources* 165: 616–619.
11 Sharma, N. and Peterson, V.K. (2013). Overcharging a Li-ion battery: effect on the Li_xC_6 negative electrode determined by in situ neutron diffraction. *Journal of Power Sources* 244: 695–701.
12 Kühne, M., Börrnert, F., Fecher, S. et al. (2018). Reversible superdense ordering of lithium between two graphene sheets. *Nature* 564: 234–239.
13 Harris, S.J., Timmons, A., Baker, D.R., and Monroe, C. (2010). Direct in situ measurements of Li transport in Li-ion battery negative electrodes. *Chemical Physics Letters* 485: 265–274.
14 Bao, W., Wan, J., Han, X. et al. (2014). Approaching the limits of transparency and conductivity in graphitic materials through lithium intercalation. *Nature Communications* 5: 4224.
15 Shi, S., Gao, J., Liu, Y. et al. (2016). Multi-scale computation methods: their applications in Li-ion battery research and development. *Chinese Physics B* 25: 018212.
16 Meng, Y.S. and Arroyo-de Dompablo, M.E. (2009). First principles computational materials design for energy storage materials in lithium ion batteries. *Energy & Environmental Science* 2: 589.
17 Car, R. and Parrinello, M. (1985). Unified approach for molecular dynamics and density-functional theory. *Physical Review Letters* 55: 2471–2474.
18 Xu, Z., Lv, X., Chen, J. et al. (2016). Dispersion-corrected DFT investigation on defect chemistry and potassium migration in potassium-graphite intercalation compounds for potassium ion batteries anode materials. *Carbon* 107: 885–894.
19 Binder, K. (1997). Applications of monte carlo methods to statistical physics. *Reports on Progress in Physics* 60: 487–559.

20 Ponce, V., Galvez-Aranda, D.E., and Seminario, J.M. (2017). Analysis of a Li-ion nanobattery with graphite anode using molecular dynamics simulations. *The Journal of Physical Chemistry C* 121: 12959–12971.

21 Zuo, P. and Zhao, Y.-P. (2015). A phase field model coupling lithium diffusion and stress evolution with crack propagation and application in lithium ion batteries. *Physical Chemistry Chemical Physics* 17: 287–297.

22 Gao, T., Han, Y., Fraggedakis, D. et al. (2021). Interplay of lithium intercalation and plating on a single graphite particle. *Joule* 5: 393–414.

23 Guo, G., Long, B., Cheng, B. et al. (2010). Three-dimensional thermal finite element modeling of Li-ion battery in thermal abuse application. *Journal of Power Sources* 195: 2393–2398.

24 Iqbal, N. and Lee, S. (2018). Mechanical failure analysis of graphite anode particles with PVDF binders in Li-ion batteries. *Journal of the Electrochemical Society* 165: 1961–1970.

25 Salkind, A.J., Fennie, C., Singh, P. et al. (1999). Determination of state-of-charge and state-of-health of batteries by fuzzy logic methodology. *Journal of Power Sources* 80: 293–300.

26 Waag, W., Fleischer, C., and Sauer, D.U. (2014). Critical review of the methods for monitoring of Li-ion batteries in electric and hybrid vehicles. *Journal of Power Sources* 258: 321–339.

27 Van Duong, M., Van Tran, M., Garg, A. et al. (2020). Machine learning approach in exploring the electrolyte additives effect on cycling performance of $LiNi_{0.5}Mn_{1.5}O_4$ cathode and graphite anode-based Li-ion cell. *International Journal of Energy Research* 45: 4133–4144.

28 Gao, M., Pan, S.-Y., Chen, W.-C., and Chiang, P.-C. (2018). A cross-disciplinary overview of naturally derived materials for electrochemical energy storage. *Materials Today Energy* 7: 58–79.

29 Ashokkumar, M. and Ajayan, P.M. (2020). Materials science perspective of multifunctional materials derived from collagen. *International Materials Reviews* 66: 160–187.

30 de Wolf, F.A. (2003). Chapter V, Collagen and gelatin. *Progress in Biotechnology* 23: 133–218.

31 Niu, J., Guan, J., Dou, M. et al. (2020). Sustainable synthesis of biomass-derived carbon electrodes with hybrid energy-storage behaviors for use in high-performance Na-ion capacitors. *ACS Applied Energy Materials* 3: 2478–2489.

32 Niu, J., Shao, R., Liu, M. et al. (2019). Porous carbons derived from collagen-enriched biomass: tailored design, synthesis, and application in electrochemical energy storage and conversion. *Advanced Functional Materials* 29, 1905095.

33 Kong, W., Duan, X., Ge, Y. et al. (2016). Holey graphene hydrogel with in-plane pores for high-performance capacitive desalination. *Nano Research* 9: 2458–2466.

34 Ma, J., Wang, L., and Yu, F. (2018). Water-enhanced performance in capacitive deionization for desalination based on graphene gel as electrode material. *Electrochimica Acta* 263: 40–46.

35 Cao, J., Wang, Y., Chen, C. et al. (2018). A comparison of graphene hydrogels modified with single-walled/multi-walled carbon nanotubes as electrode materials for capacitive deionization. *Journal of Colloid and Interface Science* 518: 69–75.

36 Mi, M., Liu, X., Kong, W. et al. (2019). Hierarchical composite of N-doped carbon sphere and holey graphene hydrogel for high-performance capacitive deionization. *Desalination* 464: 18–24.

37 Ma, S., Chiu, C.P., Zhu, Y. et al. (2017). Recycled waste black polyurethane sponges for solar vapor generation and distillation. *Applied Energy* 206: 63–69.

38 Cai, W., Li, Z., Ma, C. et al. (2021). Effect of crystal defects on solar steam generation performance of black phosphorous nanosheets. *Materials Today Energy* 19 100553.

39 Sethia, G. and Sayari, A. (2016). Activated carbon with optimum pore size distribution for hydrogen storage. *Carbon* 99: 289–294.

40 Gao, A., Guo, N., Yan, M. et al. (2018). Hierarchical porous carbon activated by $CaCO_3$ from pigskin collagen for CO_2 and H_2 adsorption. *Microporous and Mesoporous Materials* 260: 172–179.

41 Guo, J., Wu, H., Liao, X., and Shi, B. (2011). Facile synthesis of size-controlled silver nanoparticles using plant tannin grafted collagen fiber as reductant and stabilizer for microwave absorption application in the whole Ku band. *The Journal of Physical Chemistry C* 115: 23688–23694.

42 Zhang, X., Thavasi, V., Mhaisalkar, S.G., and Ramakrishna, S. (2012). Novel hollow mesoporous 1D TiO_2 nanofibers as photovoltaic and photocatalytic materials. *Nanoscale* 4: 1707–1716.

43 Suzuki, Y., Pavasupree, S., Yoshikawa, S., and Kawahata, R. (2005). Natural rutile-derived titanate nanofibers prepared by direct hydrothermal processing. *Journal of Materials Research* 20: 1063–1070.

44 Xiao, G., Huang, X., Liao, X., and Shi, B. (2013). One-pot facile synthesis of cerium-doped TiO_2 mesoporous nanofibers using collagen fiber as the biotemplate and its application in visible light photocatalysis. *The Journal of Physical Chemistry C* 117: 9739–9746.

Index

Collagen-Derived Materials: Synthesis and Applications in Electrochemical Energy Storage and Conversion, First Edition. Feng Wang, Yaqin Huang, and Jin Niu.
© 2022 WILEY-VCH GmbH. Published 2022 by WILEY-VCH GmbH.

d

e